ACID DEPOSITION CAUSES AND EFFECTS

A State Assessment Model

Alex E. S. Green and Wayne H. Smith
Editors

Sponsored by

Interdisciplinary Center for Aeronomy and (other) Atmospheric Sciences (ICAAS)
and
Center for Environmental and Natural Resources Programs (CENRP)

Proceedings of a Workshop
March 23-24, 1983

Supported by

Graduate School and Division of Sponsored Research
College of Engineering
Institute of Food and Agricultural Sciences
College of Liberal Arts and Sciences
University of Florida
Gainesville

PUBLISHER'S NOTE

This book is designed to provide accurate and authoritative information on the subject matter. It is sold with the understanding that the publisher is not engaged in rendering legal, accounting, or other professional services. If legal advice or other expert assistance is needed, the services of a competent professional should be sought. (From a declaration of principles adopted jointly by a committee of the American Bar Association and a committee of publishers.

Publication of this book does not signify that the contents necessarily reflect the views and policies of the publisher.

September 1983

Published by
Government Institutes, Inc.
966 Hungerford Drive, #24
Rockville, Maryland 20850
U.S.A.

ISBN Number: 0-86587-060-8
Library of Congress Number: 83-82073

Printed and bound in the United States of America

Acknowledgments

The conference coordinators would like to express their thanks to R. A. Bryan, W. H. Chen, G. W. Hemp, M. J. Ohanian, C. F. Sidman, Jr., F. G. Stehli, and F. A. Wood for committing the financial resources that enabled us to hold this workshop and to publish its proceedings. We would also like to thank Henry Gholz and Bob G. Volk for their valuable help in organizing the program. Marybruce Dowd, Diana Kanoy, and Jerry Schwartz provided painstaking assistance on organization and administration of the conference, for which we are most appreciative. A special thanks is extended to Ruby Moore, Floy Ramey, and Belle Wigglesworth who provided typing and other support services for the workshop. Finally, we would like to thank all participants for giving of their time and expertise to make this workshop a model cooperative venture among university, governmental and industrial scientists working toward achieving a scientific assessment and consensus on an important, albeit controversial, subject.

To Marybruce Dowd, we are especially appreciative for her painstaking review of each manuscript. Her critical, careful examination improved the accuracy of the manuscripts and the structure of each presentation. In addition she ably served as liaison between the editors, among the authors, and with the publisher. Ruby Moore entered, formatted, and generated word processor copy for the intermediate reviews and the final copy for the publisher. We thank her for this important contribution to the production of these proceedings.

Alex E. S. Green　　　　Wayne H. Smith

PREFACE

The initial purposes of this Workshop were to prepare a document to guide the University of Florida in strengthening its existing programs relating to causes and effects of acid deposition and to formulate long range research and educational programs bearing upon these topics. Under "causes" we considered anthropogenic and natural sources of acid-forming substances, the transformation of these substances in the atmosphere and their transport from source site to depositing site. Under "effects" we restricted our scope to studies on monitoring of acid deposition and effects of acid deposition upon aquatic and terrestrial ecosystems. For reasons of time and the state of knowledge, other possible effects, such as direct effects upon human health, structure and materials, or socio-economic-legal impacts, were not considered.

As plans for the Workshop evolved from discussions over the past year, we decided to establish first an overview of the current state of knowledge with the assistance of nationally recognized authorities. Then, to help construct a Florida assessment and to reach a scientific concensus on Florida's research needs, we issued a university-wide call for scientific papers relevant to acid rain and for nominations of other resource scientists available in Florida. Invitations to all persons so suggested were issued. The response was very gratifying and, on the basis of the program that finally evolved, we are hopeful that the Proceedings of this Workshop might serve a broader purpose than originally envisioned. To enable the Proceedings to reflect discussions at the Workshop, we extended the deadline for final revision of contributed papers.

The organization of this Workshop and the production of its Proceedings have been undertaken as a joint venture between two centers whose interests and involvements with acid deposition overlap. ICAAS, an interdisciplinary center spanning seven colleges and twenty departments of the University of Florida, has since 1970 been primarily concerned with scientific topics considered under "causes", as well as with health, socio-economic and legal aspects of such topics. The monograph, "Coal Burning Issues", is an example of ICAAS' prior work. CENRP, since its formation in 1973, has supported and coordinated environmental and natural resources projects in 18 academic departments and 23 outstate research centers. Volumes resulting from CENRP projects include "Proceedings of International Symposium on Biological Control of Weeds", and "Sludge--Health Risks of Land Application". The CENRP supported the installation and operation of the first collector in the National Atmospheric Deposition Program and has maintained involvement in aspects relating to "effects" of atmospheric deposition.

The one-day Workshop on Acid Deposition Science was stimulating and fruitful beyond our expectations. The strong participation of the State Department of Environmental Regulation and the electrical utility industry supported the idea that bringing together physical and biological scientists involved in major studies of acid deposition could serve a valuable role in arriving at an objective statewide assessment. As we approached the heart of the various questions related to acid deposition, not only were many agreements achieved but our definitions of the key questions requiring further research were greatly improved. It also became apparent during the discussions that we had overlooked some relevant topics and, in this connection, we solicited and obtained post conference contributions on precipitation, cloud formation and lightning in Florida. We also had several of the Florida papers expanded to cover topics originally in national overview papers which encountered delays in federal review processes.

Upon further post conference reflection, we realized that with minor changes in format we could have further improved the cross-communications between biological and physical scientists. For example, we could have avoided the parallel sessions for the physical and biological sciences by running the Workshop an extra half day. Participation exceeded our expectation, so we experienced serious time allocation problems among speakers. While we should have been more specific in designating presentation times, the best solution would have been a longer conference. With these relatively minor format changes, we believe our Workshop approach and our final report might serve as a model assessment of the problem or potential problem of acid rain in any state. While Florida is unique in its geographic and climatological features, each state is unique in other respects. Our basic procedure is first to bring together physical and biological scientists and engineers to reach dispassionate assessments on scientific issues. With such a scientific base it should be possible to proceed with greater confidence to the difficult (and sometimes emotional) socio-economic, legal and political aspects of the acid rain issue. The goal is to establish a process to arrive at the optimum public policy from the viewpoint of the state's citizenry.

Papers in this Proceedings consider how statewide assessments on important public policy issues such as the acid rain issue might be achieved. The process employed in the Workshop should be easily translatable to other states faced with concerns regarding potential acid rain problems. A final result of the Workshop was a recommendation that annual update sessions be scheduled. Should this interest be sustained, we shall be organizing subsequent workshops -- initially addressing oversights in this Workshop as well as the health, socio-legal, political and economic aspects of acid deposition. This process is designed to aid in the development of public policy by interfacing multifaceted assessments of acid deposition.

Alex E. S. Green   Wayne H. Smith

TABLE OF CONTENTS

EXECUTIVE SUMMARY: REPORT ON PRESENTATIONS AND DISCUSSIONS centered upon I. SOURCES, ATMOSPHERIC TRANSFORMATIONS AND TRANSPORT OF ACIDIC SUBSTANCES IN FLORIDA

R. B. Gaither
Department of Mechanical Engineering
University of Florida, Gainesville, FL 32611

The purposes of the Acid Deposition Science Workshop -- Causes and Effects included the bringing together of a group of knowledgeable and well informed scientists, engineers and government representatives to establish an understanding of Florida's vulnerability to atmospheric pollutants and to prepare a document to guide the University of Florida and others in strengthening existing programs, formulating long range research, and developing educational programs dealing with acid deposition. The quality of the Workshop presentations, the assemblage and comparison of data, and the extensive discussions all attested to the conclusion that these purposes were achieved.

There is considerable evidence to indicate that anthropogenic acidic pollutants are being generated at rates that have the potential for further acidifying a significant number of sensitive Florida lakes. Clearly justified is a well constructed plan for research and close monitoring of the Florida atmosphere for the next ten years. Also recognized is the need for extensive research in atmospheric chemistry and transport as well as in the combustion processes involving fuels that are being burned and are expected to be burned in Florida.

National Overview - Causes

Focusing upon information catalogued in publications describing the National Acid Precipitation Assessment Program of 1980, Michael Maxwell presented evidence that manmade atmospheric pollutants, notably those in the SO_x and NO_x families, are the major contributors to acid deposition in the industrial regions of the world. In the period 1950 to 1980, the increase of SO_x discharge to the atmosphere in the United States alone rose from 20 million tons per year to 28 million tons per year, then decreased to 24 million tons per year, providing sufficient overtaxing of natural neutralizing agents in the air, water, and vegetation to cause marked rises in acid deposition. It was noted that while this overall increase of 20 percent is disturbing on a national scale, much higher percent rises appear in regions undergoing rapid population and industrial growth. In the Southeast, the increase has been 200 percent. While such a high percentage increase does not mean the Southeast has become the major producer of acidic air pollutants (Florida produced slightly over one million tons of SO_x air pollutants in 1981), it suggests that acid sensitive lakes and agricultural forested lands without natural neutralizing agents are rapidly approaching dangerous levels of acidity.

Additional statistics indicate that the major contributors to SO_x and NO_x pollution are stationary and mobile devices that burn fossil fuels. While supporting evidence is not without critics, it is reckoned that stationary power plants emit approximately two-thirds of the total national SO_x discharge to the atmosphere. Although stationary power plants are identified as providing approximately one-third of the NO_x, automotive sources hold the lead by generating well over one-half.

Control options currently available were categorized by Maxwell as follows: (1) use of low sulfur fuels, (2) removal of sulfur from fuels, (3) combustion process modification, and (4) post combustion treatment. In discussion of these options it was noted that they are applicable to SO_x generation only and that the costs associated with some of these options using present day technology are high.

After some discussion that revealed concerns about the difficulties in obtaining reliable and correlatable data on sources and effects of NO_x and SO_x, the session proceeded with two more presentations focused upon sources, transformation, and transport of acid forming substances.

In reviewing the natural sources of acidity in precipitation, William Chameides documented observations that rainfall with a pH below 5.6 (the value attained by water in equilibrium with atmospheric CO_2) is a common occurrence in remote regions as well as industrial areas. Several natural sources of atmospheric acidity, as well as various explanations of the chemistry involved, were reviewed. Chameides concluded with the presentation of evidence that anthropogenic sources are now beginning to overtake natural sources of SO_x and NO_x in the global view. In local areas, it is clear that enhanced gaseous SO_x and NO_x levels as well as highly acidic rainfall are found in regions affected by anthropogenic activities.

A major problem in actually locating the sources of pollution affecting particular geographical areas is the absence of reliable and correlatable data concerning the role played by large scale winds. In his presentation entitled, "Transport of Pollutants on the Scale of Hundreds of Kilometers", Carl Kreitzberg noted that here we are confronted with a difficult inverse problem. It is relatively easy to trace the dispersion of pollutants from a known source. Conversely, it is extremely difficult to trace a deposition back to its sources -- especially when macroscale weather is involved. Kreitzberg reviewed the nature of different-scale storm systems and their influence on transport. After an explanation of the techniques currently available for determining trajectories, computing concentrations and identifying source regions, he concluded with a description of the seriousness of problems that are encountered in attempting to compare model predictions with observations. He emphasized that attempts to validate averages and model predictions with small amounts of data in atmospheric research can be expected to lead to erroneous conclusions.

Florida Overview - Causes

Relatively few Florida lakes have become significantly more acidic over the past twenty years, in spite of the fact that Florida is known to have many lakes sensitive to acidification (i.e., provided with minimal buffering capabilities). With that observation, Patrick Brezonik provided an introductory focus for five presentations in the second session of the Workshop. Brezonik went on to identify several serious trends. Today, Florida is the tenth largest SO_2 producing state in the nation, generating considerably more of this air pollutant than is necessary to account for all of the acid deposition that is occurring in the state.

Handicapped to some extent by a paucity of data, scientists have not yet been able to identify long term trends in the acidification of Florida lakes. However, results of recent studies indicate that the acidity of rainfall in Florida is quite seasonal (summer rains being more acidic) and more pronounced in the northern counties than in the southern. Annual pH values of less than 4.7 have been reported to occur over the northern two-thirds of Florida, with slightly less acidic precipitation occurring in the southern third.

In his presentation entitled, "Atmospheric Deposition Patterns in North Central Florida", Hans Riekerk offered data supporting the conclusion that over the past five years, the Bradford Forest Watershed in North Central Florida has experienced increasing acidity of rainwater at the rate of 0.2 pH units per year. This has apparently had no detectable effect upon watershed runoff acidity, however. After providing additional data indicating the appearance of a significant increase of rainwater acidity in regions proximate to new power plants, Riekerk made a strong case for increased long range monitoring of Florida lakes.

John Schert presented data recorded from field tests at 14 locations in Florida for the period September 1981 through September 1982, which appear generally to corroborate laboratory analyzed data of others (i.e., that northern Florida received rainfall that is more acidic than southern Florida). Concentrations of cations have also been noted as more numerous in the north than the south, indicating the transition from a continental to marine environment.

Barry Hartigan described the methodology used by the University of Florida Libraries' Energy Committee in the retrieval and utilization of reference sources. After providing advice on data gathering and interpretating data base information, Hartigan concluded with the announcement that a literature search on acid rain in Florida had recently been completed and was available.

In concluding Session II, Thomas A. Lynch provided a provocative presentation entitled, "The Scientific Needs of Environmental Economists". After identifying the economic aspects of scientific research and proposals to avert ecological damage, Lynch reviewed the alternatives available. Included in his review were the identification of secondary economic effects, such as the cost of government regulation and the description of several scenarios that could ensue when specific measures were taken to effect a reduction of emissions believed to be major sources of acid rainfall.

Sources, Atmospheric Transformations, and Transport of Acidic Substances in Florida

In opening this session, Hamilton Oven provided a general frame for the presentations to follow by identifying the major questions confronting Florida's Department of Environmental Regulation. These included the need to identify accurately and in quantifiable terms the sources of acidic substances in the atmosphere, trends in acid deposition levels, and the sources of acidic materials that are realistically controllable. After presenting a considerable amount of codified data on emission sources for SO_x and NO_x compounds, Oven provided an analysis of correlations between anthropogenic levels of sulfate and nitrate generation and the pH of acidic depositions. He concluded with the speculation that the state can maintain a status quo through such efforts as bringing older power plants up to current standards for emission (many are currently exempt from such standards), developing better combustion techniques, and calling for the introduction of relatively inexpensive control methods. Oven noted that by doing so, Florida has an opportunity now to avoid irreversible pollution.

Charles Proctor demonstrated considerable insight to combustion chemistry in the burning of "dirty" oils. Trace species of such substances as vanadium were shown to effect the production of SO_x through secondary and catalytic processes. Proctor concluded with a conviction that considerably more information and understanding of atmospheric chemistry, as well as weather conditions besides rainfall, are needed.

Krishma Pamidimukkala presented an impressive model that he and Alex Green had developed for the devolitalization and combustion of coal. The model was utilized on comparative simulations of burning coal and coal blends (gas/coal, coal/water, and gas/coal/water) in air for the purpose of oil backout and for combustion process control that could reduce NO_x and SO_x emissions. Green concluded with the outline of a research and development program that could be useful in facilitating the use of coal while reducing the emissions causing acid rain.

In a presentation entitled, "Biomass Gasification as a Source of Acidic Substances", K. M. Eoff discussed wood fuels as potential air polluters. At present, the availability of data and/or analysis is quite limited.

Reporting on several studies completed in coastal salt marsh and freshwater marsh areas, Dian Hitchcock showed that emissions from natural sources are extremely variable in both kinds of habitats, in spite of the fact that release rates are quite similar. She also agreed with Brezonik's observation that summer rains are more acidic than winter rains. Hitchcock concluded with two observations, namely: (1) Florida's lakes and wetlands are probably significant sources of atmospheric sulfur compounds; and (2) human activities in these areas may promote release of biological sulphur to the atmosphere.

An analytic Gaussian plume model that allows for predicting pollution concentration patterns was described by J. M. Schwartz. Developed by Green, Schwartz, Singhal and Bolch, an analytic three-dimensional joint frequency distribution (wind rose) map of Florida was shown to permit the assignment of average SO_2 concentrations arising from any source in Florida. The possibility of developing a simple and easy to use transport function map of Florida was discussed.

Elisa Hernandez and Karl Taylor followed by describing a procedure using the analytic Gaussian plume model to find average pollution concentration patterns in regions of Florida where wind distribution data are sparce. Their technique provides an alternate pathway for extending the present limited body of wind data to arrive at a mesoscale climatology in Florida as it bears on acid deposition.

In the final paper in the "Causes" Workshop, Ellen Chen used measurements made with the GOES satellite in connection with their Frost Warning Program to infer air movements within Florida. Her work provided impressive evidence to support a conclusion that sea breezes constitute the dominant mechanism for pollution transport and deposition in Florida.

Note Added by Editors:

Papers by Henry on "Precipitation and Cloud Climatologies of Florida" and by Green and Uman on "Lightning in Florida as a Source of Acid Rain" were added after the conference. The paper by Green and Pamidimukkala on "Oil Backout and Acid Rain" was expanded from their original version to incorporate topics originally intended for Michael A. Maxwell's paper, "Stationary Combustion Sources of Acid Forming Substances and Control", which encountered delays in federal review.

EXECUTIVE SUMMARY: REPORT ON PRESENTATIONS AND DISCUSSIONS centered upon II. TERRESTRIAL AND AQUATIC EFFECTS OF ACID PRECIPITATION IN FLORIDA

Frank G. Nordlie
Department of Zoology
University of Florida, Gainesville, FL 32611

Presentations dealing with effects of acid deposition were given in the general sessions of the program as well as in the session dealing with site-specific situations. They will be considered sequentially in this summary.

National Overview - Effects

J. W. Wilson, United States Military Academy, who gave the lead-in paper in this session, discussed in detail the efforts being made to develop a unified national network of precipitation monitoring stations. His paper described the design and implementation of The National Trends Network, and discussed, in particular, problems of choosing appropriate sites in the Southeast and in Florida. He pointed out that the implementation of the network is a dynamic process. Visitation of existing and proposed sites is currently being conducted by staff members of the Science Research Laboratory, U.S. Military Academy, with adjustments made to recommendations of the basic design as needed. Final approval for all sites rests with the Deposition Monitoring Task Group and the Task Force.

The effects of acid rain on forest productivity cannot be discussed without reference to site-specific factors such as soil type, nutrient status, etc., according to Dale W. Johnson of the Oak Ridge National Laboratory. Under certain conditions, acid rain can be shown to have a beneficial effect as well as an adverse one. Perhaps because of the negative implications of the term "acid rain", initial speculations about acid rain effects on forest productivity devoted little attention to the potential benefits of nitrate and sulfate deposition on forests deficient in S or N. In cases when nutrients are abundant and S and N are at deficiency levels, moderate amounts of acid rain would probably cause increased forest growth. However, acid rain in excessive amounts would probably reduce productivity as well as cause other damaging effects. In any case, it is clear that the costs and benefits of acid rain are highly dependent upon the quantities of nutrient capital and acid inputs as well as the susceptibility of individual members of the ecosystem. By far, the greatest amount of information on acid rain and forest productivity concerns these nutritional effects, according to Johnson.

George R. Hendrey of the Terrestrial and Aquatic Ecology Division, Brookhaven National Laboratory, opened his presentation by pointing out that acidification of surface waters has occurred in several areas of the world. Thousands of water bodies have been altered. In many individual cases the relationship to acid deposition is not conclusively proven so debate continues concerning the cause. He further indicated that it is difficult to impossible to prove to everyone's satisfaction that no factor other than acid deposition is responsible for acidification of a particular water body. A critical approach to assessing the impacts of acid deposition by evaluating each study individually can usually find one or more flaws in chemical analysis, or in the logic and assumptions of interpretations linking acid deposition to acidification at a particular site. The danger in such an approach, however, is that the overall body of evidence is not examined as a whole--the forest is not seen through all of the trees.

An alternative approach, according to Hendrey, is to evaluate cause-effect relationships of acidification on a regional basis. There is a striking coincidence between regions in which acidification is observed and annual average acidity of wet deposition expressed as pH of 4.7 or less (or wet sulfate deposition of 3 g m^{-2} yr^{-1} or more). Within acidified waters sulfate is the major anion and it dominates the chemistry of acidified waters. Finally, he stated, the historical record of acidification from fisheries, paleolimnology and aquatic chemistry points to a recent and simultaneous acidification of waters in geologically sensitive areas. While many hypotheses have been advanced to explain all of these phenomena in a piecemeal fashion, none explains them all as well as does "acid deposition".

Florida Overview - Effects

This session dealt largely with causes of acidic deposition in Florida as well as with such problems as keeping track of the rapidly expanding literature on the subject of acid deposition and with the ultimate economic and political questions regarding managing its impacts. However, Patrick L. Brezonik, now of the University of Minnesota but formerly Professor of Environmental Engineering Sciences at the University of Florida, presented research results indicating that nearly a third of Florida's 7800 lakes are potentially sensitive to acidification from atmospheric deposition, as based on a criterion of 100 µeq/L or less for alkalinity in sensitive lakes. A variety of trends can be seen in the chemistry and biological composition of softwater lakes as a function of pH, but determination of causes awaits controlled experiments. Observed trends include lower nutrient concentrations and lower population densities of phytoplankton and zooplankton in acidic lakes; whether the trends are merely correlative or are caused by pH variations is unknown. Acidic lakes also have higher concentrations of aluminum than non-acidic lakes, but concentrations are below levels generally regarded as toxic (>200 µg/L) even in the most acidic lakes.

Recent studies indicate, according to Brezonik, that sediment-water interactions are a key determinant affecting the way that seepage lakes respond to inputs of acidity. Sediments play an important (but still poorly quantified) role in neutralizing acid inputs; the likely mechanisms are cation exchange and biological sulfate reduction. Sediments also release aluminum (and possibly other toxic metals) to overlying water as the system becomes more acidic.

Terrestrial and Aquatic Effects of Acid Deposition in Florida

In his introduction, David R. Barker, the session chairman, referred to acid precipitation as a part of the general air pollution problem, but called attention to a present lack of evidence of degradative effects of acid rainfall in Florida as contrasted with such areas as northeastern United States. He noted that the most obvious and immediate impacts of acid precipitation have been on aquatic environments. He pointed out that scientists must approach the acid precipitation problem with proper concern for establishing credibility, based on the most careful gathering and interpretation of relevant information. Also, he pointed to the need for cooperation and coordination among representatives of government, industry and academia in identifying problems, designing studies, evaluating coverage, and ultimately in providing solutions to existing and potential problems. Barker commented on the problems of getting studies underway and, once underway, keeping them in operation; however, he felt that with the team available in Florida this should be no problem.

Curtis D. Pollman, the second speaker, in identifying gaps in our present knowledge of acid precipitation effects, pointed out that a serious problem exists in assessing potentially sensitive ecosystems because commonly used indices are not directly relevant to Florida. Pollman described their program of studies on lake and soil buffering capabilities as part of an integrated multifaceted effort covering ecological effects of acid deposition in Florida.

Certain fruit crops of more temperate areas show differential sensitivities to experimentally applied acid precipitation, based on studies discussed by Walter Kender. However, these effects were produced at pH's that are below present ambient levels. Also, despite some obvious tissue damage, no significant reduction in fruit yield was found. R. H. Biggs reiterated the lack of information on effects of acid precipitation on fruit crops such as citrus and addressed the need to look at the continuum in nature in designing studies and to look for areas of physiological sensitivity to assess such effects.

Henry L. Gholz pointed up the lack of evidence of detrimental effects of acid precipitation on forested ecosystems in Florida. He emphasized the need to study such problems at an ecosystem level, utilizing modeling techniques to evaluate possible interactions of stresses. R. S. Mansell also called for developing and improving models to study influences of acid rainfall on the leaching of soil cations. He pointed to the need for use of the most advanced technology and more complex models to assess various interactions.

B. G. Volk demonstrated the utility of an integrated approach in combined field and laboratory evaluations of ionic movements in soil solutions, to the water table, and down slope to an adjacent lake. He showed differential mobilization of various soil ions, including that of aluminum at pH's below 4.5. He concluded that there is a relatively rapid recovery in the soil from acid rain effects and stressed the need to characterize the types of rain events in evaluating possible results. Gabriel Bitton stressed the need for long-term studies and cognizance of differences in responses on different soil types in summarizing his data on acid effects on soil microbes. Some of his results to date showed that nitrogen fixation in one soil was reduced at a pH of 3.0 and that soil nitrification was reduced at pH's below 3.7.

D. E. Canfield cited as a significant problem in the assessment of potential detrimental effects the difficulty in defining management goals for lakes and presented his work in this context. He pointed out that while Florida lakes may respond chemically to acid precipitation as predicted by sensitivity indices, these criteria may not apply for the biota. He identified the problem of separating the results of acid stress from those of low nutrient levels. Thomas Crisman agreed and noted also the possible disparity in responses between clear and colored lakes. He urged that in assessing possible detrimental effects, a single component or species should not be evaluated without considering the impacts to the ecosystem of which it is a component. Also, he stressed that existing data show that local aquatic biotic responses to lake acidity are not identical to those of other areas. He cited data from _in situ_ experiments in which obvious deleterious effects were noted on aquatic organisms at pH's below 3.8.

The following is a series of general conclusions which appear supported by the presentations.

1. Applying sensitivity criteria for terrestrial or aquatic systems to Florida situations based on work done elsewhere should be approached with great caution.

2. There is at present a lack of evidence of detrimental effects of acid precipitation on either terrestrial or aquatic systems in Florida. This does not necessarily mean that such do not exist or might not come to pass, but simply that at present there is no definitive evidence of such.

3. It may be more difficult to assess potential detrimental effects in terrestrial systems than in aquatic systems. However, it is difficult in aquatic systems to separate possible influences of acidity, for example on the release of aluminum, from that of low nutrient levels or other collateral effects.

4. Experimental data suggest that obvious detrimental effects would be produced in aquatic systems at pH's slightly below 4.0 and in terrestrial systems at pH's below 3.0.

5. There is a serious need for taking systems approaches and using modeling in carrying out assessments of possible influences -- direct and secondary -- of acid precipitation on either terrestrial or aquatic systems.

6. There must be close cooperation among representatives of government, industry and academia to identify problems and needs in order to deal with existing and potential acid deposition effects. It would seem desirable to have an independent and dispassionate group of scientists, with the capability of evaluating such multifaceted studies, cooperate in research activities in this effort to avoid duplication, to avoid gaps in

coverage and to take advantage of opportunities to get integrated studies underway. Such an effort would obviously require multiple funding sources.

7. It is heartening to see that Florida is responding to this challenge before the results are disasterous and potentially irreparable.

In the general discussion following the rapporteur summaries, Patrick L. Brezonik made the very significant observations that aquatic systems in Florida may be feeling the stresses of increased acidification in ways that we presently cannot identify, and that the pH levels at which immediately obvious effects are produced may be well below levels at which more subtle and/or long term detrimental influences may be felt. Our present conclusion regarding the lack of evidence of detrimental effects of acid precipitation in Florida ecosystems should be considered in the context of the present limited body of information regarding such effects.

Note Added by Editors:

The paper presented in this volume for Johnson only includes the remarks he made on effects of acid deposition on forested ecosystems. The paper by Volk and Byers was expanded after the Workshop to include a review of the effects on agricultural crops. This modification was necessary to avoid publication delays relating to agency clearances. A presentation entitled "Effect of Acid Precipitation on Soil Microbial Processes" was given by G. Bitton, D. A. Graetz and B. G. Volk. This paper is not included in the Proceedings.

THE NATURAL SOURCES OF ACIDITY IN PRECIPITATION

William L. Chameides
School of Geophysical Sciences
Georgia Institute of Technology
Atlanta, Georgia 30332

Abstract

Observations have shown that precipitation with a pH below 5.6, the value attained by water in equilibrium with atmospheric CO_2, is a common occurrence throughout much of the remote atmosphere. The major acidic species have been found to be H_2SO_4, HNO_3, and HCOOH, with dissolved NH_3 playing a minor role as a neutralizing, basic species. These observations can be qualitatively explained in terms of the dissolution of acidic sulfate aerosols, nitric acid vapor, and gaseous ammonia as well as the dissolution and aqueous-phase oxidation of gaseous SO_2 and HCHO. The background levels of SO_2 and $SO_4^=$ aerosol arise primarily from the oxidation of naturally-produced reduced sulfur species, while that of HNO_3 arises from the oxidation of NO_x produced from lightning, soils, and ocean chemistry. Atmospheric HCHO can arise from the oxidation of CH_4 as well as a variety of natural non-methane hydrocarbon species. In both the global atmospheric SO_x and NO_x budgets, anthropogenic sources are beginning to overtake the natural source rates and as a result enhanced gaseous SO_x and NO_x levels, as well as highly acidic rainfall, are commonly found in regions affected by anthropogenic activities.

Introduction

Scientific interest in the chemical composition of rainwater may be documented back to the eighteenth century studies of Marggraf and Bergman, as noted in the review of Miller (1905). In fact, the term 'acid rain' (used to denote precipitation with a pH below 5.6, the value attained by water in equilibrium with atmospheric CO_2) was apparently first coined in 1858 by Smith who found evidence for increased acidity in rain in urban areas. While studies of rainwater composition have continued since this time, interest in studying the composition of rainwater has grown markedly in recent years as a result of numerous measurements documenting the occurrence of acid rain over large portions of the globe and the deleterious effects of this phenomenon upon terrestrial and aquatic ecosystems (cf., Eriksson, 1952a, b; Barret and Brodin, 1955; Junge, 1958; Jung and Gustafson, 1956; Cogbill and Likens, 1974; Wilson et al., 1980; Oden, 1976). It is now recognized that, for the most part, acid rain is caused by the presence of large quantities of dissolved H_2SO_4 and HNO_3 in precipitation (Granat, 1972). While the presence of high concentrations of H_2SO_4 and HNO_3 in rain is often the result of the oxidation and incorporation in rainwater of anthropogenically produced nitrogen oxides (NO_x) and sulfur oxides (SO_x) the observations of low-pH rain in remote locations (Sequira, 1981; Galloway et al., 1981) suggest that natural sources of these materials can add significantly to the acidity of rain in the remote atmosphere.

Clearly any program designed to alleviate the serious acid rain problem caused by anthropogenic SO_x and NO_x emissions will require an understanding of precipitation chemistry in regions relatively unaffected by anthropogenic emissions. In this work, the processes which control the acidity of rain and cloudwater in remote regions of the troposphere are examined.

Precipitation Chemistry in the Remote Troposphere

It has often been assumed that in areas removed from anthropogenic influences the pH of rain would be about 5.6, the value of water in equilibrium with atmospheric CO_2. Contrary to this assumption, however, recent observations have indicated that even in some of the most remote regions of the globe, the pH of rain is typically below 5.6 and often below 5.0 (c.f. Sequira, 1981; Galloway et al., 1982). The chemical analyses of Galloway et al. of rainwater samples collected on Amsterdam Island in the Indian Ocean is illustrative of this point. As summarized in Table 1, the average pH observed from these samples was about 4.9, in spite of the fact that the site is located thousands of kilometers from any urban source region. The acidity of this rain was found to arise primarily from three species: sulfuric acid (61%), nitric acid (12%), and formic and other organic acids (28%). As illustrated in the following section, these observations can be qualitatively explained in terms of the dissolution into cloud droplets of aerosol sulfate and nitric acid vapor as well as the dissolution and aqueous-phase oxidation of gaseous SO_2 and HCHO, all of these species being present at levels typically encountered in the background troposphere.

Table 1. Concentrations of Key Acidic and Basic Species Observed in Rainwater and Calculated in Cloudwater

Species	Volume-Weighted-Mean Concentrations in Rainwater Samples Collected on Amsterdam Island (Galloway et al., 1982)	Model-Calculated Cloudwater Concentrations for Remote Marine Troposphere	Model Calculated Cloudwater Concentrations with NO_x, HNO_3, and SO_2 levels increased by a factor of 10
$SO_4^{=*}$	4.4×10^{-6} M	9.9×10^{-6} M	42×10^{-6} M
NO_3^-	1.7×10^{-6} M	1.6×10^{-6} M	20×10^{-6} M
$(HCOO^- + RCOO^-)$	$\sim 4 \times 10^{-6}$ M	1.5×10^{-6} M	0.4×10^{-6} M
NH_4^+	2.1×10^{-6} M	6×10^{-6} M	5.9×10^{-6} M
H^+	12×10^{-6} M	17×10^{-6} M	100×10^{-6}
	(pH=4.92)	(pH=4.77)	(pH=4.0)

$SO_4^{=*}$ (=excess sulfate) is the concentration of sulfate in excess of that contributed by seasalt.

Model Calculations

In order to identify the primary chemical pathways by which acids are generated within cloudwater, a time-dependent model was developed which simulates many of the important characteristics of the coupled gas- and aqueous-phase, free radical chemistry of a stratoform cloud. Since the chemical composition of precipitation has been found to closely resemble that of the cloud from which it was formed (Lazrus et al., 1982), the model calculations should give a rough indication of the processes controlling the acidity of rainfall as well as cloudwater.

The gas-phase chemical mechanism included in the model is based on the well-known CH_4-CO-O_3-H_xO_y-NO_x-SO_x system (c.f. Chameides and Tan, 1980) with rate constants taken from the review of Demore et al. (1981). The aqueous-phase solubility constants, equilibria, and bi-molecular rate constants are largely based upon those previously reported by Chameides and Davis (1982; 1983).

The mass balance equation describing the temporal variations of a gas-phase species X with number density n is given by

$$\frac{\partial n(X)}{\partial t} = P_g(X) - D_g(X) - N\phi_a(X) \quad (1)$$

where P_g and D_g are the gas-phase photochemical production and destruction rates. The term on the right-hand side of (1) represents the rate of loss of species X via heterogeneous scavenging to a monodisperse distribution of N droplets of radius a, where ϕ_a is the rate of loss to a single cloud droplet. Following Fuchs and Sutugin (1972)

$$\phi_a(X) = 4/3\pi\ell \left(\frac{8KT}{m(X)\pi}\right)^{\frac{1}{2}} \frac{a}{1 + \left(0.7 + \frac{4(1-\alpha)}{3\alpha}\right)\frac{\ell}{a}} (n(X) - [X]/H(X)/R/T) \quad (2)$$

where ℓ is the mean free path, K is the universal gas constant, m(X) is the molecular weight of X, α is the accommodation coefficient assumed to be 0.01 for all soluble species (see Chameides and Davis (1982) for further discussion), [X] is the aqueous-phase concentration of species X in the cloud droplet in M(moles/liter), H(X) is the Henry's Law constant for X in M atm^{-1}, and $R=1.36 \times 10^{-22}$ atm-cm^3/molec/°K.

Assuming all drops to be homogeneously mixed, the mass balance equation for an aqueous species X is then given by

$$\frac{d[X]}{dt} = P_a(X) - D_a(X) + S_a(X) \quad (3)$$

where P_a and D_a are the aqueous-phase chemical-production and destruction rates of species X in a droplet of radius a and S is the source of X to the droplet via scavenging from the gas-phase. S is given by

$$S_a(x) = \frac{\phi_a(x)}{4/3\pi\, a^3\, 6.02 \times 10^{23}} \quad (4)$$

The initial conditions adopted in the calculation are outlined in Table 2. The initial gas-phase concentrations correspond to those that might be encountered in a remote region of the marine boundary layer. Superimposed upon this initial gaseous mixture is a monodisperse distribution of droplets having a radius of 12μ, a concentration of 70 cm^{-3}, and a liquid water content of 0.5 g/m^{-3}. The initial concentration of Cl^-, NH_4^+, $SO_4^{=*}$ (excess sulfate) and H^+ in the droplets were assumed to arise from the nucleation scavenging of a typical marine-boundary-layer aerosol composition. Adopting this initial composition the concentrations of all gas-phase and aqueous-phase species as a function of time (t>0) were obtained by integrating the coupled set of equations of the form of (1) and (3) using a standard Gear routine. The results of these calculations are discussed below.

Figure 1 illustrates the calculated aqueous-phase concentrations of the key acidic and basic species. As summarized in Table 1, the levels of $[H^+]$, $[SO_4^=]$, $[NO_3^-]$, $[HCOO^-]$, and $[NH_4^+]$ are qualitatively consistent with the levels observed in rainwater on Amsterdam Island. Similar to the rainwater data the calculated acidity (pH=4.77) was found to be generated by three acidic compounds (sulfuric, nitric, and formic acids) with NH_4^+ playing a minor role as a neutralizing basic species.

About 60% of model-calculated 1×10^{-5}M of $[SO_4^=]$ arose from the nucleation scavenging of acidic $SO_4^=$ aerosol. As noted by Charlson and Rodhe (1982), this process alone can lower the pH of precipitation significantly below 5.6; in this calculation the scavenging of acidic $SO_4^=$ aerosol leads to an initial pH of 5.05 (see Table 2). The remaining 40% of the $[SO_4^=]$ calculated in our model was generated from the dissolution and hydration of gaseous SO_2

(R1) $(SO_2)_g \rightleftharpoons (SO_2)_{aq}$

(R2) $(SO_2)_{aq} + H_2O \rightleftharpoons HSO_3^- + H^+$

(R3) $HSO_3^- \rightleftharpoons H^+ + SO_3^=$

followed by its oxidation in the aqueous phase. Of the myriad of processes capable of oxidizing SIV ($=(SO_2)_{aq} + HSO_3^- + SO_3^=$) to $SO_4^=$, two of the most important within clouds are

(R4) $(SO_2)_{aq} + H_2O_2 \longrightarrow SO_4^= + 2H^+$

(R5) $HSO_3^- + O_3 \longrightarrow SO_4^= + H^+ + O_2$

In both of these mechanisms the dissolved H_2O_2 and O_3 arise from the dissolution of the gaseous forms of these species.

In addition to the mechanisms described above, aqueous-phase, free radical species can play a significant role in the generation of $SO_4^=$. As recently suggested by Chameides and Davis (1982), the scavenging of the gaseous, free radical species OH and HO_2 by cloud droplets can constitute a large aqueous-phase, source of free radicals to cloud water. Once in the aqueous-phase these two radicals initiate a highly complex series of chemical reactions involving a large array of reactive radical and ionic species. The details of this chemistry have been presented by

Table 2. Initial Conditions Assumed for Remote, Marine, Stratoform Cloud

---

A. General Parameters

Solar zenith angle = 30°

T = 295°K

Relative Humidity = 100%

ε = cloud transmissivity = 0.5

α = accommodation coefficient = 0.01

N = concentration of droplets = 70 cm^{-3}

a = radius of drops = 12μ

W_L = liquid water content = 0.5 g/m^3

B. Gas-Phase Initial Concentrations*

O_3 = 3.10^{-8} v/v

CH_4 = 1.65×10^{-6} v/v

NO_x = $NO + NO_2$ = 2.0×10^{-11} v/v

HNO_3 = 2.0×10^{-11} v/v

H_2O_2 = 1.5×10^{-9} v/v

SO_2 = 5.5×10^{-11} v/v

NH_3 = 4×10^{-11} v/v

H_2CO = 3.8×10^{-10} v/v

CO_2 = 3.3×10^{-4} v/v

C. Aqueous-Phase Initial Conditions**

$[H^+]$ = $9 \times 10^{-6}M$

$[Cl^-]$ = $1 \times 10^{-4}M$

$[Na^+]$ = $1 \times 10^{-4}M$

$[SO_4^{=*}]$ = excess sulfate = $6 \times 10^{-6}M$

$[NH_4^+]$ = $3 \times 10^{-6}M$

---

* Initial gas-phase concentrations correspond to those that might typically be encountered in the remote, marine boundary layer under cloud-free conditions.

** Initial aqueous-phase conditions correspond to those that would result from nucleation scavenging of "seasalt" aerosols containing 1.8μg/m^3 Cl^- and 1.1μg/m^3 Na^+ with a scavenging efficiency of 1, and the nucleation scavenging of submicron aerosol particles containing 0.5μg/m^3 $SO_4^=$ and 0.045μg/m^3 NH_4^+ with a scavenging efficiency of 0.6.

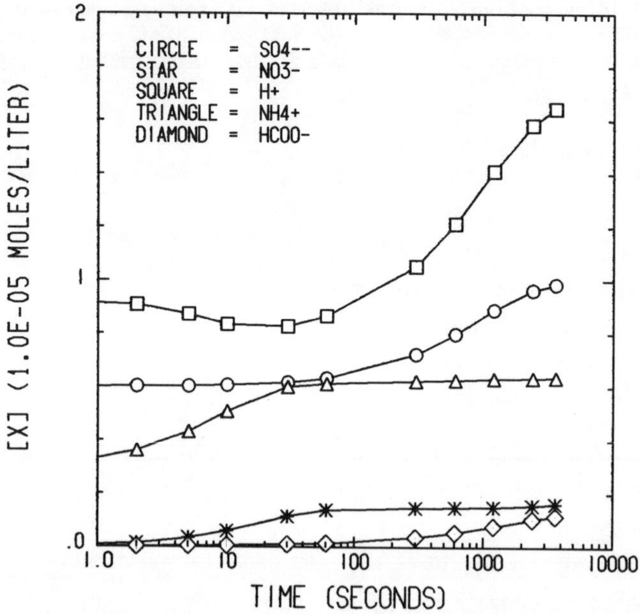

Fig. 1 The calculated temporal variation of the key acidic and basic species in remote marine cloudwater.

Chameides and Davis (1982); the salient points are illustrated in Figure 2. The OH and HO_2 radicals within cloud droplets can oxidize SIV via two mechanisms. The first mechanism involves the generation of aqueous-phase H_2O_2 from HO_2

(R6) $HO_2 + HO_2 \longrightarrow H_2O_2 + O_2$

followed by the oxidation of $(SO_2)_{aq}$ by this H_2O_2 via (R4). The other mechanism (R7) involves the reaction of OH with HSO_3^-

(R7) $OH + HSO_3^- \longrightarrow H_2O + SO_3^-$

followed by the rapid conversion of SO_3^- to $SO_4^=$.

The relative importance of each of these processes in converting SO_2 to $SO_4^=$ is illustrated in Figure 3. After one hour, virtually all the SO_2 has been converted to $SO_4^=$. Roughly 64% of the conversion was caused by the dissolution of H_2O_2 followed by (R4), 20% by OH via (R7), 7% by aqueous-phase generated H_2O_2 followed by (R4), and 8% by O_3 via (R3).

In contrast to $SO_4^=$ virtually all the 1.6×10^{-6}M $[NO_3^-]$ calculated in the cloudwater arose from the dissolution of gaseous HNO_3.

(R8) $(HNO_3)_g \rightleftharpoons H^+ + NO_3^-$

This process is quite rapid and virtually all the gas-phase HNO_3 is transferred to the droplets within the first minute of the integration (see Figure 1). Formic acid, on the other hand, is generated completely within the aqueous-phase by the dissolution and hydration of HCHO

$$\text{(R9)} \quad (HCHO)g + H_2O \rightleftharpoons ((CH_2(OH_2)_2)_{aq}$$

followed by the oxidation of the hydrated formaldehyde by aqueous-phase OH radicals

$$\text{(R10)} \quad CH_2(OH)_2 + OH \xrightarrow{O_2} HCOO^- + H_2O + H^+ + HO_2.$$

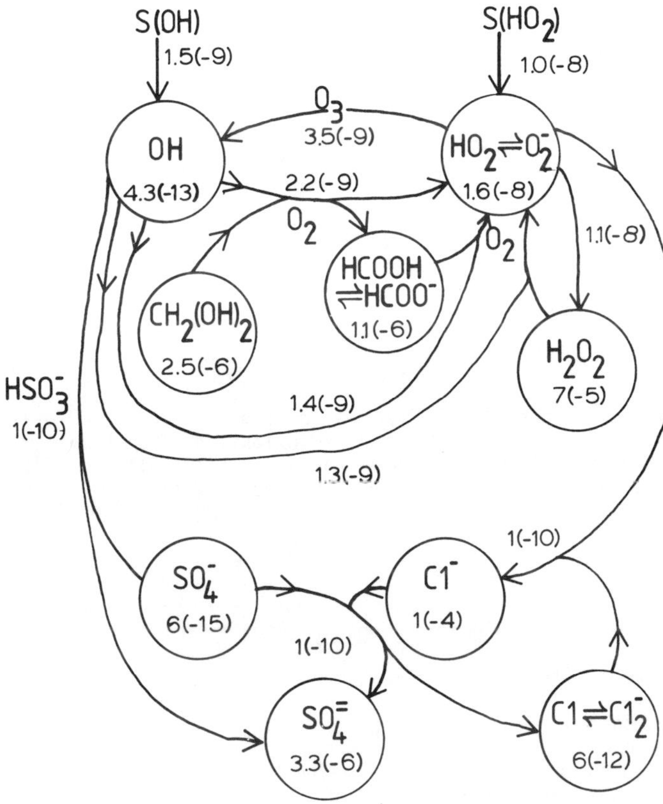

Fig. 2 The key reactive pathways for aqueous-phase OH and HO_2 radicals in remote marine cloudwater. S(OH) and S(HO_2) represent sources of OH and HO_2 to cloudwater from the gas phase. The numbers in the circles represent calculated concentrations in M and the numbers by the arrows represent calculated reaction rates in M s^{-1} at t = 15 minutes.

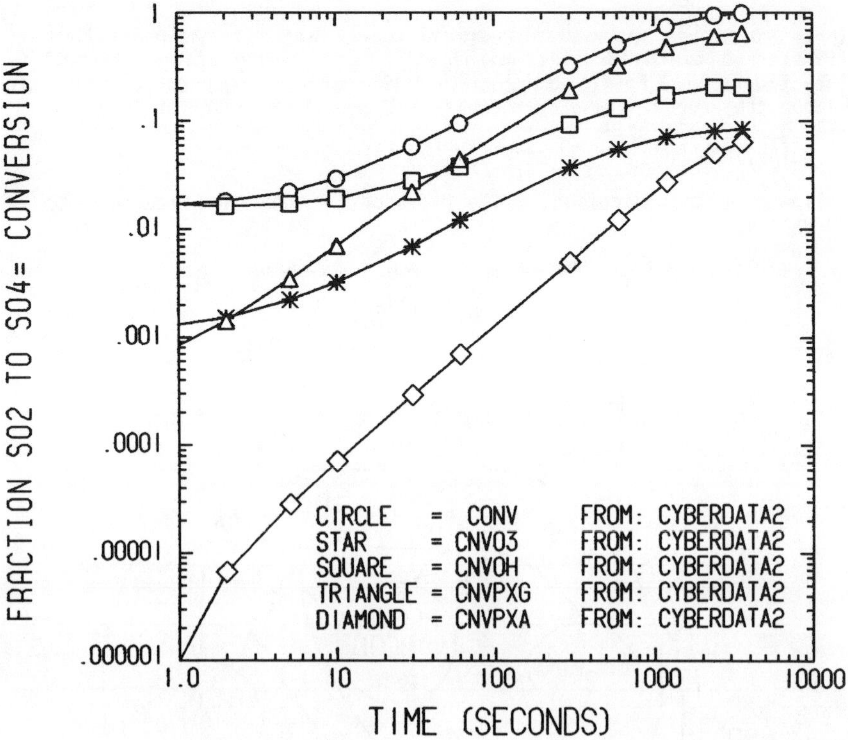

Fig. 3 Calculated percent SO_2 conversion to dissolved $SO_4^=$. Total conversion is given by circles; the oxidation of S IV by H_2O_2 from the gas-phase is given by triangles; the oxidation of S IV by dissolved OH is given by squares; the oxidation of S IV by dissolved ozone is given by stars; and the oxidation of S IV by H_2O_2 produced in the aqueous-phase is given by diamonds.

Global SO_x and NO_x Budgets

Consistent with recent rainwater analyses model calculations indicate that the pH of precipitation in the remote marine atmosphere can be expected to fall below 5.0. Apparently this acidity in precipitation can be supported by the background levels of aerosol $SO_4^=$, SO_2, and HNO_3 typically observed in the remote marine troposphere. Tables 3 and 4 summarize our current understanding of atmospheric SO_x and NO_x budgets. In the remote marine troposphere, the major natural source of SO_2 is the oxidation by OH of reduced sulfur species (principally CH_3SCH_3) emitted from the oceans; i.e.

$$\text{(R11)}\quad CH_3S\,CH_3 + OH \longrightarrow H_2O\,CH_2\,SCH_3 \xrightarrow[\text{STEPS}]{\text{MULTI}} SO_2$$

Table 3. Global SO_2 - $SO_4^{=*}$ Budget

Sources	Rate (10^{12}g yr^{-1})
Oxidation of reduced sulfur	
a. Emitted from soils (H_2S, CS_2, OCS, etc)	60
b. Emitted from ocean (CH_3SCH_3, etc)	30
Volcanoes	10
Anthropogenic emissions	65
TOTAL	165
Sinks	
Wet deposition as $SO_4^=$ produced from:	
a. Gas-phase homogeneous oxidation of SO_2	20
b. Heterogeneous (cloud) oxidation	80
Dry deposition	65
TOTAL	165

Table 4. Global NO_x - HNO_3 Budget

Sources	Rate (10^{12}g yr^{-1})
Lightning	10
Soil emission	10
Ocean emission	0.3
NH_3 oxidation	1
Stratospheric N_2O oxidation	0.5
Fossil fuel combustion	20
Biomass burning	10
TOTAL	50
Sinks	
Wet deposition of NO_3^- produced from	
a. Gas-phase homogeneous oxidation of NO_x	40
b. Heterogeneous oxidation	≤ 1
Dry deposition of HNO_3	10
TOTAL	50

To a first approximation it appears that this natural source is strong enough to maintain the background levels of SO_2 and $SO_4^=$ aerosol typically found in the remote marine troposphere (c.f. Andreae, 1982; Kritz, 1982). (Nevertheless, the possibility that some fraction of the SO_x burden in the remote marine atmosphere arises from the long-range transport of anthropogenic S cannot be completely ruled out.) In the case of the nitrogen oxides, the major natural source appears to be the fixation of N_2 by lightning, although the oxidation of NH_3 by OH may also make a significant contribution. NO_x produced from lightning is then converted to HNO_3 vapor by OH

$$\text{(R12)}\quad NO_2 + OH \xrightarrow{M} HNO_3$$

which can then be dissolved in cloud water to contribute to its free acidity.

Interestingly, it now appears probable that the aqueous-phase oxidation of aldehydes (e.g. HCHO) can also contribute to the acidity of rainwater by producing formic acid and other organic acids. Thus HCHO, like SO_x and NO_x, must be considered to be an acid precursor. In our model the initial level of 0.4 ppbv assumed for HCHO was generated from the oxidation of CH_4, i.e.

$$\text{(R13)}\quad CH_4 + OH \rightarrow CH_3 + H_2O \xrightarrow[\text{STEP}]{\text{MULTI}} HCHO$$

However, it is possible that the oxidation of other non-methane hydrocarbons (e.g. C_3H_6, C_3H_8) typically found in the remote, marine troposphere could lead to further enhancements in the HCHO level above 0.4 ppbv and thus higher levels of HCOOH in cloud water.

However, while natural SO_x and NO_x sources can generate significant levels of acidity in clouds and rainwater it does not mean that anthropogenic processes are not responsible for the very high levels of acidity in precipitation, with pH below 4.0, often found in regions influenced by anthropogenic emissions. Note in Tables 3 and 4, that in both the SO_x and NO_x budgets anthropogenic sources make up a significant fraction of the total source strength. The importance of these anthropogenic sources is reflected in the fact that background levels of SO_x and NO_x over the continental U.S. and Europe are typically an order of magnitude or more larger than the levels found in the remote, marine troposphere (c.f. Huebert and Lazrus, 1980; Maroulis et al., 1980). As indicated in Table 1, the effect of these higher SO_x and NO_x levels upon cloud (and rainwater) acidity can be quite substantial.

Conclusion

Model calculations appear to confirm recent observations indicating that natural levels of $SO_4^=$ aerosol, SO_2, and HNO_3 in the remote marine troposphere are sufficiently large to cause the pH of cloudwater (and rainwater) to fall below 5.0. These sulfur oxide and nitrogen oxide levels are probably maintained by the emissions of reduced sulfur species from the ocean and N_2 fixation in lightning. In regions affected by anthropogenic SO_x and NO_x emissions, however, the levels of sulfur and nitrogen oxides are typically enhanced by a factor of 10 or more and can thereby cause an order of magnitude increase in the acidity of precipitation.

Literature Cited

Andreae, M. O., Marine aerosol chemistry at Cape Grim, Tasmania, and Townsville, Queensland, J. Geophys. Res., 87, 8875-8886, 1982.

Barret, E., and G. Brodin, The acidity of Scandanavian precipitation, Tellus, 7, 251-257, 1955.

Chameides, W. L., and A. Tan, The two-dimensional diagnostic model for tropospheric OH: An uncertainty analysis, J. Geophys. Res., 86, 5209-5223, 1981.

Chameides, W. L., and D. D. Davis, The free radical chemistry of cloud droplets and its impact upon the composition of rain., J. Geophys. Res., 87, 4863-4877, 1982.

Chameides, W. L., and D. D. Davis, Cloud chemistry: An aqueous-phase source of formic acid, Nature, submitted, 1983.

Charlson, R. J., and H. Rodhe, Factors controlling the acidity of natural rainwater, Nature, 295, 683-685, 1982.

Cogbill, C. V., and G. E. Likens, Acid precipitation in the Northeastern United States, Water Resour. Res., 10, 1133-1137, 1974.

Demore, W. B., D. M. Golden, R. F. Hampson, M. Kurylo, C. J. Howard, M. J. Molina, A. R. Ravishankara, Chemical kinetics and photochemical data for use in stratospheric modelling, JPL Publication 82-57, 186 pp., 1982.

Eriksson, E., Composition of atmospheric precipitation I. Nitrogen compounds, Tellus, 4, 215-232, 1952.

Fuchs, N. A., and H. G. Sutugin, High-dispersed aerosols, (in Intern Reviews of Aerosol Physics and Chemistry, 2, edited by G. M. Hidy and J. R. Brock), Pergamon Press, N. Y., 1-60, 1971.

Galloway, J. N., G. E. Likens, W. C. Keene, and J. M. Miller, The composition of precipitation in remote areas of the world, J. Geophys. Res., 87, 8771-8786, 1982.

Granat, L., On the relation between pH and the chemical composition in atmospheric precipitation, Tellus, 24, 550-560, 1972.

Huebert, B. J., and A. L. Lazrus, Tropospheric gas-phase and particulate nitrate measurements, J. Geophys. Res., 85, 7322-7328, 1980.

Junge, C. E., and T. A. Ryan, Study of the SO_2 oxidation in solution and its role in atmospheric chemistry. Quart. J. Roy. Met. Soc., 84, 46-55, 1958.

Kritz, M. A., Exchange of sulfur between the free troposphere, marine boundary layer, and the sea surface, J. Geophys. Res., 87, 8795-8804, 1982.

Maroulis, P. J., A. L. Torres, A. B. Goldberg, and A. R. Bandy, Atmospheric SO_2 measurements on Project GAMETAG, J. Geophys. Res., 85, 7345-7349, 1981.

Oden, S., The acidity problem - An outline of concepts, Water, Air, Soil Pollution, 6, 137-166, 1976.

Wilson, J., Volker Mohnen, and J. Kadlecek, Wet Deposition in the Northeastern United States, ASRC Publication 796, 1980.

TRANSPORT OF POLLUTANTS ON THE SCALE OF HUNDREDS OF KILOMETERS

Carl W. Kreitzberg
Department of Physics and Atmospheric Science
Drexel University, Philadelphia, PA 19104

Abstract

The role of the large-scale wind field in determining gross trajectories and the role of turbulence in determining dilution are discussed briefly. Attention is focused on the role of intermediate-scale circulations in stretching pollutants into filaments that increase the rate of dilution and expand the region in which patches of high concentration may be encountered.

The nature of different scale storm systems as they influence transport is reviewed. Techniques of determining trajectories, computing concentrations and discriminating source regions are explained. Problems encountered in comparing model predictions with observations are discussed.

While much of the discussion deals with midlatitude conditions in general, a concluding section points out the very special nature of problems in Florida.

The Large-Scale Wind Field

The National Weather Service expends considerable resources to determine the large-scale wind field. A network of rawinsondes, ballonborne sensors tracked by radio, includes about 100 sites at 300 km spacing in the United States that make observations daily near sunrise and sunset (7am and 7pm EST). Satellite cloud observations are used to derive "cloud track winds" over oceans and satellite sounders measure the average temperature in deep layers from which general wind patterns can be deduced from approximate physical relations (1). Some commercial aircraft automatically report flight-level winds. Numerous land and ship surface weather reports can be used to infer information on the low-level wind and stability.

The National Meterological Center, NMC, of the National Weather Service, uses sophisticated statistical and physical relationships and the largest available computers to combine the above observations and numerical weather predictions to specify the large-scale wind field and the stability on a regular grid in six layers every six hours (2). Admittedly, these observations and gridded analyses are too coarse to depict deviations in these fields that have characteristic scales less than about 500 km (i.e., sinusoidal wave lengths of 2000 km). Moreover the depth of the surface mixing layer at mid-day and during the night is neither observed nor is it of substantial concern in the design of the dynamical models at NMC. Nevertheless, NMC gridded analyses are the best routinely available information on conditions off-shore and are the starting point of air pollution transport models that must deal with coastal regions.

The NMC gridded analyses will be substantially upgraded in the next year or two as modern computers come on-line at NMC. Since it costs on the order of $5,000 per day to perform such analyses, it is impractical to obtain high quality large-scale analyses in coastal regions for the five-year period one would want for air transport climatology studies using past data. The hope is that NMC will be charged with responsibility for producing such data sets during the mid-1980s as part of their overall mission. In the meantime, it is feasible to carry out case studies for tens of days using the high quality techniques that can soon be implemented routinely at NMC.

Large-Scale Trajectories

Given gridded large-scale analyses at six-hour intervals, large scale trajectories can be calculated in a straight-forward manner. Essentially, the displacement from minus to plus three hours of the analysis is the velocity times six hours. This trajectory represents the layer-mean transport and the layers are about 1 km deep in the lower atmosphere. Vertical displacement of large-scale trajectories derived from large-scale gridded analyses are poorly defined and this leads to errors in horizontal displacement when the winds change substantially with height. These errors accumulate with time and become important beyond 24 hours.

Limitations in accuracy of large-scale trajectories are likely not to be a major problem as long as it is clearly understood that they do not provide information on motions produced by smaller-scale circulations, such as are produced by heavy rain, terrain and coastal effects. What is somewhat unnerving is that effective means of quantifying the adequacy of large-scale trajectories for different applications is generally lacking. When the new generation of NMC analysis and prediction models come on-line, it will be important to establish some measure of trajectory accuracy.

Turbulent Mixing and Dilution

There are five mechanisms by which pollution concentration decrease during atmospheric transport; four are real and the fifth is the very important aspect of averaging. First, air can mix with less polluted air producing an intermediate concentration. Turbulence produces this mixing and dilution. Second, precipitation can remove pollution, particularly if the pollutant enters cloud droplets that are then collected by the much larger precipitating particles (3). Some cleansing occurs as precipitation falls through the air below cloud base. Third, pollutants can be removed by the ground through dry deposition. This mechanism is clear for particulates but more complex for gases. The fourth mechanism is chemical transformation, such that some pollutants increase as others decrease. Averaging over space or time will reduce peak concentrations by definition.

Turbulent mixing and dilution is very effective where substantial gradients of concentration exist on scales of hundreds of meters, as in the "well-mixed" surface layer that varies in depth with time of day.

The concept of plume concentration varying inversely with wind speed presumes that the plume does not merely get stretch out in high winds but that these high winds imply greater mechanical mixing so the plume cross-sectional area remains large. The concentration gradient can remain on a scale small enough for effective dilution through two mechanisms, wind shear and stretching.

If there is a significant change in the wind direction with height, then the plume fans out over a broad region thereby producing a large area that has substantial gradients. Similarly, plumes are systematically stretched by circulations on scales larger than those that are effective in producing dilution directly. This stretching produces filaments of high concentration that again have a large ratio of surface area to volume so that rapid dilution can continue.

Perhaps even more important is the contribution of this stretching and the attendant meandering of the filaments to the decrease of the average concentration as the averaging time and space increases. The more these plume elements meander, the more the average plume concentration falls below that of the concentration within the filaments.

Gaussian plume models are very simple to treat analytically and numerically. They are effective on two very different scales. Over short distances and travel times, tens of minutes and a few kilometers, turbulent dilution dominates and Gaussian-like plumes develop. Conventional and generally accepted Gaussian plume models do at least provide a standard reference for comparing concentrations to be expected in different climates out to 100 km. If substantial time, space or event (ensemble) averaging is performed, the average plume will be Gaussian-like for hours and hundreds of kilometers, depending on the extent of the averaging.

Beyond 100 km, wind direction changes must be considered and concentration patterns are usually determined from the composite of puffs released at frequent intervals. Techniques for treating the vertical shear are available but rates of lateral dispersion are not a function of wind field properties that relate to stretching, such as vorticity, because these properties are hard to estimate on the scale required.

The state-of-the-art in pollution transport and dilution modeling using large-scale wind data is reasonably advanced with three possible exceptions. The vertical resolution of current large-scale models is too coarse and the lack of mid-day and nighttime observations of surface mixing layer heights means that wind shear and mixing depth are poorly represented. Absence of observations off-shore above the surface make it very difficult to routinely deduce maritime boundary layer depth and wind shear conditions.

Mesoscale Circulations

Whereas large-scale circulations are reasonably well observed and dynamically predicted between observation times, mesoscale circulations are not treated routinely. Six aspects of mesoscale circulations will be discussed briefly: storm induced circulations, stratiform cloud impact on mixing depth, coastal circulations, terrain circulations, filament producing circulations and boundary layer rolls.

Condensation releases heat that produces circulations on scales very different than those that produce condensation initially. Showers are accompanied by vertical motions two orders of magnitude larger than large-scale vertical motions; both updrafts and downdrafts can occur. Downdrafts are produced by evaporative cooling and so are dependent on the dryness of the air around the shower and on the wind shear.

Oftentimes showers become organized into mesoscale convective disturbances ranging from rainbands and cumulus clusters to squall lines and mesoscale convective complexes. Certainly these circulations can not be ignored in treating problems such as acid rain.

Stratiform clouds and precipitation have significantly less impact on circulations so they do less to reduce the value of large-scale flow fields and trajectories. On the other hand, these cloud systems substantially alter daytime surface heating and low clouds alter nighttime surface cooling. Therefore, even stratiform clouds substantially reduce the value of mixing layer information that can be treated by routine large-scale analysis and predictions systems.

Coastal circulations are important when the land/sea temperature gradient that drives these circulations is large and when the large scale wind field is weak or parallel to the coast to provide time for the forces to generate the sea-breeze and showers on the sea-breeze front.

Terrain-induced circulations depend upon both the large-scale wind field and the terrain so they can be very difficult to treat. With light winds and strong radiation, circulations develop that can be modeled and understood and that have complexities limited to the complexity of the terrain. Simple air pollution models using conventional weather data do not provide reliable estimates of ground level concentration in irregular terrain. Dynamical models can be expected to find widespread application to this problem as computing costs plunge and as increases in skill can be demonstrated.

Boundary layer rolls and stretching will be considered together for two reasons. First, they are poorly observed, understood and modeled. Second, they significantly impact short-term concentrations but are likely to get averaged out when only long-term concentrations are of interest.

The rolls are very effective in circulating pollutants from ground level to the top of the mixed layer and in bringing plumes aloft down to the surface. Stretching is important to understanding how dilution can continue at a significant rate after plumes would otherwise be too large for turbulent mixing to be effective. A plume with a characteristic width (sigma) of several kilometers cannot be diluted efficiently by circulations on scales of less than one hundred meters.

High resolution sampling of plumes 100 km from their source will not reveal a structure that relates directly to the average concentration. With introduction of accurate, continuous wind profiling radars, the statistics will become available for scientists with the largest computers to develop a firm basis for relating concentrations in the filaments to averaged concentrations.

Comparison of Models With Observations

Validation of pollution transport and dispersion models is essential because of the variety of assumptions and approximations they contain that can be justified only if they can be verified. When chemical transformations add to the uncertainties, the validation problem beyond 100 km is nearly hopeless in many cases. It is really necessary to separate the chemical model validation from the transport and dispersion model validation.

Similarly, if the potential sources of the pollutant are not known reliably, it is nearly impossible to backtrack from receptor data and deduce contributions from sources beyond 100 km. An unknown nearby source can easily obscure the information on contributions from remote sources.

When precipitation contamination is being studied, the problem of sampling period can be acute. If the sampling period is determined by the start and stop of rainfall, rather than a fixed time period, it is referred to as event sampling. Event sampling isolates the period in which the precipitation occured to more easily identify the air that was involved. Whether the pollutant came from air in the boundary layer or from an airstream at a higher level is not obvious unless the potential sources of the pollutant are well identified beforehand. Short period rainfalls, such as from a single cluster or band of showers, make it that much easier to identify the important air. A rainfall sample collected from a series of showers or from an extended period of precipitation could have become polluted from air exposed to an extremely large variety of sources. The initial rainfall could have very different contamination levels than subsequent rainfall from the same storm; this would depend on details of the storm and the chemistry.

The problems discussed above have led to careful tracer experiments where there is a single source of known strength, reasonably simple meteorology without precipitation, no uncertainty in chemical transformations and quality-controlled receptor data. A variety of air pollution models were tested in this best of all worlds and results were still disappointing. Either much better specification of the meteorology is required or the models all require substantial improvement or else extensive averaging is required to obtain a well enough defined solution (less randomness) that the models have a chance of anticipating it.

This less than satisfactory state of pollution transport and dispersion modeling is attracting considerable attention. There are a variety of new techniques for remotely detecting winds and turbulence and special tracer studies with airborne and ground samplers are being conducted. Computing power per dollar is continuing to grow at an ever increasing rate. Progress can certainly be expected even if a simple solution never emerges. Since there are so many critical problems that require improvement in transport and dispersion modeling, the massive effort that is required will be justified as soon as it can be coordinated among the interested parties.

Considerations Unique to Florida

Florida has a far more tropical climate and far more exposure to maritime and coastal influences that any other part of the contiguous United States. The extent of land covered by water and the rate of evapotranspiration by the vegetation not only varies from season to season but also from year to year.

The wind and stability aloft that impact strongly on power plant plume dispersion are observed only at four locations in Florida and only at sunrise and sunset. This state is surrounded by data sparse waters over which few useful observations are made of wind and stability aloft and these fields can be expected to differ substantially from conditions produced over the land.

The lack of surrounding upper-air data means that Florida will obtain less information from the NMC large scale analyses than other states. Florida's level topography and strong coastal influences make this state ideally suited to mesoscale numerical modeling, and successful results have already been shown in case studies (4,5). It could be that Florida will find that its own state mesoscale model would be so useful that it could justify the expense of running it operationally before other states are ready to do so. This idea was impractical a few years ago but may become practical very soon as the cost of computing drops rapidly.

Literature Cited

1. Brandli, H.W., 1976: Satellite Meteorology (Wind, 105-121), Air Weather Service, AWS-TR-76-264, United States Air Force, Scott AFB, Ill. 62225.

2. Shuman, F., and Hovermale, J., 1968: An operational six layer primitive equation model, J. App. Meteor., 7:(525-547).

3. Fowler, D., 1980: Removal of sulphur and nitrogen compounds from the atmosphere in rain and by dry deposition (22-32), in Ecological Impact of Acid Precipitation, Proceedings of an international conference, Sandefjord, Norway.

4. Pielke, R.A., 1974: A three-dimensional numerical model of the sea breezes over South Florida, Mon. Wea. Rev., 102:(115-139).

5. Tripoli, G.J., and Cotton, W.R., 1980: A numerical investigation of several factors contributing to the observed variable intensity of deep convection over South Florida, J. Appl. Meteor., 19:(1037-1063).

THE NATIONAL TRENDS NETWORK: A UNIFIED EFFORT FOR MONITORING ACIDIC DEPOSITION

Jerre W. Wilson and John K. Robertson*
Science Research Laboratory
United States Military Academy
West Point, New York 10996

Abstract

The Deposition Monitoring Task Group of the Interagency Task Force on Acid Precipitation is charged with the responsibility of establishing a national network to monitor long-term trends in atmospheric deposition. This paper describes the design of the National Trends Network with particular emphasis on Florida and the Southeast.

Introduction

Interest in acidic deposition has spawned the development of a variety of monitoring networks. Wisniewski and Kinsman (1) recently listed over fifty networks in operation across the United States. While these organizations differ tremendously in size, expenditures, coverage, and protocol, they all share a common mission of monitoring ionic concentration in precipitation and acidic deposition over a specified area. The specific goal of each network will determine sampling protocol and data utilization. Research sites tend to operate on a daily or event schedule while general trends can be monitored utilizing a weekly or monthly collection routine.

Network Concept

The National Acid Precipitation Assessment Plan (2) provided a mandate for establishing an Interagency Task Force on Acid Precipitation. The task force is chaired jointly by the Department of Agriculture, Environmental Protection Agency, and the National Oceanic and Atmospheric Administration and is subdivided into ten task groups. The Deposition Monitoring Task Group is charged with the responsibility of developing a national network for measuring atmospheric deposition. The network to be developed by the task group is termed the National Trends Network (NTN) and is designed to consist of 150 sites and operate over a period of 10 to 20 years. Its purpose is to monitor long-term trends and detect chemical differences due to general source types. The sampling frequency will be weekly with the sites located relatively remote from identifiable point source influences.

It is recognized that existing networks have different siting criteria, different analytical laboratories, and different methods of analysis. However, the task group decided to attempt to integrate the sites

---

*The views expressed in this presentation are those of the authors and do not reflect the official policy or position of the Army, the Department of Defense, or the United States Government.

of the National Atmospheric Deposition Program (NADP), Multistate Atmospheric Power Production Pollution Study (MAP3S), Reference Climatological Stations (RCS), Tennessee Valley Authority (TVA), and the Utility Acid Precipitation Sampling Program (UAPSP) whenever possible. Maximum utilization of these sites will preserve historical data bases, stabilize existing sites, and provide experienced personnel. The UAPSP has subsequently decided not to participate and those sites initially chosen are being replaced. Negotiations are currently being conducted with the networks not operating on a weekly sampling protocol to delineate what additional equipment and manpower will be necessary for integration into NTN. Figure 1 shows the distribution of the sites considered.

Network Design

The NTN design developed by Robertson and Wilson (3) is based on an ecosystem concept designed to insure all regions of the United States are adequately covered. Bailey's ecoregions description (4) was chosen as the regional discriminator. It combines biotic and abiotic factors into one central classification system, and generalized properties over broad areas, thus producing a reasonable number of regions for consideration. Bailey's scheme divides the country into domains, divisions, provinces, and sections. Domains are based on broad climatic regimes, divisions are subdivisions of domains, and provinces delineate divisions into uniform regional climates and zonal soils. An example of this classification scheme is shown in Figure 2. The 2000-series represents the humid temperate domain, the 3000-series the dry domain, and the 4000-series shows the humid tropical domain. Within this broad definition, these domains are subdivided with most of the Southeast designated as a subtropical division. A further classification which encompasses most of Florida is the outer coastal plain forest province as shown in Figure 3. The southernmost portion of Florida is a humid tropical domain, a Savanna division, and an Everglades province.

Based on guidance from the Deposition Monitoring Task Group approximately 95 sites were selected on an ecosystem equal-area basis. Additional selection guidance raised the network total to 150 sites and will be discussed later. The area of each province was compared to the total area of the United States and the resulting percentage used to determine the number of stations each province should have for representative coverage. All existing sites of the NADP, TVA, MAP3S, RCS, and UAPSP networks were examined within each province and compared to the siting criteria established by the task group for local influences. The MAP3S Emission Inventory (5) was utilized to identify local point sources, compute distance and direction, and obtain estimates of the five criteria pollutants emitted by point sources within 50 km of the site. These considerations, in addition to maximization of spatial resolution, were utilized to select proposed existing sites for inclusion in the NTN. In the outer coastal plain forest province, Figure 3, the existing sites AR02, R17, U18, and FL03 were selected. Austin Cary, FL00, is currently an inactive NADP site.

As shown in Figure 1, most existing sites are located in the northeastern United States so that some ecoregions require new sites to be

EXISTING SITES CONSIDERED

Fig. 1 Existing sites considered for inclusion in the National Trends Network

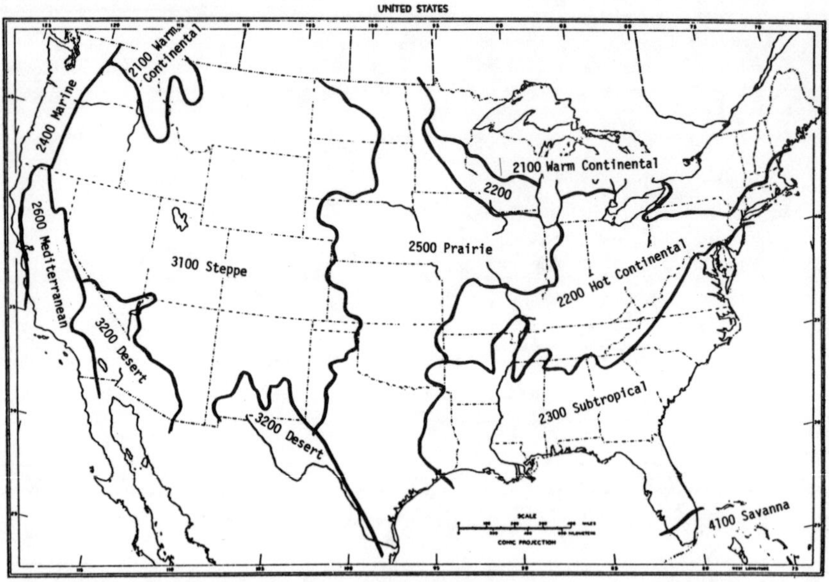

Fig. 2 Ecosystem divisional boundaries, after Bailey (4)

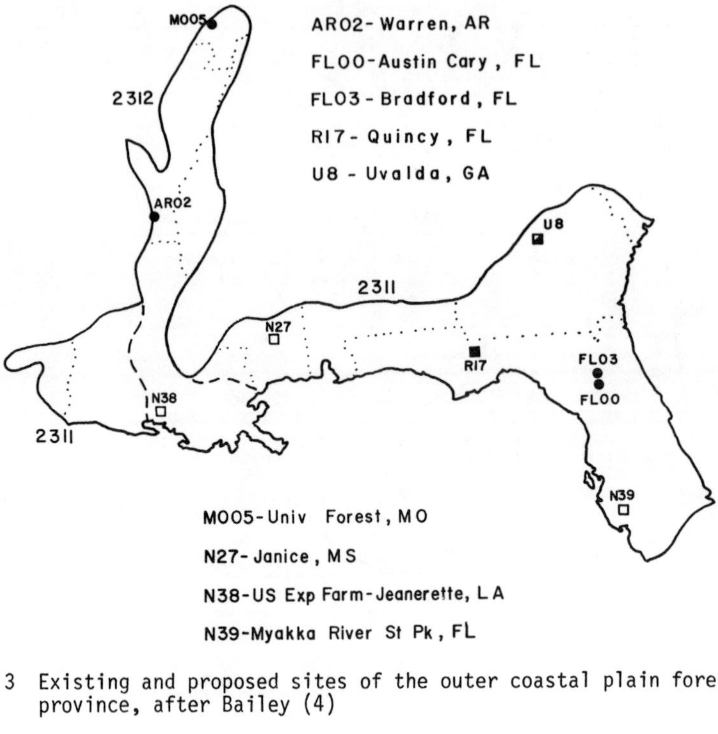

Fig. 3 Existing and proposed sites of the outer coastal plain forest province, after Bailey (4)

established. The siting criteria for local influences and emission inventory plots were used to identify acceptable areas where land use and ownership would be relatively stable for a number of years. Federal and state land were prime candidates in this respect. In Figure 3, N27 was identified as a potential new site for the equal-area basing concept. Figure 4 illustrates the 95 sites chosen for each of the ecosystem provinces. The open squares identify new sites to be established.

The task group further specified that approximately 25 sites should be selected utilizing existing deposition gradients, 15 using sensitivity analyses, and 15 supplementary sites selected. Deposition analyses of Semonin et al. (6) were used to identify areas not covered by ecosystem equal-area sites that needed additional sites to define strong gradients of sulfate, nitrate, hydrogen, and ammonium ion deposition. Omernik and Powers' analysis of the sensitivity of surface waters (7) was used to provide monitoring in sensitive areas not covered by ecosystem and gradient sites and 15 supplementary sites were added for continuity. These supplementary sites included the three joint US-Canadian monitoring sites in Canada, five urban sites, and those needed to provide uniform spatial coverage. The local criteria and emission inventory analysis was also applied to existing and proposed sites in these categories. Figure 5 shows the final proposed NTN design of all classes of sites. This network design provides good spatial resolution while placing the emphasis in sensitive areas and regions of high gradients in acidic deposition.

Summary

The National Trends Network will provide a unified effort in monitoring acidic deposition. The design is constructed so that adequate spatial coverage is achieved using an ecosystem-basing concept with additional emphasis placed in areas of high deposition gradients, sensitive areas, and special interest areas.

The implementation of the network is a dynamic process. Visitation of existing and proposed sites is currently being conducted by staff members of the Science Research Laboratory, U.S. Military Academy, with adjustments made to recommendations of the basic design as needed. Final approval for all sites rests with the Deposition Monitoring Task Group and the Task Force.

Acknowledgements

This work is sponsored under U.S. Geological Grant #4000-2519-82. A special thanks to Miss Sue Romano and Mrs. Shirley Bonsell for preparation of the manuscript.

ECOSYSTEM EQUAL-AREA SITES

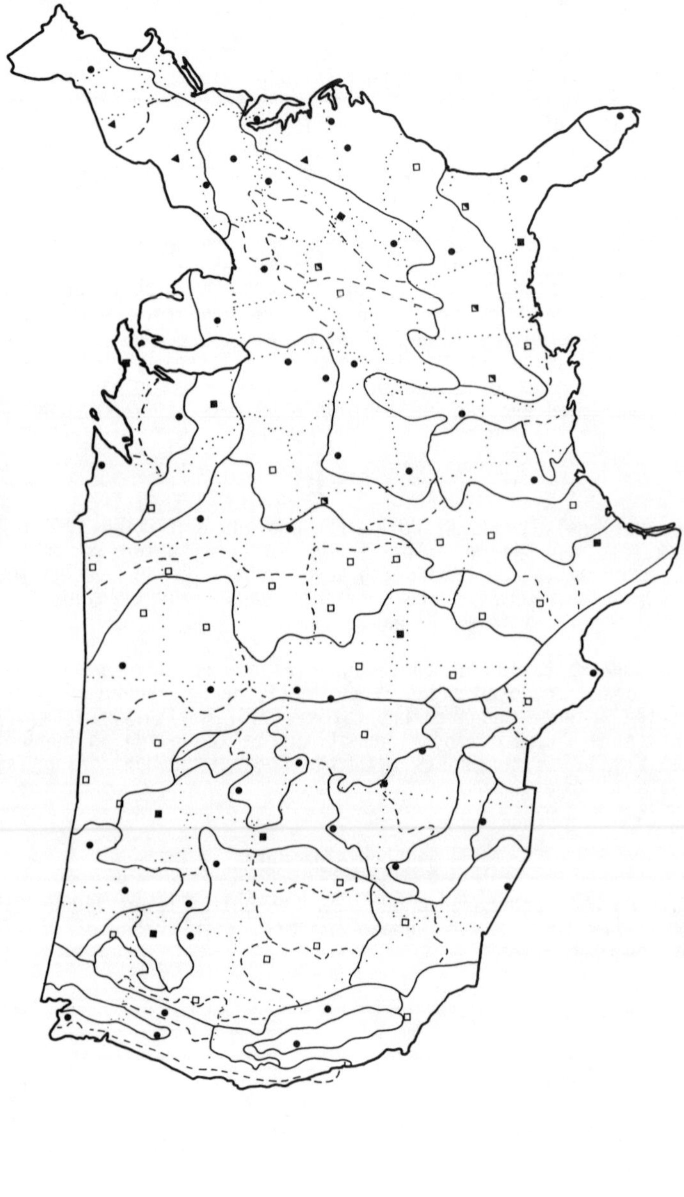

Fig. 4 Ecosystem equal-area sites proposed for the National Trends Network

PROPOSED NTN SITES

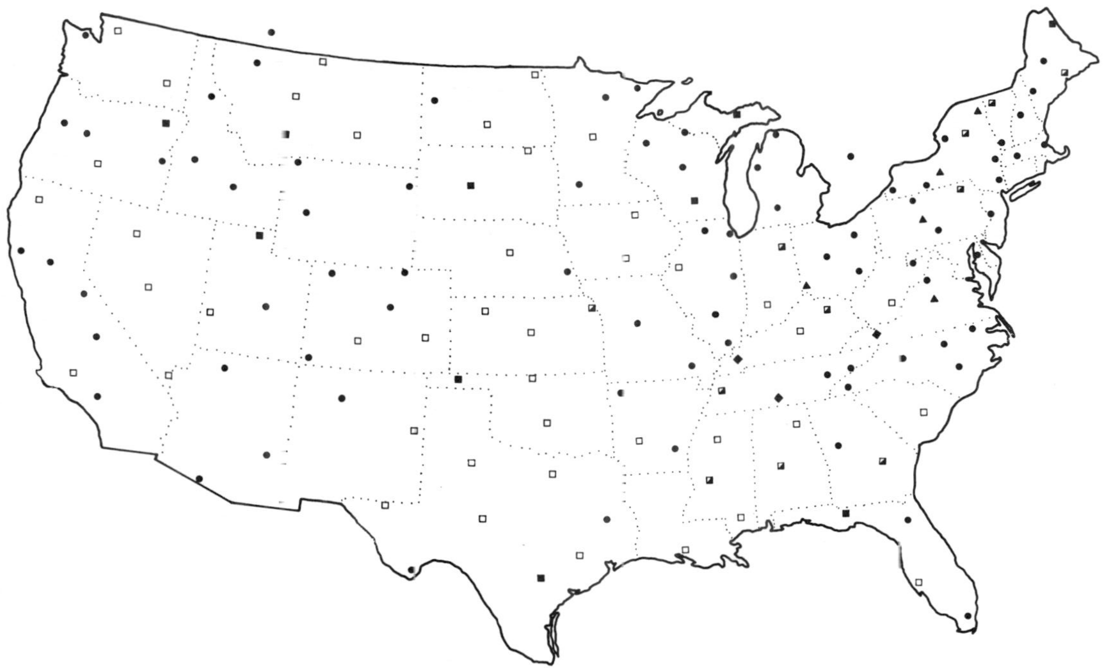

Fig. 5 Proposed ecosystem equal-area, deposition gradient, sensitivity, and supplementary sites of the National Trends Network

Literature Cited

1. Wisniewski, J., and Kinsman, J., 1982. An Overview of Acid Rain Monitoring Activities in North America. Bulletin of the American Meteorological Society, Vol. 63 (6):598-618.

2. The National Acid Precipitation Assessment Plan, Interagency Task Force on Acid Precipitation, Washington, D.C., June, 1982.

3. Robertson, J., and Wilson, J., 1983. The National Acid Precipitation Trend Network: A Design and an Evaluation of Current Precipitation Monitoring Sites. U.S. Military Academy, prepared for the U.S. Geological Survey.

4. Bailey, R.G., 1980. Description of the Ecoregions of the United States. USDA, Misc. Pub. No. 1391.

5. Benkovitz, C., 1980. The MAP3S Emission Inventory. Brookhaven National Laboratory.

6. Semonin, R., Bowersox, V., Gatz, D., Peden, M., and Stensland, G., 1981. Study of Atmospheric Pollution Scavenging. Illinois State Water Survey, Report No. COO-119-63.

7. Omernik, J., and Powers, C., 1982. Total Alkalinity of Surface Waters - A National Map. Environmental Protection Agency, Corvallis, OR, EPA-600/D-82-333.

ACID RAIN AND FOREST PRODUCTIVITY[1,2]

Dale W. Johnson
Environmental Sciences Division
Oak Ridge National Laboratory
Oak Ridge, Tennessee 37830

Abstract

Acid rain can cause increases or decreases in forest productivity depending on site nutrient status and the duration and rate of inputs. Some acid irrigation studies have shown short-term growth increases due to increased N availability, yet long-term growth reductions remain theoretically possible because of cation depletion and toxic aluminum accumulations in soils.

It cannot be overemphasized that the problem of assessing acid rain effects on forest productivity is one of quantification. Generalizations about acid rain effects without reference to the amount of acidic input and site nutrient conditions are hazardous.

Introduction

Only recently has acid rain been recognized as a potential cause of increases as well as decreases in forest productivity (Abrahamsen 1980b, Cowling and Dochinger 1980). Perhaps because of the negative implications of the term "acid rain" itself, initial speculations about acid rain effects on forest productivity devoted little or no attention to the potential benefits of concurrent sulfate and nitrate deposition on forests deficient in S or N (Johnson and Sundberg 1972). Potential forest growth reductions due to acid rain are very real, but so are potential growth increases. The net effect of acid rain on forest growth depends upon a number of site-specific factors such as soil type, nutrient status, and amount of atmospheric acid input. In cases where cation nutrients are abundant and sulfur or nitrogen are at deficiency levels, moderate inputs of acid rain will likely cause increased forest growth. At the other extreme (perhaps more familiar to readers of the acid rain literature), acid rain in sufficient amounts can reduce productivity in sites with adequate N and S but deficient in cation nutrients. In either case, however, the productivity costs and benefits of acid rain are highly dependent upon the quantities of nutrient capital and acid inputs involved.

---

[1] Research sponsored by contract RP-1813-1 with the Electric Power Research Institute for the Office of Health and Environmental Research, U.S. Department of Energy, under contract W-7405-eng-26 with Union Carbide Corporation. Publication No. 1717, Environmental Sciences Division, ORNL.

[2] Reprinted from the Proceedings of the XVII IUFRO World Congress, Japan 1981 with permission of the International Union of Forest Research Organizations.

It should be made clear at the outset that the terms "positive" and "negative" or "beneficial" and "adverse" in the following discussion refer to increases and decreases in productivity, respectively. These terms are consistent when applied to forests managed for wood products, but perhaps not when applied to forests managed as wilderness or recreation areas. The ultimate value judgment as to acid rain effects on a forest ecosystem must be made in the context of the desired use of that ecosystem.

The concepts of deficiency, sufficiency, and toxicity are familiar to students of both plant and animal nutrition, and they will provide the framework for the following discussion of acid rain and forest productivity. These concepts can be applied to H^+ as well because, although H^+ is not normally considered a nutrient, plants have optimal pH ranges for growth, and sufficient deviations from this optimum in either direction will cause growth reduction.

By far the greatest amount of information on acid rain and forest productivity concerns nutritional effects. Abrahamsen (1980b, p. 14) in his very thorough review of the topic, states that, "Apart from possible direct effects of acid precipitation on forest trees, the effects on forest growth can be considered as a nutrition problem". Nonetheless, both speculation and some empirical evidence indicate that acid rain can have "direct" or nonnutritional effects on forests, also (e.g., Tamm and Cowling 1977). Therefore, the following discussion is divided into two major sections, nutritional and nonnutritional effects. The latter section, although conceptually very broad, is much shorter than the former, reflecting a relative lack of information and perhaps a general research need.

Nutritional Effects

Acid rain, in the forms of H_2SO_4 or HNO_3, is only part of a broad spectrum of pollutant inputs which can affect forest nutrition either positively or negatively (Cowling and Dochinger 1980). For any given site, evaluating the effects of acid rain in isolation from the frequently associated inputs of NH_4^+, for instance (Abrahamsen 1980b), would not only be unwise but totally impractical. However, the kinds and amounts of associated pollutant inputs vary considerably, and adequate discussion of pollutant effects on forest nutrition in general is beyond the scope of this paper. Therefore, only a "pure" system of H_2SO_4 or HNO_3 input will be considered in the following discussion, with the full realization that these may be but a part of total nutrient deposition at many sites.

Effects of H^+ Inputs

Deficiencies of H^+ (i.e., excessively alkaline soils) are virtually unknown in forests, but estimations of excessive or toxic levels of H^+ input must be based upon the natural H^+ production within forest ecosystems. Within forest soils, H^+ is produced by carbonic and organic acids, either of which can play a major role in cation leaching under natural conditions (McColl and Cole 1968, Johnson et al. 1977). Organic acids also leach from the canopies of some unpolluted subalpine and northern forest ecosystems, substantially reducing the pH of "clean" rain (5.65) passing through the forest canopy (Table 1). Oxidation of S, N, and P following litter decomposition also produces H^+, and a very significant mechanism of H^+ production in many forests is H^+ release by plant roots while taking up cations (Sollins et al. 1980, Andersson et al. 1980).

Table 1. pH, SO_4^{2-}, HCO_3^-, Cl^-, color total cation, and cation-anion balance of precipitation and throughfall from Findley Lake, Washington, and Petersburg, Alaska (Johnson 1975)

	pH	HCO_3^-	SO_4^{2-}	Cl^-	Color[a] Units	Σ cations	Σ cations- Σ anions[b]
	μeq/L						
				Findley Lake, Washington			
Precipitation	5.6	66	35	<10	0	106	5
Throughfall	4.4	0	64	<10	44	116	52
				Petersburg, Alaska (Wadleigh Site)			
Precipitation	5.5	72	<10	25	0	56	-41
Throughfall	5.0	51	23	51	44	133	88

[a] Alpha-platinum standard

[b] Positive values imply the presence of unmeasured anion, probably organic (Johnson et al. 1977)

There are basically two ways to assess the roles of internal vs atmospheric H^+ inputs to forest ecosystems. One approach involves a detailed assessment of H^+ production and consumption within the ecosystem, and generally indicates that atmospheric inputs comprise relatively small contributions to the total H^+ budget (Ulrich 1980, Sollins et al. 1980, Andersson et al. 1980). The other involves an input-output assessment only, where the net effects of atmospheric vs internal H^+ inputs on cation nutrient export are considered (Ulrich 1980, Cronan et al. 1978, Cole and Johnson 1977).

While the latter approach is more integrative and direct, it does not distinguish between cation leaching and soil acidification. Although these processes are often referred to together, they differ in several important respects. Specifically, soils can be leached of cations without being acidified if the rate of weathering from primary minerals to exchange sites exceeds the leaching rate. On the other hand, soils can be acidified without being leached of cations if the anion of the acid is immobilized in the soil (Johnson and Cole 1977). Under these circumstances, both the incoming H^+ and associated anion (usually SO_4^{2-}) are retained in the soil, causing acidification but no net cation loss. An example of this situation is given in Table 2, where H_2SO_4 was applied to a forest soil which retained both SO_4^{2-} and H^+.

The effect of acid rain on soil cation leaching is highly dependent upon the mobility of the anion(s) associated with the acid (Johnson and Cole 1980). This is due to the requirement for charge balance in soil solution, a requirement that precludes the leaching of cations without associated mobile anions. In the case of SO_4^{2-}, adsorption to free ion and aluminum oxides can be a major factor restricting mobility in highly-weathered soils (Johnson 1980) whereas biological immobilization is more prevalent for NO_3^- (Vitousek et al. 1979). Soils low in free Fe and Al or high in organic matter (the latter of which appears to block sulfate adsorption sites; Johnson et al. 1979, 1980) are therefore generally susceptible to leaching by H_2SO_4 (e.g., Cronan et al. 1978), whereas ecosystems rich in N (i.e., where biological immobilization of NO_3^- may be minimal) are probably susceptible to leaching by HNO_3.

Table 2. Effects of H_2SO_4 irrigation on cation leaching and soil pH in a sulfate-adsorbing soil (from Johnson and Cole 1977)[a]

Horizon	Soil pH		Cation budget for 4 months following acid application	
	AT	CT	H^+	TC
			------(eq/ha)------	
Acid application	-	-	35,520	0
A (0-12 cm)	4.6	5.1	0.2	11,200
B (12-50 cm)	5.5	5.5	0.2	200

[a] AT = acid treated, CT = control, TC = $Ca^{2+} + Mg^{2+} + K^+ + Na^+$

When cation leaching is accelerated by acid inputs under controlled irrigation experiments, Ca^{2+} and Mg^{2+} appear to be the most affected and K^+ the least affected (Abrahamsen 1980a,b; Hovland et al. 1980; Ogner and Teigen, in press; Wood and Bormann 1976). To some extent, this may reflect the relative amounts of these cations on soil exchange sites, but the rate of increase in K^+ depletion seems to be consistently below that for Ca^{2+} or Mg^{2+} under acid irrigation as well (Abrahamsen 1980a; Ogner and Teigen, in press; Wood and Bormann 1976). This relative lack of response in K^+ may be due to the greater plant requirements for K^+ as opposed to those for Ca^{2+} or Mg^{2+} and possibly also due to fixation of K^+ in 2:1 clays.

The effect of acid rain on cation nutrient availability will depend on the net effects of leaching vs weathering (Abrahamsen 1980b, Ulrich et al. 1980). Weathering might be stimulated by additional H^+ input, offsetting the leaching of exchangeable cations to some extent. However, most acid irrigation studies (Abrahamsen 1980b) and one study under ambient acid input conditions (Ulrich et al. 1980) indicate a net decline in exchangeable cations with time.

Acid rain may indirectly affect the availability of N and P in forest soils, also. Under sufficiently acid conditions (generally pH < 4.0), soil Al is mobilized (Cronan and Schofield 1979, Ulrich et al. 1980), leading possibly to Al toxicity and immobilization of P (Abrahamsen 1980b). Tamm (1976) predicted short-term increases in N availability (due to decreased decomposition and microbiological N immobilization) and tree growth due to acid rain, but long-term declines in both N status and growth due to net N losses from the ecosystem. With regard to decomposition, empirical results have been variable. Some irrigation studies note decreased decomposition or CO_2 evolution with sulfuric acid inputs (Tamm et al. 1976, Bååth et al. 1980), whereas others show little or no effect (Abrahamsen 1980a, Roberts et al. 1980, Hovland et al. 1980). However, several independent studies show increased N availability at least during the initial stages of sulfuric acid input (Roberts et al. 1980, Abrahamsen 1980a, Ogner and Teigen, in press), and not surprisingly, this increased availability produces growth increases in situations where N is limiting (Tveite and Abrahamsen 1980a, b; Tamm and Wiklander 1980). Whether this increase in N availability is due to changes in microbial activity or to the acid-catalyzed hydrolysis of labile soil N is unknown as yet, however, and deserves further investigation. In either event, the results of the Norwegian studies in which both N availability and nitrate leaching were stimulated by H_2SO_4 inputs strongly suggest that, contrary to earlier predictions that nitrification would be inhibited by acid inputs (Tamm 1976), nitrification can be stimulated by acid inputs. The positive response of nitrifiers to increased ammonium supplies appears to outweigh the negative effects of H^+ inputs on nitrifier activity in such cases. This definitely has negative long-term implications for forest N and cation status if nitrate production exceeds plant uptake, resulting in net ecosystem N and cation loss.

Ulrich et al. (1980) forecast negative long-term effects of acid rain on the nutrition of a *Fagus sylvatica* stand at the Solling site due to soil acidification and toxic accumulations of soluble Al. Resampling soils from the site over a 13-year period showed as predicted by Tamm (1976), significant accumulations of N-poor organic matter in the forest floor (which are thought to promote internal H^+ production as well as to tie up N) and very large increases in soluble Al. As with all field studies of ambient acid rain effects, no control sites were available to show to what extent these changes would have occurred without the influence of acid rain. Accumulations of N-poor organic matter in the forest floor have been noted during stand development in relatively unpolluted Douglas-fir forests, for example (Turner 1975). Nonetheless, these findings deserve careful consideration because the only alternative to uncontrolled pre- and post-treatment assessments of this nature is to artificially add acid to the ecosystem, perhaps inducing effects that would not occur under ambient loadings (Bache 1980).

Johnson et al. (1981) present circumstantial evidence suggesting forest growth loss in the New Jersey pine barrens due to acid rain. That is, a decrease in the growth of pines (as indicated by tree ring analysis) beginning in the mid 1950's coincides closely with a decrease in stream pH. Because the soils are sandy, shallow, and very poorly buffered, the acid rain effect is quite possible although as yet unproven.

Effects of SO_4^{2-} Inputs

Deficiencies of S were indicated in forests remote from pollutant inputs in eastern Australia (Humphreys et al. 1975) and the northwestern United States (Youngberg and Dyrness 1966, Will and Youngberg 1978). Humphreys et al. (1975) strongly suggest that pollutant inputs from power plants are beneficial in S-deficient Australian forests, particularly when the soils have little SO_4^{2-} adsorption capacity. Fertilization with S in such cases is very inefficient because SO_4^{2-} rapidly leaches away, necessitating continual refertilization to maintain adequate S supply. In these situations, the continual input of moderate amounts of atmospheric sulfur may be an ideal, if inadvertant, source of fertilizer.

At the other extreme, continual atmospheric S inputs may help alleviate sub-optimal sulfate availability in sulfate "fixing" soils such as those of the tropics that are rich in hydrated Fe and Al oxides. Although adsorbed, water-insoluble sulfate is thought to be available to plants in the long run, the intensity or rate of supply to soil solution can be less than that required by plants, effecting an S limitation (Hasan et al. 1970).

It is also clear that fertilization with N, a common practice in many forested regions of the world, results in the rapid utilization of ecosystem S supplies, possibly leading to S limitations (Humphreys et al. 1975, Turner et al. 1980). In fact, it has been strongly suggested that forest N and S status must be evaluated together because of the closely related roles of these elements in the synthesis of protein (Kelly and Lambert 1972, Turner and Lambert 1980, Turner et al. 1980). In relatively unpolluted regions of the northwestern United States, there is evidence that lack of Douglas-fir growth response to N is due to marginal S status (Turner et al. 1977). Thus, it seems evident that moderate amounts of acid rain as H_2SO_4 could especially beneift forests undergoing fertilization with N. In the United States, this currently involves a total of about 10^6 ha of forest lands, primarily in the northwest and southeast (Bengston 1979).

On the other hand, excessive inputs of S lead to large accumulations of SO_4^{2-} in both plant tissue and soils (Hesse 1957, Leaf 1968, Turner et al. 1980, Johnson et al. 1979, Johnson and Henderson 1979). Sulfur supplied in excess of that which can be utilized for protein synthesis, i.e., that present in excess of a set S:N ratio of 0.03 on a gram-atom basis, accumulates as SO_4^{2-} in plant tissue (Leaf 1968, Kelly and Lambert 1972, Turner and Lambert 1980). In fact, foliar SO_4^{2-} content has proven to be an excellent diagnostic criterion for plant S status (Leaf 1968, Kelly and Lambert 1972, Turner and Lambert 1980) and as an indicator of S pollution (Bieberdorf et al. 1958). For *Pinus radiata*, foliar SO_4-S levels less that 80 ppm indicate S deficiency; levels in excess of 400 ppm indicate excess SO_4^{2-} and probable N deficiency (Kelly and Lambert 1972). Toxic levels of SO_4^{2-} are not well defined (Turner and Lambert 1980). Linzon et al. (1979) point out that acute effects of excess S, such as during an SO_2 fumigation, must be considered separately from chronic effects in which S builds up in plants to threshold levels for injury, and furthermore, that such threshold levels may vary with site conditions.

Hesse (1957) and Turner et al. (1980) found that excess S not only accumulates as SO_4^{2-} in plant tissues but also cycles as SO_4^{2-} within forest ecosystems. This is not to say that individual SO_4^{2-} ions necessarily remain intact throughout uptake, return, and release by decomposition, but, because large SO_4^{2-} components were found in every component and transfer process within the cycle, this possibility cannot be excluded. Why large amounts of SO_4 cycling occur under conditions of excess S input is not clear; there is no evidence to indicate that the cycling of SO_4^{2-} causes harm, and it may, in fact, help maintain charge balance in plant roots as they take up cations, precluding the necessity for H^+ release from roots to some extent.

By far the greatest sink for excess SO_4^{2-} in many forest ecosystems is the soil (Johnson et al. 1979, 1980). On Walker Branch Watershed, Tennessee, for instance, 48% of the total S input accumulates in the soil whereas only 13% accumulates in vegetation (Table 3; Shriner and Henderson 1978, Johnson and Henderson 1979). As discussed previously, sulfate accumulates in soils by inorganic adsorption to free Fe and Al oxides. The effects of this accumulation are beneficial in two ways: (1) prevention of cation leaching by H_2SO_4 by immobilizing the sulfate anion, and (2) creation of new cation exchange sites (Wiklander 1980, Rajan 1978).

Table 3. Content and annual accumulation of S in soils and vegetation of Walker Branch Watershed (from Shriner and Henderson 1978)

Component	Content ($kg \cdot ha^{-1}$)	Accumulation ($kg \cdot ha^{-1} \cdot year^{-1}$)
Input = 18 $kg \cdot ha^{-1} \cdot year^{-1}$		
Vegetation	73	+2.3
Soil	1300	+8.6

Effects of NO_3^- Inputs

Deficiencies of N are very common in forests throughout the world (Abrahamsen 1980b) and inputs of NO_3^- (as well NH_4^+ and other forms of N) are very likely to improve forest productivity in many cases. For instance, Wood and Bormann (1976) found that additions of artificial acid rain containing NO_3^- stimulated the growth and foliar N concentrations of white pine (*Pinus strobus*) seedlings in a greenhouse study. The authors assumed a fertilization effect by NO_3^- additions, but the results of Tveite and Abrahamsen (1980) and Ogner and Teigen (in press) suggest that accelerated N mineralization or reduced heterotrophic N uptake may have been a factor.

The benefits of NO_3^- deposition on forest N status may depend upon the species involved. Some tree species are better adapted to NH_4^+ than NO_3^- as a source of N, some species are better adapted to NO_3^-, and some appear to require both (Krajina et al. 1973, McFee and Stone 1968). It is also noteworthy that NO_3^- uptake does not require exudation of H^+ uptake from tree roots as NH_4^+ uptake does (Van den Driessche 1978). Thus it is conceivable that HNO_3 deposition could cause reductions in internal H^+ production.

Nearly all forest ecosystems for which nutrient budgets are available appear to accumulate NO_3^- as well as other forms of N (i.e., inputs > outputs; Abrahamsen 1980b). Because NO_3^- is very poorly adsorbed to most soils (Johnson and Cole 1980), this accumulation is probably due to biological uptake (although denitrification may be a factor). In some cases where N is particularly deficient, NO_3^- uptake may occur even within the forest canopy (Table 4 from Johnson 1975). The inhibiting effect of NO_3^- immobilization on the leaching potential of HNO_3 is the same as that of SO_4 immobilization on the leaching potential of H_2SO_4 even though the mechanisms of immobilization for those two anions are entirely different.

Table 4. pH, SO_4^{2-}, and NO_3^- of precipitation, throughfall, and soil solution following a rain event on June 25-July 3, 1973 (Johnson, unpubl. data), and average annual fluxes of SO_4 and NO_3^- for 1973-1975 (Johnson 1975) at the Thompson Site, Washington

Sampling level	pH	SO_4^{2-}	NO_3^-
		June 25-July 3, 1973	
	-----	μeq/L	-----
Precipitation	4.2	120	352
Throughfall	5.6	94	< 5
Soil solution (50 cm)	6.3	35	< 5
	Average annual fluxes ($eq \cdot ha^{-1} year^{-1}$)		
Precipitation		260	38
Throughfall		320	< 1
Soil solution (50 cm)		390	< 1

Because forest N requirements are relatively high compared to S, there are very few instances of excessive NO_3^- inputs except in fertilization and wastewater or sludge application studies. Even in these situations, however, it is often impossible to separate the effects of NO_3^- inputs from those of NH_4^+ or organic N inputs. In any case, a major negative impact of excessive N input (defined here as being in excess of the ability of biota to utilize it) is excessive nitrate and associated cation leaching (Overrein 1971, Breuer et al. 1979, Hook and Kardos 1978,

Cole and Scheiss 1980). Forest ecosystems with adequate or excessive amounts of N (e.g., N-fixing species) are probably very susceptible to HNO_3 leaching, as mentioned previously.

Biological utilization of atmospherically deposited NO_3^- may ultimately improve forest growth and N status to the point where other nutrients become limiting (possibly including S in some cases). While this can hardly be construed as a negative impact overall (since deficiencies in the most limiting nutrient, N, will have been overcome), it is a factor worthy of consideration, especially with respect to cation nutrients that may be concurrently depleted by acid rain-accelerated leaching. Indeed, there are indications that after an initial growth increase due to N, Mg became limiting and growth decreased following the cessation of H_2SO_4 irrigation in the Norway spruce experiment of the Norwegian SNSF Project (Tveite 1980, Abrahamsen 1980b). Further observations must be made before any final conclusions are drawn on the cause of the growth decline, however.

Excess N can accumulate in tree tissues (e.g., Turner 1977), but unlike S it apparently does so largely in organic forms (accumulation of NO_3^- and NH_4^+ in plants is usually less than 10-20% of total N; Epstein 1972).

Nonnutritional Effects

Tamm and Cowling (1977) list several possible effects acid rain might have on forest trees, including damage to cuticle structure, interference with guard cell functioning, direct poisoning of plant cells after acid diffusion through stomata or cuticle, decreased photosynthetic efficiency, alteration of leaf or root exudation processes, interference with reproductive processes, deleterious interactions with other pollutants and environmental factors, increased susceptibility to drought and other environmental stresses, and alteration of symbiotic and host-parasite relationships. Of this list of possible effects, even though it is not complete in itself, only a small number have been addressed in empirical studies.

Seedling germination and growth have been both stimulated and inhibited in a variety of studies involving artificial H_2SO_4 irrigation (Lee and Weber 1979, Abrahamsen et al. 1976, Matziris and Nakos 1977). Abrahamsen (1980b) reviews these studies and concludes that pH has a moderate effect on seedling germination and gives an overall optimum value of 4.7. He also reviews studies of seedling establishment and suggests that this process is more sensitive to pH effects than germination. The readers are referred to that paper for details which will not be repeated here.

Jacobsen (1980) reviews the state of knowledge regarding the phytotoxicity of acid precipitation, and some highlights are worth repeating here. it appears that foliar tissue damage is due mainly to free H^+ rather than to other ionic constituents of acid rain. In general, damage does not occur to any large extent until pH reaches values of 4.0 or below, although this will vary with species, stage of plant development, duration and frequency of exposure, and a number of other environmental factors (Wood and Bormann 1976, Evans et al. 1978, Jacobsen 1980). Deciduous leaves appear to be more sensitive to acid rain damage than coniferous leaves, and the most vulnerable stage of plant development appears to be just prior to maximum leaf enlargement (Jacobsen 1980, Evans et al. 1977). There is little empirical evidence that acid rain affects plant physiological processes, but the possibility remains real and studies are currently underway. Further research is also needed to determine the effects of leaf structure and wettability on susceptibility to acid rain damage (Jacobsen 1980).

Shriner and Cowling (1980) review the effects of acid rain on plant pathogens, and only a brief summary is presented here. Acid rain can have either positive or negative impacts upon plant diseases by affecting the microenvironment of infection courts on leaf surfaces. It can promote infection by eroding protective cuticular barriers, interfering with guard cell functioning, creating lesions through which pathogens can invade, and leaching spore-germination inhibitors from plant tissues. On the other hand, there is empirical evidence that sulfuric or nitric acid applications reduce the incidence of some diseases, apparently by creating unfavorably acidic conditions for the disease organisms themselves.

Summary

The effects of acid rain on forest productivity have been thought of and researched primarily in the sense of a nutritional problem, probably because of the initial concerns for Ca^{2+} depletion in Scandinavian forests (Engstrom et al. 1971). Results to data have clearly shown that nutritional effects can be positive (at least initially) by the addition of S and N to forests deficient in these nutrients and possibly also by the stimulation of a net increase in soil N mineralization. Although the fertilization effects of acid rain are totally inadvertent, they are no less real than very purposeful fertilization programs, and there is no scientifically justifiable reason for casting one in a negative light and one in a positive light. On the other hand, the possibilities that acid rain may cause forest growth reductions over time are also very real. Alleviation of N or S deficiencies will create greater demands for other nutrients. The cation leaching potential of acid rain may eventually be realized as soil SO_4^{2-} adsorption sites fill up and biological demand for NO_3^- is reduced, allowing greater mobility of these anions. The net result of these changes in nutrient flux and status could be the depletion of cations to deficiency levels, acidification of soils, concomitant solubilization of Al, and reduction in available P.

The accumulation and cycling of excess S is probably the least understood potential long-term effect of acid rain inputs. The plateau between S sufficiency and toxicity in forest ecosystems appears to be quite broad, since excess S often cycles without apparent harm as SO_4^{2-}. Inputs of S usually constitute a more significant increment to the natural sulfur flux within forest ecosystems than equivalent inputs of H^+ do to the natural flux of H^+. Therefore, it would appear that further emphasis ought to be given to the effects of the SO_4^{2-} component of acid rain. Similarly, further emphasis ought to be given to the effects of N inputs, because such inputs appear to be increasing (Abrahamsen 1980b) and N is the most commonly limiting nutrient in forest ecosystems.

Whether SO_4 accumulates to toxic levels or cations become limiting or some other factor such as host-parasite interactions or cell poisoning becomes critical first, excessive inputs of acid rain must ultimately lead to forest growth reduction. Surely there has never been any question of this; the question is one of quantification not only of the "toxicity" thresholds but also of the deficiency-sufficiency thresholds. The entire problem of acid rain and forest productivity was unknowingly summed up very nicely by the 16th-century Swiss physician Paracelus: "All substances are poisons; there is none which is not a poison; the right dose differentiates a poison from a remedy."

Literature Cited

Abrahamsen, G. 1980a: Impact of atmospheric sulfur deposition on forest ecosystems. pp. 397-416. In: Shriner, D. S., C. R. Richmond, and S. E. Lindberg (eds.), Atmospheric Sulfur: Environmental Impact and Health Effects. Ann Arbor Science, Ann Arbor, Michigan.

Abrahamsen, G. 1980b: Acid precipitation, plant nutrients, and forest growth. pp. 58-63. In: Drabløs, D., and A. Tollan (eds.), Ecological Impact of Acid Precipitation. The Norwegian Interdisciplinary Research Programme - Acid Precipitation - Effects on Forest and Fish, Oslo-As, Norway.

Andersson, F., Fagerstrom, T., and Nilsson, S. I. 1980: Forest ecosystem responses to acid deposition-hydrogen ion budget and nitrogen/tree growth model approaches. pp. 319-334. In: Hutchinson, T. C., and M. Havas (eds.), Effects of Acid Precipitation on Terrestrial Ecosystems. Plenum Press, New York.

Bååth, E., Berg, B., Lohm, U., Lundgren, B., Lundkvist, H., Rosswall, T., Soderstrom, B., and Wiren, A. 1980: Effects of experimental acidification and liming on soil organisms and decomposition in a Scots pine forest. Pedobiologia, vol. 20, pp. 85-100.

Bache, B. W. 1980: The acidification of soils. pp. 183-202. In: Hutchinson, T. C., and M. Havas (eds.), Effects of Acid Precipitation on Terrestrial Ecosystems. Plenum Press, New York.

Bengston, G. W. 1979: Forest fertilization in the United States: Progress and outlook. J. For., vol. 78, pp. 222-229.

Bieberdorf, F. W., Shrewsbury, C. L., McKee, H. C., and Krough, H. L. 1958: Vegetation as a measure indicator of air pollution. Part I. The pine (*Pinus taeda*). Bull. Torrey. Bot. Club, vol. 85, pp. 197-200.

Breuer, D. W., Cole, D. W., and Scheiss, P. 1979: Nitrogen transformation and leaching associated with wastewater irrigation in Douglas-fir, poplar, grass, and unvegetated systems. pp. 19-33. In: Sopper, W. E. (ed.), Municipal Wastewater and Sludge Recycling on Forest Land and Disturbed Land. The Pennsylvania State Univ. Press, University Park.

Cole, D. W. and Johnson, D. W. 1977: Atmospheric sulfate additions and cation leaching in a Douglas-fir ecosystem. Water Resour. Res., vol. 13, pp. 313-317.

Cole, D. W. and Scheiss, P. 1980: Renovation of wastewater and response of forest ecosystems: The Pack Forest Study. pp. 323-331. In: McKim, H. L. (ed.), The State of Knowledge of Land Treatment of Wastewater, Vol. I. U.S. Army Corps of Engineers, Cold Regions Research and Engineering Lab, Hanover, New Hampshire.

Cowling, E. B., and Dochinger, L. S. 1980: Effects of acidic precipitation on health and productivity of forests. pp. 165-173. In: Miller, P. R. (Technical Coordinator), Effects of Air Pollutants on Mediterranean and Temperate Forest Ecosystems. U.S. Forest Service Gen. Tech. Rep., PSW-43, Upper Darby, Pennsylvania.

Cronan, C. S., Reiners, W. A., Reynolds, R. L., and Lang, G. E. 1978: Forest floor leaching: Contributions from mineral, organic, and carbonic acids in New Hampshire subalpine forests. Science, vol. 200, pp. 309-311.

Cronan, C. S., and Schofield, C. L. 1979: Aluminum leaching response to acid precipitation: Effects on high elevation watersheds in the Northeast. Science, vol. 204, pp. 304-306.

Engstrom, A., Backstrand, G., and Stenram, H. 1971: Air pollution across national boundaries. The impact on the environment of sulfur in air and precipitation. Sweden's case study for the United Nations conference on the human environment. Royal Ministry for Foreign Affairs, Royal Ministry of Agriculture, Stockholm, Sweden.

Epstein, E. 1972: Mineral Nutrition of Plants. John Wiley and Sons, New York. pp. 288-290.

Evans, L. S., Gmur, N. F., and Kelsch, J. J. 1977: Perturbations of upper leaf surface structures by acid rain. Environ. Exp. Bot., vol. 17, pp. 145-149.

Evans, L. S., Gmur, N. F., and Da Costa, F. 1978: Foliar response of six clones of hybrid poplar. Phytopathology, vol 68, pp. 847-856.

Hasan, S. M., Fox, R. L., and Boyd, C. C. 1970: Solubility and availability of sorbed sulfate in Hawaiian soils. Soil Sci. Soc. Am. Proc., vol. 34, pp. 897-901.

Hesse, P. R. 1957: Sulphur and nitrogen changes in forest soils of East Africa. Plant Soil, vol. 19, pp. 86-96.

Hook, J. E., and Kardos, L. T. 1978: Nitrate leaching during long-term spray irrigation for treatment of secondary sewage effluent on woodland sites. J. Environ. Qual., vol 7, pp. 30-34.

Hovland, J., Abrahamsen, G., and Ogner, G. 1980: Effects of artificial acid rain on decomposition of spruce needles and mobilization and leaching of elements. Plant Soil, vol 56, pp. 365-378.

Humphreys, F. R., Lambert, M. J., and Kelly, J. 1975: The occurrence of sulphur deficiency in forests. pp. 154-162. In: McLachlan, K. D. (ed.), Sulphur in Australasian Agriculture. Sydney Univ. Press, Sydney, Australia.

Jacobsen, J. S. 1980: Experimental studies on the phytotoxicity of acidic precipitation: The United States experience. pp. 151-160. In: Hutchinson, T. C., and M. Havas (eds.), Effects of Acid Precipitation on Terrestrial Ecosystems. Plenum Press, New York.

Johnson, A. H., Siccama, T. G., Wang, D., Barringer, T. H., and Turner, R. S. 1981: Decreases in stream pH and tree growth rates in the New Jersey pinelands: A possible link to acid rain. Agronomy Abstr., 1980, p. 202. J. Environ. Qual. (in press).

Johnson, B., and Sundberg, R. 1972: Has the acidification by atmospheric pollution caused a growth reduction in Swedish forests? Department of Forest Yield Research, Royal College of Forestry, Stockholm, Sweden. Nr. 20.

Johnson, D. W. 1975: Processes of elemental leaching in some tropical, temperate, alpine and northern forest soils: Factors influencing the availability and mobility of major leaching agents. Ph.D. Thesis, University of Washington, Seattle, Washington. 165 pp.

Johnson, D. W. 1980: Site susceptibility to leaching by H_2SO_4 in acid rainfall. pp. 525-536. In: Hutchinson, T. C., and M. Havas (eds.), Effects of Acidic Precipitation on Terrestrial Ecosystems. Plenum Press, New York.

Johnson, D. W., and Cole, D. W. 1977: Sulfate mobility in an outwash soil in western Washington. Water Air Soil Pollut., vol. 7, pp. 489-495.

Johnson, D. W., and Cole, D. W. 1980: Anion mobility in soils: Relevance to nutrient transport from terrestrial ecosystems. Environ. Int., vol. 3, pp. 79-90.

Johnson, D. W., Cole, D. W., and Gessel, S. P. 1979: Acid precipitation and soil sulfate adsorption properties in a tropical and temperate forest soil. Biotropica, vol. 11, pp. 38-42.

Johnson, D. W., Cole, D. W., Gessel, S. P., Singer, M. J., and Minden, R. V. 1977: Carbonic acid leaching in a tropical, temperate, subalpine and northern forest soil. Arct. Alpine Res., vol. 9, pp. 329-343.

Johnson, D. W., and Henderson, G. S. 1979: Sulfate adsorption and sulfur fractions in a highly weathered soil under a mixed deciduous forest. Soil Sci., vol. 128, pp. 34-40.

Johnson, D. W., Hornbeck, J. W., Kelly, J. M., Shank, W. T., and Todd, D. E. 1980: Regional patterns of soil sulfate accumulation: Relevance to ecosystem sulfur budgets. pp. 507-520. In: Shriner, D. S., C. R. Richmond, and S. E. Lindberg (eds.), Atmospheric Sulfur Deposition: Environmental Impact and Health Effects. Ann Arbor Science, Ann Arbor, Michigan.

Kelly, J., and Lambert, M. J. 1972: The relationship between sulphur and nitrogen in the foliage of *Pinus radiata*. *Plant Soil*, vol. 37, pp. 395-408.

Krajina, V. J., Modoc-Jones, S., and Mellor, G. 1973: Ammonium and nitrate in the nitrogen economy of some conifers growing in Douglas-fir communities of the Pacific Northwest of America. *Soil Biol. Biochem.*, vol. 5, pp. 143-147.

Leaf, A. L. 1968: K. Mg, and S deficiencies in forest trees. pp. 88-102. In: Forest Fertilization: Theory and Practice. Tennessee Valley Authority, Muscle Shoals, Alabama.

Lee, J. J., and Weber, D. E. 1979: The effect of simulated acid rain on seedling emergence and growth of eleven woody species. *For. Sci.*, vol. 25, pp. 393-398.

Linzon, S. N., Temple, P. J., and Pearson, R. G. 1979: Sulfur concentrations in plant foliage and related effects. *J. Air Pollut. Control Assoc.*, vol. 29, pp. 520-525.

Matziris, D. I., and Nakos, G. 1977: Effect of simulated "acid rain" on juvenile characteristics of Aleppo pine (*Pinus halepensis* Mill.). *For. Ecol. Manage.*, vol. 1, pp. 267-272.

McColl, J. G., and Cole, D. W. 1968: A mechanism of cation transport in a forest soil. *Northwest Sci.*, vol. 42, pp. 134-140.

McFee, W. W. and Stone, E. L. 1968: Ammonia and nitrate as nitrogen sources for *Pinus radiata* and *Picea glauca*. *Soil Sci. Soc. Am. Proc.*, vol. 32, pp. 879-884.

Ogner, G., and Teigen, O. Effects of acid irrigation and liming on two clones of Norway spruce. *Plant Soil* (in press).

Overrein, L. N. 1971: Isotope studies on nitrogen in forest soil. I. Relative losses of nitrogen through leaching during a period of forty months. *Medd. Nor. Skogforsoeksves. Veg.*, vol. 29, pp. 261-280.

Rajan, S. S. S. 1978: Sulfate adsorbed on hydrous alumina, ligands displaced, and changes in surface charge. *Soil Sci. Soc. Am. J.*, vol. 42, pp. 39-44.

Roberts, T. M., Clarke, T. A., Ineson, D., and Gray, T. R. 1980: Effects of sulphur deposition on litter decomposition and nutrient leaching in coniferous forest soils. pp. 381-394. In: Hutchinson, T. C., and M. Havas (eds.), Effects of Acid Precipitation on Terrestrial Ecosystems. Plenum Press, New York.

Shriner, D. S., and Cowling, E. B. 1980: Effects of rainfall acidification on plant pathogens. pp. 435-442. In: Hutchinson, T. C., and M. Havas (eds.), Effects of Acid Precipitation on Terrestrial Ecosystems. Plenum Press, New York.

Shriner, D. S., and Henderson, G. S. 1978: Sulfur distribution and cycling in a deciduous forest watershed. J. Environ. Qual., vol 7, pp. 392-397.

Sollins, P., Grier, C. C., McCorison, F. M., Cromack, K., Jr., Fogel, R., and Fredriksen, R. L. 1980: The internal element cycles of an old-growth Douglas-fir ecosystem in western Oregon. Ecol. Monogr., vol. 50, pp. 261-285.

Tamm, C. O. 1976: Acid precipitation: Biological effects in soil and on forest vegetation. Ambio, vol. 5, pp. 235-238.

Tamm, C. O., and Cowling, E. B. 1977: Acidic precipitation and forest vegetation. Water Air Soil Pollut., vol. 7, pp. 503-511.

Tamm, C. O., and Wiklander, G. 1980: Effects of artificial acidification with sulfuric acid on tree growth in Scots pine forest. pp. 188-189. In: Drabløs, D. and A. Tollan (eds.), Ecological Impact of Acid Precipitation. The Norwegian Interdisciplinary Research Programme - Acid Precipitation - Effects on Forest and Fish. Oslo-As, Norway.

Tamm, C. O., Wiklander, G., and Popovic, B. 1976: Effects of application of sulfuric acid to poor pine forests. pp. 1011-1024. In: Proc., First International Symposium on Acid Precipitation and the Forest Ecosystem. U.S. Forest Service Gen. Tech. Rep. NE-23, Upper Darby, Pennsylvania.

Turner, J. 1975: Nutrient cycling in a Douglas-fir ecosystem with respect to age and nutrient status. Ph.D. Thesis. University of Washington, Seattle, Washington.

Turner, J. 1977: Effect of nitrogen availability on nitrogen cycling in a Douglas-fir stand. For. Sci., vol. 23, pp. 307-316.

Turner, J., Johnson, D. W., and Lambert, M. J. 1980: Sulphur cycling in a Douglas-fir forest and its modification by nitrogen applications. Oecol. Plant., vol. 15, pp. 27-35.

Turner, J., Lambert, M. J., and Gessel, S. P. 1977: Use of foliage sulphate concentration to predict response to area application by Douglas-fir. Can. J. For. Res., vol. 7, pp. 476-480.

Turner, J., and Lambert, M. J. 1980: Sulfur nutrition of forests. pp. 321-333. In: Shriner, D. S., C. R. Richmond, and S. E. Lindberg, (eds.), Atmospheric Sulfur Deposition: Environmental Impact and Health Effects. Ann Arbor Science, Ann Arbor, Michigan.

Tveite, B. 1980: Effects of acid precipitation on soil and forest. 9. Tree growth in field experiments. pp. 206-207. In: Drabløs, D., and A. Tollen (eds.), Ecological Impact of Acid Precipitation. The Norwegian Interdisciplinary Research Programme - Acid Precipitation - Effects on Forest and Fish. Oslo-As, Norway.

Tveite, B., and Abrahamsen, G. 1980: Effects of artificial acid rain on the growth and nutrient status of trees. pp. 305-318. In: Hutchinson, T. C., and M. Havas (eds.), Effects of Acid Precipitation on Terrestrial Ecosystems. Plenum Press, New York.

Ulrich, B. 1980: Production and consumption of hydrogen ions in the ecosphere. pp. 255-282. In: Hutchinson, T. C., and M. Havas (eds.), Effects of Acid Precipitation on Terrestrial Ecosystems. Plenum Press, New York.

Ulrich, B., Mayer, R., and Khanna, P. K. 1980: Chemical changes due to acid precipitation in a loess-derived soil in central Europe. Soil Sci., vol. 130, pp. 193-199.

Van den Driessche, R. 1978: Response of Douglas-fir seedlings to nitrate and ammonium nitrogen sources at different levels of pH and iron supply. Plant Soil, vol. 49, pp. 607-623.

Vitousek, P. M., Gosz, J. R., Grier, C. C., Melillo, J. M., Reiners, W. A., and Todd, R. L. 1979: Nitrate losses from disturbed ecosystems. Science, vol. 204, pp. 469-474.

Wiklander, L. 1980: Interaction between cations and anions influencing adsorption and leaching. pp. 239-254. In: Hutchinson, T. C., and M. Havas (eds.), Effects of Acid Precipitation on Terrestrial Ecosystems. Plenum Press, New York.

Will, G. M., and Youngberg, C. T. 1978: Sulfur status of some central Oregon pumice soils. Soil Sci. Soc. Am. J., vol. 42, pp. 132-134.

Wood, T., and Bormann, F. H. 1976: Short-term effects of a simulated acid rain upon the growth and nutrient relations of Pinus strobus, L. pp. 815-826. In: Proc., of the First International Symposium on Acid Precipitation and the Forest Ecosystem. U.S. Forest Service Gen. Tech. Rep. NE-23, Upper Darby, Pennsylvania.

Youngberg, C. T., and Dyrness, C. T. 1966: Biological assay of pumice soil fertility. Soil Sci. Soc. Am. Proc., vol. 29, pp. 182-187.

A REGIONAL PERSPECTIVE ON AQUATIC EFFECTS OF ACID DEPOSITION

George R. Hendrey
Terrestrial and Aquatic Ecology Division
Brookhaven National Laboratory
Upton, NY 11973

Abstract

Acidification of surface waters has occurred in several areas of the world. Thousands of water bodies have been altered. In many individual cases the relationship to acid deposition is not conclusively proven so debate continues concerning the cause. It is difficult, if not impossible, to prove to everyone's satisfaction that no factor other than acid deposition is responsible for acidification of a particular water body. A critical approach to assessing the impacts of acid deposition by evaluating each study individually can usually find one or more flaws in chemical analysis, or in the logic and assumptions of interpretations linking acid deposition to acidification at a particular site. The danger in such an approach, however, is that the overall body of evidence is not examined as a whole: the forest is not seen through all of the trees.

An alternative approach is to evaluate cause-effect relationships of acidification on a regional basis. There is a striking coincidence between regions in which acidification is observed and annual average acidity of wet deposition expressed as pH of 4.7 or less (or wet sulfate deposition of 2 g m^{-2} yr^{-1} or more). While wet deposition pH and wet sulfate deposition represent a large fraction of total acids or acid forming substances deposited ("acid deposition"), they appear to be reasonable surrogates for total deposition of acids and acid forming substances (SO_x, NO_x, NH_4). Within acidified waters sulfate is the major anion and it dominates the chemistry of acidified waters. In most cases no source for sulfate ions other than atmospheric deposition can be demonstrated. Nitrate correlates well with peak acidity in areas having persistent snow pack but it can be shown that episodes of high acidity would be much less severe if sulfate were absent.

Finally, the historical record of acidification from fisheries, paleolimnology and aquatic chemistry point to a recent and simultaneous acidification of waters in geologically sensitive areas. While many hypotheses have been advanced to explain all of these phenomena in a piecemeal fashion, none explain them all as well as does "acid deposition."

Introduction

An overview of acid deposition science, which is the topic of today's presentations, gives us the opportunity to look at all of the evidence concerning this man-made problem from a broad perspective. Instead of a microscopic approach to examining cause-effect relationships in which scientists invariably find more questions than answers, the problem can be examined in toto to see what sense can be made of all the information currently available. Because so much of what is known was determined in Scandinavia and Canada, it is essential to include studies from these areas in any reasonable overview. With respect to aquatic environments, a general pattern of cause-effect relationships has been available for several years (see Ambio, 1976*; Braekke, 1976). Stated briefly, lakes and streams in several areas of the world have lost fish populations as a consequence of acidification; this acidification has been caused by increased concentrations of $SO_4^=$, primarily, and of NO_3^- for episodic peaks in acidity; the sources of increased concentrations of these strong mineral-acid anions is atmospheric deposition of SO_x and NO_x. More recent studies have provided results which are consistent with this pattern.

In this presentation I will first discuss briefly overall biological changes associated with acidification, then review some of the evidence concerning the degree and extent of fisheries losses due to acidification of surface waters, and finally I will discuss some of the evidence which links acidic clear water lakes and streams to the deposition of SO_x and NO_x.

General Biological Changes

Effects of acidification of surface waters on biota other than fish have been reviewed elsewhere (Evans et al., 1981; Hendrey et al., 1980b). Biological differences associated with acidity among lakes, and changes in acidified lakes, are very marked as pH decreases below 6. In general, the number of species becomes progressively smaller as lake acidity increases. A few populations of plants and invertebrates may become very abundant. Litter decomposition appears to decrease while benthic primary production increases. Plant matter accumulates to an unusual degree, even though productivity per unit of biomass may be lower. Apparently, removal mechanisms such as invertebrate grazing and microbial activity are retarded.

Before fish are completely eliminated, the quality of fish food items (invertebrates) is altered and the total biomass may decrease. Following removal of fish predators, predator-prey relationships are drastically altered, with a few species of large, acid-tolerant invertebrates becoming the top carnivores. Several studies have found invertebrate shredders, collectors and scrapers to be reduced more than raptorial species. Factors governing community structure are poorly understood with respect to acidification. For example, the consequences of removing some species of predators because of their sensitivity to acidic conditions, particularly loss of fish species, may profoundly influence both trophic structure and nutrient dynamics.

The biotic variability among aquatic habitats is so great that it is probably not useful at this time to discuss the details of community structure in different locations. There is considerably more interest in fisheries than in other biota, however, so a brief review of that topic is in order.

---

* Information on acid deposition up to 1976 has been summarized in the series of articles contained in the journal Ambio, vol. 5, no. 5-6, 1976.

Acidification and Fisheries Losses

The first reports linking acid deposition to surface water acidification came from Scandinavia. In Norway, fish hatcheries in the southern part of the country experienced episodic fish kills in the early 1920s. The kills were associated with acidity peaks in the rivers from which the hatcheries drew their water and passing the water through a bed of crushed oyster shells was a common solution to the problem. Over the decades to the 1950s, Norwegian fishermen noted that more and more lakes became fishless. In 1959, Fisheries Inspector G. Dannevig noted that rain in southern Norway was extremely acidic and that the fishless lakes and streams were acidic. He put one and one together and came up with an explanation which has withstood scrutiny at every conceivable level.

Fisheries losses in North America, while less extensively documented, are parallel to losses in Scandinavia. The pattern and timing of losses indicate a common mechanism. Futhermore, lakes in which these losses have been documented have similar chemistries with respect to sulfate and aluminum concentrations. In lakes which formerly supported good fisheries, concentrations of aluminum now often exceed 0.2 mg ℓ^{-1}, which is sufficiently toxic to prevent successful reproduction for many fish species.

Fisheries Losses in Scandinavia

In Norway, over an area of 33,000 km^2 which receives strongly acidic atmospheric deposition, thousands of lakes have been acidified and fish populations have become virtually extinct in an area of 13,000 km^2. Atlantic salmon has been completely eliminated from seven rivers which were formerly important fisheries in southern Norway. This loss occurred despite attempts to manage the fishery. Similar rivers in Norway which do not lie in areas receiving very acidic precipitation have not suffered such losses (Leivestad et al., 1976). Historical records show that brown trout populations have been lost from over 2,500 lakes in this impacted area and remaining populations are declining. These losses occurred since 1940 and were particularly severe after 1960 (Sevaldrud et al., 1980; Muniz and Leivestad, 1980). Similar observations have been made in Sweden where acidification has altered approximately 18,000 lakes (Swedish Ministry of Agriculture, 1982).

Fisheries Losses in Canada

In Canada, fishery records are not as extensive as in Norway and assessment of fisheries impacts at a national level is not yet complete. A series of provincial studies, however, reveals significant fisheries losses. Examination of data on fisheries in 37 Nova Scotia rivers, dating as far back as the early 1900s, shows that at least 7 rivers have lost their Atlantic salmon, and runs have declined significantly in at least 8 other rivers. For 12 rivers with pH >5.0, angling success does not show a long-term trend, while angling success has decreased significantly in 10 rivers with pH <5.0 (U.S./Canada Work Group, 1981). In Ontario, 140 lakes have lost all fish population (OME, 1979), and an estimated 11,400 lakes (over 300,000 ha) are currently stressed by acidic deposition (Kelso and Minns, 1982). Approximately 40 native lake trout populations have become extinct in acidified lakes in the Sudbury area, including some of the largest acidified lakes known, e.g., Chiniguchi (1,511 ha), Matagamasi (1,343 ha), and Florence (1,007 ha) (Kelso and Gunn, in press).

Fisheries Losses in the United States

In the United States, documentation of fisheries losses is clearest for the Adirondack region of New York State. At least 180 lakes and ponds have lost their brook trout populations. Comparison of recent water chemistry data to historic measurements indicates that surface water acidification has occurred in the Adirondacks (Pfeiffer and Festa, 1980). Similar comparisons show historic acidification of surface waters in Maine, New Hampshire, New Jersey, and North Carolina (Davis et al., 1978; Johnson, 1979; Hendrey et al., 1980a).

Linking Acidification to Acid Deposition

While there is abundant evidence from several countries that acidification has occurred, in many of these individual cases the relationship to acid deposition is not conclusively proven so that debate still persists as to the causes. In fact, it is difficult, or even impossible, to prove that acid deposition, and not some other factor, is responsible for acidification of a particular water body. A critical approach to assessing the impacts of acid deposition by evaluating each separate study can usually find one or more flaws in analysis or in the assumptions or logic of interpretations linking acid deposition to acidification. The danger in such a critical approach, however, is that the overall body of evidence is not examined: the forest is not seen because of all the trees.

An alternative approach to evaluating acid deposition's cause-effect relationships is to examine the evidence on a regional basis. That is, one should look for an explanation of regional acidification in each of the three areas of the world discussed above. At the same time, one can see whether or not a common mechanism accounts for the simultaneity and similarity of acidification in these geographically large regions.

Regional Trends in Acidification

Acidification of freshwaters is defined as the reduction of alkalinity due to acid inputs to aquatic systems. Logically, an increase in hydrogen ion concentration of natural waters will occur only if the principal H^+ buffer, which is bicarbonate for most unacidified, clear surface waters, is decreased. Waters which have low alkalinities have little chemical buffering capacity and are thus sensitive to substantial inputs of acids. Alkalinity levels below 200 microequivalents per liter are considered indicative of moderate to high sensitivity to acidification, while values above 500 microequivalents per liter are considered to indicate non-sensitivity.

Determining the rates at which waters have been acidified may be accomplished by three means: by statistical inference from a set of sample stations representative of a larger population of stations sampled at two times separated by a sufficient interval; by trend analysis for specific sites; or by means of models describing some time-dependent variable associated with site specific or regional acidification. Because the modeling approach is not yet very well advanced, just the first two methods will be discussed.

Several studies have examined sets of data collected from lakes and/or streams at two different points in time. This is often accomplished by sampling a large population of waters in a short period, and the method is called synoptic sampling. For example, Hendrey et al. (1980a) sampled 38 streams of North Carolina and 68 streams and lakes in New Hampshire in the summer of 1979. The pH and alkalinity values of those samples were compared to pH and alkalinity values reported for samples collected in the summers of 1960-64 and 1937-39, respectively. This comparison found that in North Carolina the mean hydrogen ion concentration of the population of waters sampled increased by 82%, with 79% of the stations increasing and 21% decreasing. The population mean alkalinity value (conservatively corrected for earlier analytical methodological errors by subtracting 40 microequivalents from the alkalinity values) decreased by 31% with 71% of the stations decreasing and 19% increasing. The New Hampshire stations had a mean hydrogen ion increase for the population sampled of 247% with 90% of the stations increasing. Similar studies have been conducted elsewhere in the U.S. and Canada (U.S./Canada Work Group I, 1981) and in Scandinavia (Drabløs and Tollan, 1980; Swedish Ministry of Agriculture, 1982), demonstrating marked increases in acidity (or decreased alkalinity) for modern data compared to historical data.

In surveys such as these, one may not properly discuss the significance of change in any one particular lake or stream on the basis of a sampling which has only two data points. This is because the values of pH and alkalinity undergo large fluctuations seasonally, and in some cases in response to single storms (Hendrey et al., 1980a). When only two samples are available as end members over a long period of time, the data base is insufficient to conduct statistical analyses on a single sampling station. But, remembering that in a synoptic sampling program a large number of samples are taken from a population of waters, the change in values observed for that population can be described and applied to similar waters of the region investigated.

The other method of evaluating trends in acidification is by time series or trend analysis for each site. To identify which particular stations have been acidified requires a reasonably large number of samples for any station of interest. It is desirable to have enough data so that the short-term (daily to seasonal) variations can be statistically eliminated to identify the existence of long-term trends. This certainly will reduce the number of stations included in the analyses but one can be statistically certain of the trend analysis for the station of interest. This approach has been used for Nova Scotia rivers (Thompson et al., 1980) which had a five-fold increase in acidity between 1954 and 1974. In New Jersey, Johnson (1979) found significant long-term decreases in the pH of two streams.

In general, use of long-term records to evaluate acidification trends on a national basis rather than just for specific sites, has not yet progressed very far. This is due to a lack of sufficiently well collated long-term data records for many waters in acidified regions, and to concentration changes associated with stream discharge fluctuations which make stream data "noisy." Discharge compensation techniques are available and will eventually be applied to acidification studies. This type of analysis is in progress and may prove to be very useful for both site-specific and regional trend analyses.

Use of older data has been criticized by Kramer and Tessier (1982). They noted that possible errors in measurement could result from use of color indicators for alkalinity titration and for pH; use of soft glass or metal sampling vessels; improper sample storage, etc. It is true that such problems may exist in older data, but that does not prove the older data to be beyond correction. Kramer's critique points out both the possible uncertainty in use of such data and the need for other methods of evaluating acidification.

Linking Regional Acidification and Acid Deposition

Natural ecosystems are complex. The routes taken by rain water and dry deposition through a forest canopy, shrubs, litter layer, soil horizons, streams, and lakes are often tortuous. It seems unlikely that all of the conceivable interactions along these routes can ever be known. It is, however, not necessary to quantify them all in order to show that acidification is due to acid deposition. Several other types of evidence are available.

Whether regional acidification is shown by synoptic surveys in which old and new data are compared, or by analysis of more or less long-term continuous data for individual sites, results indicating widespread acidification of surface waters not associated with mine drainage are only reported for regions receiving highly acidic depositions. If analytical problems were the cause of the apparent trends, as suggested by Kramer and Tessier (1982), then such trends should be widely observed in low alkalinity waters not receiving acid deposition; they are not.

In southern Scandinavia, southeastern Canada, and the northeastern U.S., sulfate is the dominant anion in very many acidic clear water lakes including those in which evidence indicates acidification has taken place. These areas are underlain by granite or other silicious bedrock (Wright et al., 1980). Intensive studies in these areas including mass balance studies (based on the principle of conservation of mass) show that this sulfate does not appear to be associated with the watershed mineralogy, but is derived from atmospheric deposition (Drabløs and Tollan, 1980). Electroneutrality considerations demonstrate that if these sulfate concentrations were substantially reduced or absent, either the alkalinity would have to increase to balance the sum of cations, or cation concentration would have to decrease in proportion to the reduced sulfate, or both. Reduced cation input to aquatic systems might result from decreased weathering of base cations from the watershed or decreased input of H^+ associated with H_2SO_4. These considerations, based on first principles of chemistry, strongly support the idea that regional acidification is due to acid deposition.

Summary

Debates over cause and effect relationships concerning acid deposition and surface water acidification may never be resolved to everyone's satisfaction. This is especially so when the debates are focused on fine details of processes at specific sites. On the other hand, examining the evidence as a whole on a regional basis leads to some very convincing conclusions.

Surface water acidification in Europe and North America has been demonstrated by both biological and chemical changes over the past half century. The rate of change is not easy to specify just yet, but both regional and site specific approaches can be employed in assessing temporal aspects of the problem.

Regions shown to have severe acidification problems receive rain with average acidities of pH 4.7 or less, or sulfate deposition of 2 g m^{-2} yr^{-1} or more. Sulfate is the dominant anion in impacted waters. Regionally, this sulfate is derived from atmospheric deposition, rather than from local geologic or pedologic sources. If the sulfate concentrations in these waters were lower, either alkalinity must rise, or cations (H^+ and probably aluminum) must decrease, or both, to maintain electroneutrality. In any case, conditions for fish would certainly improve.

From the regional perspective, there can be little question as to the occurrence of freshwater acidification or its cause in the deposition of SO_x and NO_x. Further studies in this area should be aimed at improving our ability to predict rates of recovery from acidification following abatement of the causes of acid deposition.

Literature Cited

Braekke, F. H., Ed., 1976. Impact of Acid Precipitation of Forest and Freshwater Ecosystems in Norway, SNSF Project FR 6/76, Aas-NLH, Norway.

Davis, R. B., Smith, M. O., Bailey, J. H., and Norton, S. A., 1978. Acidification of Maine (U.S.A.) Lakes by Acidic Precipitation, Verh. Internat. Verein, Limnol. 10:532-537.

Drabløs, D., and Tollan, A., Eds., 1980. Ecological Impacts of Acid Precipitation, Proc. International Symposium, Sandefjord, Norway, March 11-14, 1980.

Evans, L. S., Hendrey, G. R., Stensland, G. J., Johnson, D. W., and Francis, A. J., 1981. Acidic precipitation: Considerations for an air quality standard, Water Air Soil Pollut. 16:469-509.

Hendrey, G. R., Galloway, J. N., Norton, S. A., Schofield, C. L., Shaffer, P. W., and Burns, D. A., 1980a. Geological and Hydrochemical Sensitivity of the Eastern United States to Acid Precipitation, EPA-600/3-80-024. 100 pp.

Hendrey, G. R., Yan, N. D., and Baumgartner, K. J., 1980b. Responses of freshwater plants and invertebrates to acidification, pp. 457-466 in Restoration of Lakes and Inland Waters, Proc. Internat. Symp. on Inland Waters and Lake Restoration, Portland, ME. U.S.E.P.A., 552 p.

Johnson, A. H., 1979. Evidence of acidification of headwater streams in the New Jersey Pinelands. Science 106:834-836.

Kelso, J. R. M. and Gunn, J. M. (in press). Additional information on the responses of fish communities to acidic waters in Ontario, in G. R. Hendrey (Ed.), Lake Acidification: Biotic Responses in the pH Range 5 to 6, Ann Arbor Science.

Kelso, J. R. M. and Minns, C. K., 1982. Current status of lake acidification and its effect on the fishery resources of Canada. pp. 69-81 in Acid Rain/Fisheries, American Fisheries Society, Bethesda, MD. 357 p.

Kramer, J. and Tessier, A., 1982. Acidification of aquatic ecosystems: A critique of chemical approaches, Environ. Sci. Technol. 16:606A-615A.

Leivestad, H., Hendrey, G. R., Muniz, I. P., and Snekvik, E., 1976. Effects of Acid Precipitation on Freshwater Organisms, pp. 87-111 in Impact of Acid Precipitation on Forest and Freshwater Ecosystems in Norway. F. H. Braekke, Ed., SNSF-Project FR6/76, Aas-NLH, Norway.

Muniz, I. P., and Leivestad, H., 1980. Acidification--Effects on Freshwater Fish. pp. 84-92 in Drabløs and Tollan (Eds.), Ecological Impact of Acid Precipitation, Proc. International Conf., Sandefjord, Norway, March 11-14, 1980.

OME, 1979. Acid Precipitation in Ontario. Ontario Ministry of the Environment. 30 pp.

Pfeiffer, M. and Festa, P., 1980. Acidity Status of Lakes in the Adirondack Region of New York in Relation to Fish Resources. FW-P168 (10/80). New York State Department of Environmental Conservation, Albany, NY.

Sevaldrud, I. H., Muniz, I. P., and Kalvenes, S., 1980. Loss of Fish Populations in Southern Norway. Dynamics and Magnitude of the Problem. pp. 350-351 in Drabløs and Tollan (Eds.), Ecological Impact of Acid Precipitation, Proc. International Conf., Sandefjord, Norway, March 11-14, 1980.

Swedish Ministry of Agriculture, 1982. Acidification Today and Tomorrow. Ministry of Agriculture, Environment '82 Committee, Stockholm. 231 pp.

Thompson, M. E., Elder, F. E., Davis, A. R., and Whitlow, S., 1980. Evidence of Acidification of Rivers in Eastern Canada, pp. 244-245 in D. Drabløs and A. Tollan (Eds.), Ecological Impact of Acid Precipitation. Proc. Internat. Symp., Sandefjord, Norway, March 11-14, 1980.

U.S./Canada Work Group I, 1981. United States-Canada Memorandum of Intent on Transboundary Air Pollution. Phase II. Interim Working Paper.

Wright, R. F., Conroy, N., Dickson, W. T., Harriman, R., Henriksen, A., and Schofield, C. L., 1980. Acidified Lake Districts of the World, pp. 377-379 *in* Drabløs and Tollan (Eds.), *Ecological Impact of Acid Precipitation*, Proc. International Symposium, Sandefjord, Norway, March 11-14, 1980. 383 pp.

REGULATORY ASPECTS OF ACID RAIN PROBLEMS

Victoria J. Tschinkel
Department of Environmental Regulation
State of Florida
Tallahassee, Florida 32301

Abstract

Because of interstate atmospheric transport of acid rain precursors, the problem is of national (and international) concern in scope. Recent debate has brought the question into sharp focus. The vehicle for debate has been amendments (e.g. Moynihan, Stafford, Mitchell bills) to the Clean Air Act that set targets for reductions in sulfur dioxide emissions. Acid rain considerations have become the key issue in reauthorization of the Clean Air Act.

There is no doubt that much of the USA (including Florida) is experiencing significant acid rain and that it is due to anthropogenic acid gas emissions. For at least the northeast USA and parts of Canada, acid rain is a serious problem and some controls should be implemented now. But, there are many unknowns about acid rain which make the job of regulation very difficult. There are additional factors relating to location, geology, emission characteristics and growth patterns that make Florida unique and add more complications. Unfortunately, energy conservation and alternative energy sources will not help in the near-term.

The existing regulatory structure established by the Clean Air Act is inadequate to deal with acid rain. Thus, amendments will be necessary. The current proposals in Congress to control acid rain are inequitable to Florida because of the unique features noted earlier. Because of all of the uncertainties on both sides of the issue, the best near-term (i.e. next 5-10 years) strategy for Florida is to cap SO_2 emissions at the 1982 statewide level (about 1.1 million tons) and aggressively complete current studies of acid deposition and its effects on terrestrial and aquatic systems.

Background

Within the past two years the national debate on acid rain has been brought into sharp focus by proposed amendments to the Clean Air Act to deal with the problem. Although there have been some individual state initiations (there is for example, a Minnesota Acid Rain Act) it is generally recognized that, because of interstate atmospheric transport of acid rain precursors, national legislation will be necessary.

Following the release of the National Research Council Report on acid rain in the summer of 1981 (Atmosphere-Biosphere Interactions: Towards an Understanding of the Consequences of Fossil Fuel Combustion), which boldly called for a 50% reduction in SO_2 emissions, Senator Mitchell of Maine introduced legislation (S.1706) calling for a 10 million ton annual reduction in SO_2 emissions over 1980 levels for 31 states: all states east of the Mississippi River and those bordering on the west. The act established an "Acid Rain Control Region" which required each state in the region to reduce its emissions levels. Florida's share of proposed reductions was about 340,000 tons. Reductions were to occur within 10 years of enactment. Interstate trading was allowed, as well as "swaps" of NO_x for SO_2 emissions (since acid rain in the eastern U.S. is one-third nitric acid, the Mitchell bill allowed for the reduction of 2 tons of NO_x emissions to count as a one ton SO_2 reduction).

Another acid rain bill introduced in the 97th Congress was the Moynihan bill which targeted only "major sources" (greater than 50,000 tons SO_2 emitted per year) and exempted sources meeting NSPS. However, this bill never progressed very far and the action has been the Mitchell bill. The Mitchell bill was modified in the Senate Committee on Environment and Public Works (Stafford) to become S.3041. It reduced the annual reduction target from 10 to 8 million tons and extended the deadline for compliance to 1995. It further required the governors in the 31 state region to negotiate shares of the total reduction. The negotiations were binding on all states if 75% of the governors agreed. Each state's share of the reductions was based on its share of the emissions above 1.5 lbs. SO_2 per million BTU heat output. S.3041 passed Stafford's Committee in July, 1982 by a vote of 15-2 and was included in a package of Clean Air Act amendments. But the 97th Congress concluded without reauthorizing the Clean Air Act.

The bill (S.3041) has been re-introduced in the 98th Congress by Stafford. Also, Mitchell has re-introduced his bill essentially as it was introduced in the 97th Congress. Other proposals calling for a moratorium on new emissions, liming of affected lakes, and accelerated research are being considered in the House.

The acid rain issue has thus become the principal issue in the debate over reauthorization of the Clean Air Act. The electric utility industry, through the Edison Electric Institute, have made it their first priority. Many New England states, and conservation groups allied in the National Clean Air Coalition, have lined up on the other side.

Nature of the Problem

There is no doubt that much of the USA (including Florida) is experiencing rainfall significantly more acidic than natural background conditions and that it is due to anthropogenic acid gas emissions. Some groups, such as EEI and the American Coal Association, seem to be still arguing that acid rain may be due to natural causes. However, the evidence seems inescapable that: (1) we are experiencing acid rain, which is far more acidic on the average than would be expected due to natural causes (it's probably true that "natural" rainfall acidity wasn't always 5.6; it probably fluctuated around this point due to natural inputs. But rainfall in Florida, for example, is now 10 times more acidic than this on an average annual basis.); and (2) anthropogenic acid gas emissions are the primary cause (as argued in the NRC Report, there is no other logical source for acid rain).

For at least the northeast U.S. and parts of Canada, acid rain has become a serious problem and something should be done now. Again, although many groups are still debating the issue, there does appear to be convincing evidence that some lakes in the northeast USA (e.g. lakes in the Adirondacks) and Canada have been acidified and sterilized by acid rain. Due to their low buffering capability, many more thousands of lakes in the northeast USA and Canada appear to be endangered. If this much damage were being caused by a known source (such as a sewage plant) steps would have already been taken to stop it. Unfortunately, what to do about acid rain is the dilemma.

Regulation Constraints

There are many unknowns about acid rain which make the job of regulation very difficult. Is long range transport the primary source of acid rain or are local sources equally or even more important? What is the role of NO_x in acid rain? Is it fruitless to control SO_2 without reductions in NO_x?

Some aquatic systems are quite acidic yet are healthy systems. For example, the Upper Suwannee River in Florida can have pH values below 3.5 as it drains from the Okefenokee Swamp but has a healthy fishery. The prevailing view for northern USA lakes is that when the pH goes below 5, fish cannot survive.

There are many more questions, but it is clear that we don't know nearly as much as we should about atmospheric transport, chemical transformations, ecological effects and other aspects of acid rain as we would like to. Regulatory agencies are used to dealing with incomplete data -- it's a fact of life. But the story seems especially incomplete with acid rain.

Regulatory Situation in Florida

There are additional factors unique to Florida which add even more complications to the regulatory picture. Florida is a peninsula, surrounded by water on three sides. With our current knowledge, there seems to be little likelihood that our emissions are affecting other states. Florida is fairly remote from the center of the acid rain controversy in New England and Canada. The statewide average rate of about 1.9 lbs SO_2 per million BTU is already quite low compared to other states (especially mid-western states like Illinois, Indiana, Ohio). Many of Florida's waters are already naturally quite acidic, yet healthy; the factors causing fish mortality in New England and Canadian lakes don't seem to hold for Florida. Alternately, many of the state's lakes and soils are very well buffered and don't appear to be susceptible to acid rain.

Florida is a high growth state. More generating capacity is inevitable. Controlling SO_2 and NO_x emissions is much tougher for Florida than for low or no growth states. The state is very dependent on imported oil (the most dependent of all states). We need to convert to coal which may mean even more SO_2 emissions without additional costly controls.

Energy Conservation and Alternative Energy Sources

Energy conservation and alternative energy sources won't help in the near term. Conservation brought about by high prices and public awareness has been a very significant factor in reducing Florida's demand for energy. Energy use across the board is below earlier projections. Each barrel of oil or ton of coal not burned means less emissions, fewer acid rain precursors and presumably, less acid rain. However, conservation also means forestalling the construction of new plants which must meet stringent New Source Performance Standards (NSPS). Thus, existing oil-fired plants, which emit two or three times more SO_2 per Mm BTU heat output, continue to be used as base-load units (as opposed to use as intermittant peaking units if new plants burning cheaper coal were base-loaded) which means more emissions. On balance then, conservation does not help, and may even exacerbate, the acid rain situation.

Alternative energy sources (especially solar thermal, photovoltaics, wind) look very promising in the long term -- that is, beyond the year 2000. The Energy Research Advisory Board Solar R & D Panel concluded that these sources will provide between 5-11% of national energy needs by 2000. But for the next 20 years, they probably won't help the acid rain problem. The "transition fuel" to the future will continue to be largely coal.

Existing Regulatory Structure

The existing regulatory structure established by the Clean Air Act (CAA) is inadequate to deal with acid rain and amendments will be necessary. It is generally recognized that the CAA is inadequate to deal with interstate air pollution problems. Two 1977 amendments to the act -- Sections 110(a)(2)(E) and 126 -- require that State Implementation Plans (SIP's) prohibit a major stationary source from preventing another state from attaining ambient standards, and allow states to petition EPA to find such sources in violation of that state's SIP. But EPA has never reviewed SIP's for this provision and petitions filed by a number of states (especially New England area states) have languished in EPA.

The CAA is primarily aimed at controlling specific sources and is not well suited to deal with cumulative or general effects. The Act also seems weak in dealing with international air pollution problems.

Another problem is that acid rain results from air pollution, but is of primary concern because of aquatic effects. Thus, emissions from one state may not violate another state's SIP, but still cause acid rain and aquatic damage in the receiving state. Finally, it even appears that the CAA may have worsened the acid rain problem by encouraging tall stacks to meet local ambient standards. Tall stacks inject pollutants higher in the atmosphere and seem to be causing greater long-range transport.

Clean Air Act Amendments and Florida

Current proposals in Congress to control acid rain (Mitchell and Stafford Bills) are inequitable for Florida and should be modified. The

Mitchell and Stafford bills especially hurt a high growth state like Florida. The state already has a very low average state emission rate compared to most states in the proposed acid rain control region.

Florida has very high utility rates compared to high emitting states. Even with full implementation of Mitchell or Stafford bills, which hit the high emitting states the hardest (e.g. Ohio, Illinois, Indiana, etc.), the high emitting states will still have lower average utility bills than Florida.

It's not at all clear that Florida is even contributing to the acid rain problems of other states -- our problems are most likely from our own plants and imports from the north. The Florida Acid Deposition Study of the Florida Electric Power Coordinating Group (FCG) is the most comprehensive state level effort underway in the USA. It will be completed in 1985. Thus, it's premature to move now on controls.

Another problem with these bills is their 10 year implementation time-frame. The CAA is on a 5 year reauthorization cycle. There is concern that we'll start down the road required by these bills, only to have a whole new Congress -- with a different political agenda -- substantially modify requirements.

Near-Term Florida Strategy

The best near-term strategy for Florida on acid rain is to cap SO_2 emissions at their 1982 levels and aggressively complete the FCG study. Because of all of the uncertainties on both sides of the issue, it appears best for the next 5-10 years to cap SO_2 emissions at the 1982 statewide level (about 1.1 million tons) while the FCG study is being completed.

The cap on SO_2 emissions is achievable. The bringing on line of new coal-fired plants meeting NSPS emission rates alone will almost achieve the cap (some oil-fired plants, now burning an average of 2.2% sulfur, would become peaking units). If there is a delay in bringing some of these new plants on line due to decreased demand, we can make up the difference by burning low sulfur fuels for existing plants. The Department has done some preliminary calculations on capping emissions and the results (Figures 1 and 2) are encouraging.

The costs of the cap seem very moderate. Since NSPS are required for all new units anyway, there is no net cost. Even if all existing units are required to switch to low sulfur fuels, the costs for the average residential customer (approximately 1000 kwh/month) are estimated to be 60¢/month. Even if we're off by 100%, the average increase would still be only about $1.00/month. This seems like a reasonable investment given the potentially very high environmental costs of acid rain.

Finally, we need to aggressively pursue the FCG study, and make sure the study is finished on time (1985). In addition, the study process should be opened up in its final phases to state agencies, university faculty and conservation groups, so the study findings have broad consensus if we have to implement some level of additional controls.

Table 1. Annual Statewide Analysis of Cost of 1% Oil and Coal*

Alternative and Cost	Year		
	1983	1986	1990
Oil $ (millions)	$32.74	$14.73	$22.65
Cost/1000kwh per customer per month	$ 0.29	$ 0.18	$ 0.24
Coal $ (millions)	$34.18	$34.18	$34.18
Cost/1000kwh per customer per month	$ 0.31	$ 0.28	$ 0.10
Total $ (millions)	$66.92	$48.91	$56.83
Cost/1000kwh per customer per month	$ 0.60	$ 0.46	$ 0.34

* Current Spot Market Differential Between 1% and 2.5% Sulfur Oil at $1.03 per Barrel

Table 2. Annual Statewide Analysis of Cost of 1% Oil and Coal*

Alternative and Cost	Year		
	1983	1986	1990
Oil $ (millions)	49.11	22.10	33.98
$/1000kwh	.44	.27	.36
Coal $ (millions)	34.18	34.18	34.18
$/1000kwh	.31	.28	.10
Total $ (millions)	83.31	56.28	68.16
$/1000kwh	.75	.35	.46

* Most Representative Historic Differential Between 1% and 2.5% Sulfur Oil at $1.50 per Barrel. No Change in Coal Differential.

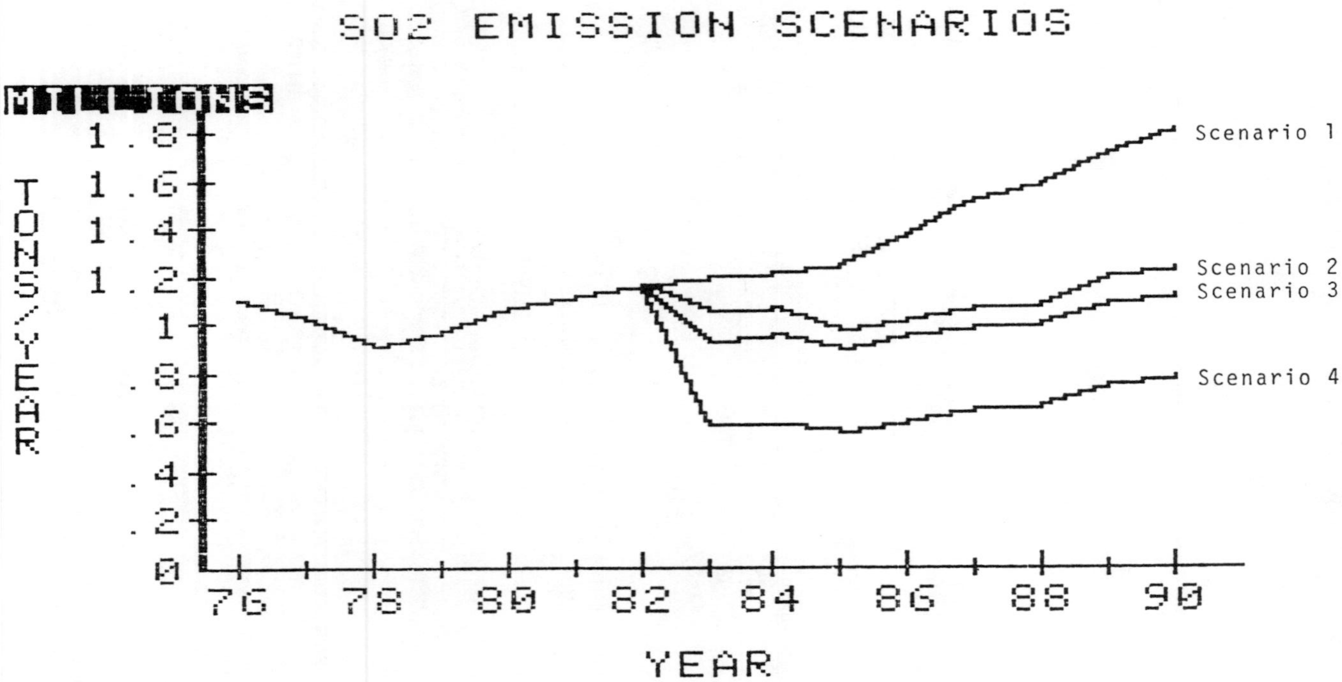

Fig. 1 SO_2 Emission Scenarios

Scenario 1: Total Statewide SO_2 Emissions Uncontrolled (non-NSPS)
Scenario 2: Implementation of New Source Performance Standards for New Sources (NSPS)
Scenario 3: NSPS and 1% Sulfur Content Heavy Oil for Existing Oil-fired Units
Scenario 4: NSPS, 1% Heavy Oil, and 1% Sulfur Coal for Existing Coal-fired Units

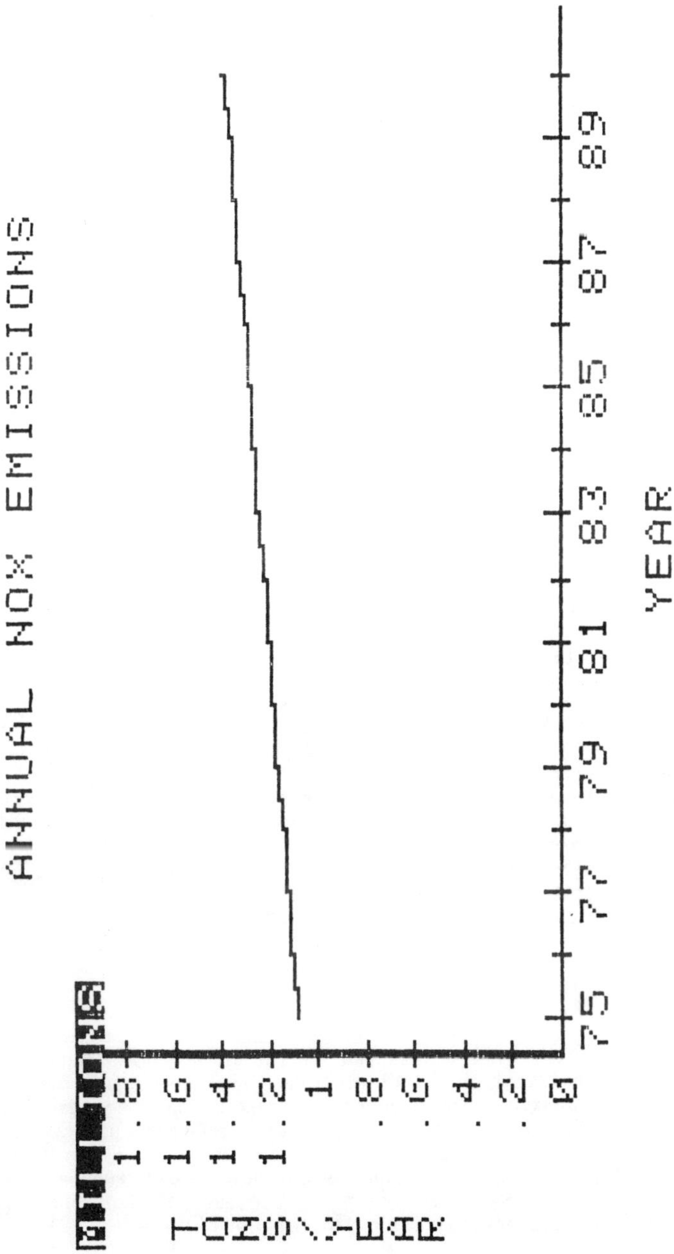

Fig. 2 Annual NO_x Emissions

ACID PRECIPITATION AND ITS EFFECTS ON FLORIDA LAKES: WHERE DO WE STAND AFTER SEVEN YEARS OF RESEARCH?

Patrick L. Brezonik
Department of Civil and Mineral Engineering
University of Minnesota, Minneapolis, MN 55455

Introduction

The role of the atmosphere in the biogeochemical cycles of several important elements (C, N, S) has been recognized for many years. Over the past decade scientists also have attained a growing appreciation of the role of the atmosphere in the biospheric transport of a wide variety of anthropogenic contaminants, including heavy metals, refractory organics, and mineral acidity. Much research has been focused on the significance of atmospheric sources of these contaminants to aquatic and terrestrial ecosystems. By far the greatest research and public attention has been paid to the deposition of mineral acidity by atmospheric precipitation, but this is only the most visible symptom of a much more widespread pollution problem.

In brief, acid precipitation is caused by emissions of sulfur and nitrogen oxides (SO_2 and NO_x) to the atmosphere, primarily by fossil fuel combustion. These gases are transported advectively in the atmosphere and gradually are transformed into the strong mineral acids, sulfuric and nitric acids, which are brought back to earth by rain and snow. The general causes of precipitation acidity are well-known, but the mechanisms and rates of transformation of SO_2 and NO_x into acids within the atmosphere are still poorly understood. Quantification of linkages between specific emission sources and local deposition rates remains a complex and unresolved problem.

The phenomenon of acid precipitation was first noted in Scandinavia in the 1960s, and the first report of its occurrence in the United States was by Likens (1972). Numerous studies reported over the past decade have documented the occurrence of acid precipitation over much of the United States and Canada, and these studies show that the region receiving the greatest acid deposition occurs in the northeastern United States and southeastern Canada (New York, New England, and southern Ontario and Quebec). Consequently, research on the effects of acid precipitation on aquatic and terrestrial ecosystems has been focused in these areas. However, monitoring programs also have demonstrated that acid precipitation occurs throughout the entire eastern United States, and many areas receiving acid deposition have been shown to be potentially sensitive to its effects.

Prominent among the sensitive regions is the state of Florida, which is the focus of this review. Research on the occurrence, causes, and effects of acid precipitation in Florida has been conducted for about seven years, and a considerable body of information has been developed. This paper is not intended to be an exhaustive review of past work but will emphasize results obtained by the author's previous research group in the Department of Environmental Engineering Sciences at the University of Florida, whose work has formed the basis for much of our current understanding of the "acid precipitation problem" in Florida.

Occurrence and Geographical Distribution of Acid Precipitation in Florida

Among the first measurements of rainfall pH in Florida are those made by Bourne (1975) near Gainesville during 1974-75, as part of studies on mineral inputs to cypress swamps. He found pH levels of 4.5-5.0 in several samples of precipitation collected in "bulk samplers" (which were open to the atmosphere on a continuous basis and thus collected both rainfall and dryfall). The first intensive studies oriented toward the acidity of Florida precipitation were undertaken by Hendry (1977) as part of a continuing effort to evaluate atmospheric inputs to cypress swamps. He sampled bulk precipitation on a biweekly basis at a cypress dome site just north of Gainesville for a one year period beginning in June of 1976 and simultaneously collected "wet-only" deposition on an event basis at a site on the University of Florida campus. With some modifications in sampler design and location, event sampling of rainfall at Gainesville has been continued by Hendry and his coworkers to the present time. The initial studies of precipitation acidity showed that wet-only precipitation at Gainesville had a (volume-weighted) average pH of 4.53 in 1976-77 (Hendry and Brezonik 1980), and stoichiometric calculations (Granat 1972) indicated that about two-thirds of the acidity in the rain was caused by sulfuric acid, while nitric acid contributed most of the remaining one-third. Similar data were collected at Tallahassee at about the same time as Bourne's sampling (R. Turner, unpublished data); these results apparently have not been published separately, although they have been cited in maps showing pH trends on a national scale (e.g. Likens 1976; Likens et al. 1979).

In 1977 our work was expanded to cover the entire state of Florida with a network of precipitation collectors in a project funded by the U.S. EPA. The network consisted of 22 bulk collectors and four wet/dry collectors located in all regions of the state (from Jay in the western Panhandle near Pensacola, to Bahia Honda in the lower Keys) (see Figure 1a). The complete network was in operation from May 1978 to April 1980. Initial results were described by Brezonik et al. (1980), and more detailed analyses for the first year were given by Edgerton et al. (1981) and Hendry et al. (1981). Summaries of the entire data base and detailed discussions of the results are given by Hendry (1983) and in the final project report (Brezonik et al. 1983).

In brief, results from the rainfall network demonstrated the occurrence of acid precipitation throughout the entire state, and a distinct north-south gradient was observed (Figure 1b), with consistently higher acidity (lower pH values) in the northern half of the state. The volume-weighted annual average pH of bulk precipitation at sites in northern Florida (including the Panhandle and northern half of the peninsula) was 4.6-4.7, whereas sites in south Florida had lower acidities (average pH values of 4.9-5.4). Highest values were found at sites affected by emissions from limerock operations or by wind erosion of basic soil particles. Bulk precipitation at Bahia Honda, in the lower Keys, was near geochemical neutrality (5.56). Individual rain samples with pH < 4.0 were found at Jay (3.73) and Gainesville (3.7-3.9) on several occasions.

Stoichiometric considerations indicate that sulfuric acid accounts for about 70% of the acidity in Florida rainfall, and nitric acid accounts for the remainder. Regression analysis of the statewide data

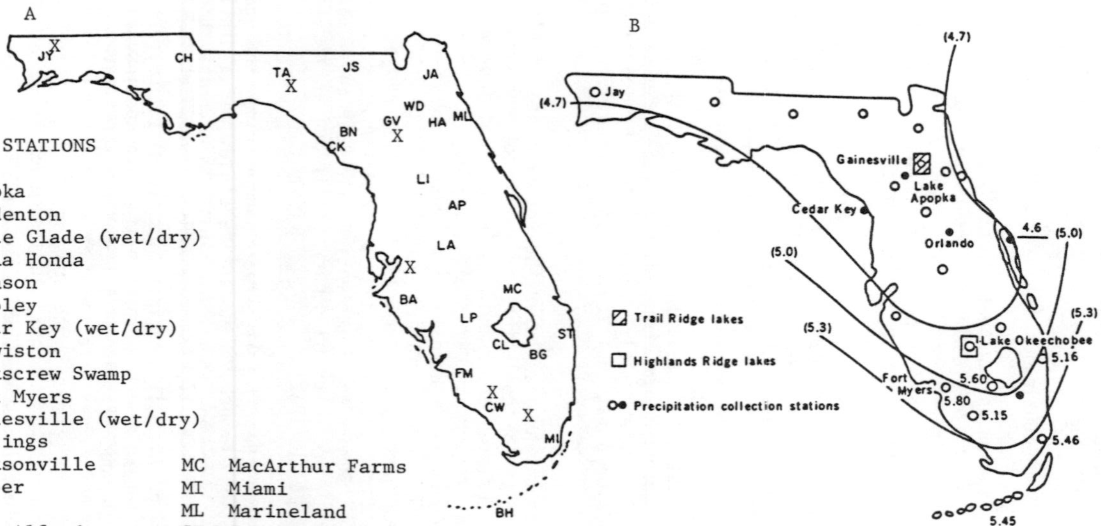

KEYS TO STATIONS

AP	Apopka		
BA	Bradenton		
BG	Belle Glade (wet/dry)		
BH	Bahia Honda		
BN	Bronson		
CH	Chipley		
CK	Cedar Key (wet/dry)		
CL	Clewiston		
CW	Corkscrew Swamp		
FM	Fort Myers		
GV	Gainesville (wet/dry)		
HA	Hastings		
JA	Jacksonville	MC	MacArthur Farms
JS	Jasper	MI	Miami
JY	Jay	ML	Marineland
LA	Lake Alfred	ST	Stuart
LP	Lake Placid	TA	Tallahassee
LI	Lisbon	WD	Waldo

Fig. 1 (A) Location of sampling stations for Florida precipitation network, 1978-79. X denotes location of wet-only sampler in 1980 network. (from Hendry et al. 1981)
(B) Isopleths of pH for bulk precipitation in state, May 1978-April 1979. (from Brezonik et al. 1980)

demonstrated that acidity (H+ concentration) could be explained by a simple two variable model as the difference between the concentration of excess sulfate (i.e. total sulfate minus sea salt sulfate) derived from SO_2 and the concentration of excess calcium:

$$[H+] = 6.1 + 0.54[SO_4^{2-}]_{xs} - 0.35[Ca^{2+}]_{xs}, \quad R^2 = .75$$

where subscript xs stands for nonmarine-derived material. The importance of calcium as a neutralizing agent of acidity in precipitation evidently derives from its association with carbonate in soil and dust particles.

Distinct seasonal differences were found in rainfall pH, with summer (May-October) rains having pH values 0.2-0.3 units lower than winter (November-April) rains over the entire state (Figure 2a). Similarly, concentrations of excess sulfate are demonstrably higher in summer rains than in winter rains (Figure 2b). These trends may reflect somewhat higher rates of power generation in Florida during summer, or they may reflect differences in the efficiencies of scavenging by intensive convective summer showers compared to winter frontal events. In addition, the warmer temperature and more intensive solar radiation in summer may accelerate production of H_2SO_4 and HNO_3 from the emitted gaseous precursors. Because about 65% of the annual rainfall in Florida occurs during the summer months, the seasonal differences in deposition of H^+ are even more pronounced than the concentration differences imply.

Consideration of seasonal patterns in climate and large-scale movements of air masses leads to the conclusion that the acid rain phenomenon in Florida is at least partially (and probably mostly) the result of acid precursors (SO_2 and NO_x) generated and emitted within the state. Peninsular Florida is isolated meteorologically from most of the continental United States during most of the summer months, and prevailing winds are from the southeast (Trade winds from the Atlantic Ocean and Caribbean Sea). Cold fronts from the north rarely penetrate the state during the months of May to September. Consequently, it is difficult to attribute summertime acidity to northern sources. A mass balance model for sulfur in the Florida atmosphere (Edgerton 1981) indicates that ample SO_2 is emitted within the state to account for the acidity of Florida rain (Table 1). Edgerton's calculations indicate that approximately 100,000 metric tonnes of excess sulfur are deposited throughout Florida by bulk precipitation each year. In comparison, an inventory of statewide sulfur oxide emissions from human activities shows that about 500,000 tonnes (as S) are emitted yearly. It may come as a surprise to many who regard Florida as relatively non-industrialized that the state is ranked tenth among all states in terms of sulfur oxide emission rates. Figure 3 shows the geographic distribution of areal emission rates within the state; the Tampa Bay area has the highest emission rates, but other major regional sources are the Pensacola area and Duval County (Jacksonville).

Based on estimates of biogenic sulfur emissions extrapolated from the literature, Edgerton concluded that biogenic sources are insufficient to account for the observed sulfate content of Florida rainfall. In general, the mass balance model supports the hypothesis that most of the acidity in Florida rain is derived from in-state sources. The accuracy and simplicity of the model are such, however, that external contributions, particularly from adjacent states in the Southeast, cannot be evaluated quantitatively at the present time.

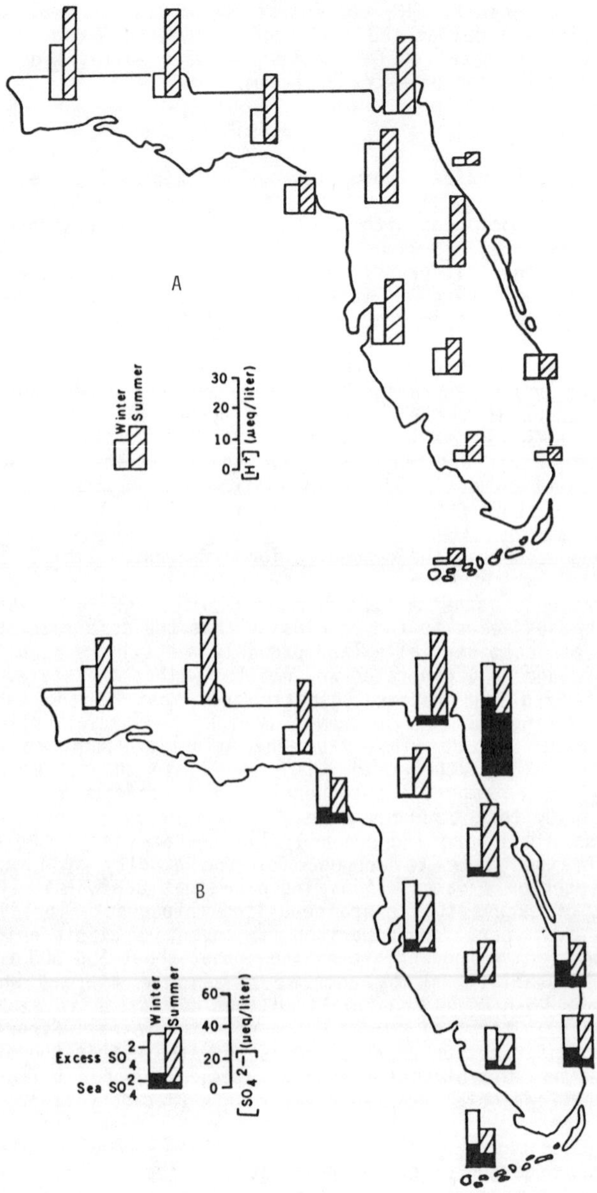

Fig. 2 Average concentrations of H^+ (A) and sulfate forms (B) in winter (Nov-April) and summer (May-Oct) for 1978-79. For clarity, not all sites are shown. (from Brezonik et al. 1980)

Table 1. Mass balance of sulfur in atmosphere over Florida (after Edgerton 1981)

Term	Magnitude ($\times 10^7$ kg-S/yr)	Magnitude (kg-S/ha)	% of Inputs
Inputs			
In-state emissions of SO_2	50.0	32.3	71-81
Biogenic emissions	0.1- 1.1	0.1-0.7	0-2
N-C Imports*	7.1-11.3	4.6-7.3	11-17
S-M Imports*	3.5- 7.0	2.3-4.6	5-11
Oceanic**	1.2- 1.5	0.8-1.0	2-3
Total	61.6-70.9	39.7-45.7	100
Internal Sinks			
Wet deposition	9.4-11.1	6.1-7.2	13-18
Dry deposition	4.6-15.5	3.0-9.9	6-25
Total	14.0-26.6	9.1-17.1	20-43
Export	35.3-56.9	22.8-36.7	57-80

* N-C = northern-continental imports of excess-sulfate
S-M = southern-marine imports of excess-sulfate

** Sea-salt (estimated from sea-salt S deposition in bulk precipitation)

Statewide Trends in "Wet-only" Precipitation

Results from wet/dry collectors located at several sites during the statewide program yielded higher rainfall acidities than found in adjacent bulk collectors (pH values were 0.2-0.3 units lower in wet-only samples). Data from these collectors also indicated that most of the calcium deposition occurred as dry fall. Because south Florida sites generally had higher rates of calcium deposition than north Florida sites, it seemed likely that the north/south gradient in acidity observed in bulk collectors was partially the result of greater dry fall or $CaCO_3$ in south Florida. To evaluate this hypothesis and obtain more detailed information on the chemistry of wet-only precipitation across the state a network of six wet/dry collectors was sampled on a weekly basis throughout 1980 (Edgerton and Brezonik 1981). Samplers were located at Jay (near Pensacola), Tallahassee, Gainesville, Tampa, Corkscrew Swamp (southeast of Fort Myers), and at the north end of Everglades National Park (see Figure 1a for locations of samplers).

Volume-weighted mean concentrations of acidity-related species at the six sites during 1980 are summarized in Table 2, and results showed similar trends to the earlier bulk precipitation network. The acidity of rainfall throughout northern Florida was nearly uniform, and the four northern sites (Jay to Tampa) had mean values in the range 4.5-4.6. Similarly, the two south Florida sites had almost identical mean pH values over the year, but the rain was less acidic (pH 4.8). Rainfall in the north thus was twice as acidic as that in the south, and the difference was significant at the 99% confidence level. Nonetheless, the results for south Florida indicate much greater acidity in wet-only precipitation than we found in bulk precipitation.

Table 2. Acid/base chemistry of wet-only precipitation throughout Florida, 1980 (from Edgerton and Brezonik 1981)

	Parameter (ueq/L)					
Site	pH	H^+	NH_4^+	Ca^{2+}	NO_3^-	xs-SO_4^{2-}
Corkscrew	4.83[b]	14.8[a]	5.1[a]	5.8	9.0[a]	14.2[a]
Everglades	4.81[b]	15.1[a]	7.6	14.6	13.4	20.0
Gainesville	4.54	28.8	9.8	5.9	12.7	28.1
Jay	4.51	30.9	10.9	3.1[a]	11.9	35.1
Tallahassee	4.59	25.7	9.9	3.1[a]	11.3	27.4
Tampa	4.53	29.5	25.8[b]	29.6[b]	20.7[b]	66.5[b]

[a] Lower than other values at 0.01 level of significance

[b] Higher than other values at 0.01 level of significance

The data in Table 2 also indicate north-south gradients in concentrations of ammonium and excess sulfate, with values generally declining from Jay to the Everglades. Values for Tampa do not conform to this trend because of large local sources in the area. Analysis of variance separated the six stations into five distinct groups based on excess sulfate concentrations: Tampa > Jay > Tallahassee and Gainesville > Everglades > Corkscrew (Table 2). The high concentrations at Tampa and Jay in part reflect the influence of local sources of SO_2, primarily from electric power plants, but some of the excess sulfate at Tampa probably reflects wind erosion of gypsum ($CaSO_4 \cdot 2H_2O$) from large mounds of this material associated with phosphate plants. The high concentrations of calcium in Tampa rain support this conclusion. The higher value at the Everglades site compared to Corkscrew Swamp probably reflects the relative proximity of the Miami-Palm Beach urban area to the former site.

On the other hand, nitrate concentrations are rather evenly distributed across the state; including all sites concentrations ranged over a factor of 2.3, and excluding Tampa, the range was only a factor of 1.5. The rather consistent concentrations of nitrate across the state probably reflect a relatively slow conversion of NO_x to nitric acid in the atmosphere (which tends to smooth out local sources), and the small range in concentrations also explains the lack of correlation between rainfall acidity and nitrate. In general, it appears that the geographical variation in rainfall acidity in Florida is explained by variations in sulfuric acid content (hence variations in SO_2 emissions), with nitric acid providing a relatively uniform background level across the state.

It is interesting to note that although the Tampa site had much higher concentrations of acidic components (sulfate and nitrate) than the other northern sites, rainfall was no more acidic there than elsewhere in the northern half of the state. This finding should not be taken to imply that Tampa area emissions do not contribute to the acidification of rainfall. The mix of emissions in the immediate area apparently is such that most of the acidity in Tampa rain is neutralized by local emissions or suspension of basic material (ammonia and calcium carbonate), and some of the excess sulfate originates as neutral $CaSO_4$.

Sensitivity of Florida Lakes to Acidification

Florida has about 7,800 lakes and reservoirs, about 3,500 of which are named (Huber et al. 1982). The lakes cover 9,270 km^2 or about 6% of the state's surface area. A major factor determining whether acid precipitation is a potentially serious environmental problem in Florida or is just an environmental curiosity is the sensitivity of its lakes to acidification by atmospheric deposition. To evaluate this aspect we recently examined the existing data on alkalinity levels in Florida lakes (Hendry 1983; Hendry and Brezonik 1983). Alkalinity is a direct analytical measure of the acid neutralizing capacity of a water. Data on alkalinity were retrieved from a recently assembled computerized data base called FLADAB (Huber et al. 1982), which contains nearly all available water quality data on Florida lakes. FLADAB contains various data on almost 800 lakes, and pH and alkalinity data for 596 lakes located in 58 of the 67 Florida counties were retrieved from the data base. Data thus are available for about 8% of the total number of lakes in the state, but the lakes with data comprise about 4,000 km^2 of total surface area, or about 43% of the total lake surface area in the state. (Excluding Lake Okeechobee from both totals, the data set represents about 28% of the remaining lake area.)

The frequency distribution of alkalinity in the 596 lakes is as follows:

Alakalinity (ueq/L)	% of total lakes
<100	31
100- 500	27
500-1000	13
1000-1500	11
1500-2000	8
>2000	10

It is apparent that most of the lakes in Florida have poorly buffered soft waters; 58% have less than 500 ueq/L of alkalinity, and 31% have less than 100 ueq/L. The latter value is commonly used as a criterion for identifying lakes considered sensitive to acidification (e.g. Gmur 1981).

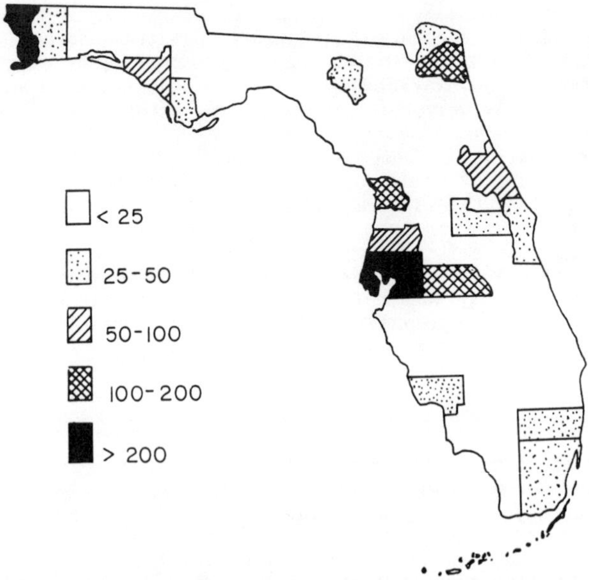

Fig. 3 SO_2 emissions by county in Florida during 1980. Units in by S/ha.yr. Total for 1980 was $5x10^5$ tonnes as S. (from Edgerton 1981)

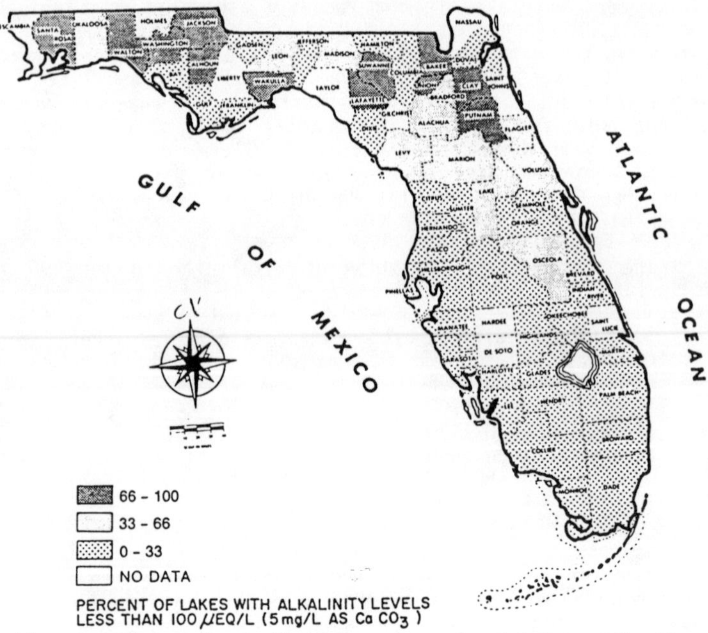

Fig. 4 Counties containing the indicated percentage of lakes with alkalinity levels less than 100 µeq/L. (from Hendry 1983)

Figure 4 shows that most of the sensitive lakes occur in the northern part of the peninsula and in the Panhandle. More than two-thirds of the (FLADAB) lakes in eleven counties of these areas have alkalinities < 100 ueq/L. Large portions of these counties contain poorly-buffered soils with cation exchange capacities near 1 meq/100 g and near zero base saturation ratios (Carlisle et al. 1978). The geographic distribution of lakes with low alkalinities are in good agreement with areas of Florida classified as having soils sensitive to acid deposition (McFee 1980; Klopatek et al. 1980).

Chemical and Biological Trends in Soft Water Lakes as a Function of pH

As a first step in evaluating the possible changes that may occur in a lake as it undergoes acidification we conducted a survey of 20 softwater lakes in north-central and south-central Florida (Brezonik et al. 1983a). Comprehensive analyses were done to characterize the chemical composition of the lakes (major and minor ions, nutrient forms, aluminum) (Hendry 1983) and their biological communities (phytoplankton, zooplankton, and benthic invertebrate populations) (Schulze 1980). A total of 20 lakes were sampled quarterly during 1978-79; 13 of the lakes are in the Trail Ridge region of north-central Florida (primarily Clay and Putnam Counties), and seven are in south-central Florida (Highlands County) in the Lake Wales region of the Central Highlands. The two lake districts are similar topographically and have similar soils (highly sandy) and native vegetation, but they differ in rates of acid deposition, the northerly group receiving about twice as much per year as the southerly group.

Analysis of variance showed that the lakes could be divided into two groups of ten based on pH: an acidic group pH <5.5 and a non-acid group pH >5.5. All of the acidic lakes are in the northern group. Table 3 summarizes limnological characteristics of the two pH groups and shows that as a group the acid lakes have lower dissolved salts (as measured by specific conductivity), lower concentrations of organic nutrient forms (C, N, P), and higher concentrations of aluminum. In addition, the acidic lakes generally have lower levels of chlorophyll and lower populations of phytoplankton and zooplankton; moderately good correlations were found between plankton population densities and pH (see Figure 5). Overall, the acid lake group exhibits a greater degree of oligotrophic conditions. These results support a hypothesis that acidification causes "oligotrophication", i.e. a decline in nutrient levels and in productivity, presumably because nutrient recycling is less efficient under acidic conditions. However, trends inferred from cross-sectional studies (surveys) are not sufficient to prove hypotheses, and controlled experiments are needed to test the hypothesis.

Alternative explanations are possible for the observed pH trends in nutrient concentrations and planktonic populations. Lakes in the acidic group have somewhat more pristine watersheds and may receive less nutrient enrichment from cultural sources. Moreover, the lower levels of dissolved solids in the acidic lakes suggest that these lakes receive a greater proportion of their water and mineral inputs directly from precipitation, and this factor may predispose the lakes not only to be more

Table 3. Means and standard deviations of water quality and chemistry of the two lake groups (from Hendry 1983)

Parameter	Acidic Group 10 Lakes(N=40)	Non-Acidic Group 10 Lakes(N=40)	Probability Means are Not Different*
Lab H^+ (ueq/L)	10.5 ± 5.0 (pH 4.98)	0.62 ± 0.62 (pH 6.21)	0.0001
Specific Conductance (uS/cm at 25°C)	41.5 ± 18.2	92.9 ± 39.8	0.0001
Na^+ (mg/L)	2.88 ± 1.27	5.92 ± 2.13	0.0001
Ca^{2+} (mg/L)	0.99 ± 0.75	3.67 ± 1.52	0.0001
Mg^{2+} (mg/L)	0.54 ± 0.27	2.15 ± 1.36	0.0001
K^+ (mg/L)	0.57 ± 0.35	2.03 ± 1.67	0.0001
Al (III) (ug/L)	58 ± 29	23 ± 15	0.0001
Cl^- (mg/L)	5.76 ± 4.01	11.10 ± 3.55	0.0001
SO_4^{2-} (mg/L)	5.51 ± 1.94	13.01 ± 6.51	0.0001
HCO_3^- (mg/L)	--	8.66 ± 5.28	0.0001
NH_4^+-N (ug/L)	72 ± 99	41 ± 48	0.0812
NO_3^--N (ug/L)	21 ± 41	16 ± 17	0.4158
TON (ug/L)	199 ± 131	405 ± 193	0.0001
TN (ug/L)	292 ± 219	462 ± 196	0.0005
SRP (ug/L)	7 ± 6	10 ± 11	0.0940
TP (ug/L)	16 ± 12	26 ± 22	0.0109
ORG P (ug/L)	9 ± 9	16 ± 19	0.0353
TOC (mg/L)	4.6 ± 3.4	7.0 ± 3.9	0.0039
SiO_2 (ug/L)	459 ± 341	420 ± 416	0.6481
Chlorophyll _a_ (ug/L)	2.74	10.0	0.005†

* Probabilities listed are from ANOVA (Barr et al. 1976) on the annual means for each parameter in each group. If $P < 0.05$, means are accepted as statistically different.

† Based upon Kruskal-Wallis Rank Sum Test (Schulze 1980)

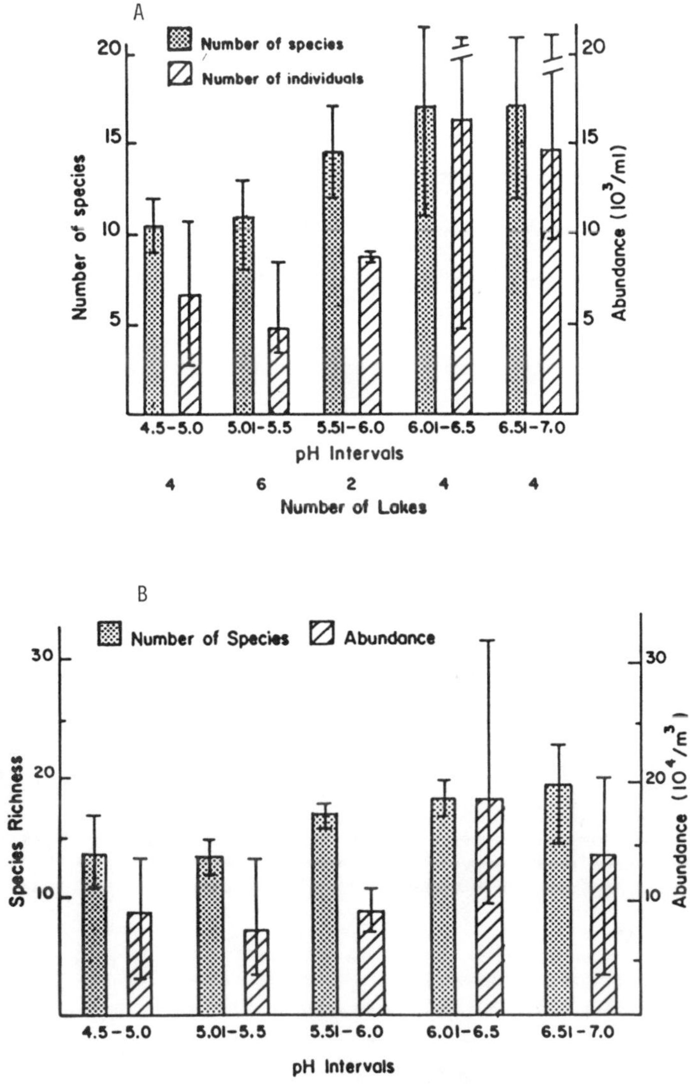

Fig. 5 Trends in number of species and abundance of individuals for phytoplankton (A) and zooplankton (B) in 20 softwater Florida lakes over five pH intervals. (from Brezonik et al. 1983)

oligotrophic, but also to be more acidic. In this sense, the trends in trophic state and acidity may be correlative rather than causative. The latter idea could be examined if accurate information were available regarding water, mineral, and nutrient budgets for the 20 lakes. Unfortunately, quantitative information is lacking, and the hydrologic characteristics of the lakes (most are seepage lakes) make such information difficult to obtain.

An intensive watershed/lake study has been conducted by our group and by researchers from the Department of Soil Science (see paper by Volk and Byers in this volume) at one of the acidic lakes (McCloud Lake) for the past two years, and the results of this project should be useful in evaluating questions posed by the earlier survey. Water and mineral budgets for the lake are being calculated at present. Results are being assembled and evaluated from controlled mesocosm experiments in which the pH of water within large enclosures of lake water (limnobags) has been modified. Conclusions from this study will be available by mid-1983.

Historical Trends in the Chemical Composition of Softwater Florida Lakes

Water chemistry data are available on some of the Trail Ridge lakes from the late 1950's, and thus it is possible to examine present conditions in a historical context. Table 4 lists data from 1957-60 (Clark et al. 1964), from 1968-69 (Shannon 1970), and from the recent survey (Brezonik et al. 1983) for pH, alkalinity, and excess sulfate concentration in some of the lakes. These parameters were selected to reflect possible changes from increased deposition of acidic precipitation. Some correction factors were applied to the alkalinity data to account for differences in analytical procedures (Hendry 1983). Comparison of the historical and recent data indicates that changes have occurred in all of the lakes during the last 20 years. All lakes show increases in excess sulfate, and several lakes experienced doubling of concentration in that time. Alkalinity decreases of > 25 ueq/L occurred in Lakes Lowery, Brooklyn, and Magnolia, and sulfate increased concommitantly by 16-34 ueq/L. These lakes (and several others listed in Table 4) also have shown decreases in pH over the historical record. In contrast, pH, alkalinity, and excess sulfate levels increased in Lakes Geneva and Kingsley over the historical record. These lakes are the largest Trail Ridge lakes included in the survey, and they have moderate amounts of residential development in their watersheds. The observed increases likely reflect increased cultural influences on water chemistry over the past 20 years. Historical data for lakes in south central Florida (in the Lake Wales region) show little change in pH and alkalinity over the past 10 years (Hendry 1983). These lakes presently are receiving lower loadings of acidity from the atmosphere compared to their northern counterparts, but it also is possible that any temporal trends have been overwhelmed by variations in loadings from their watersheds. Moreover, the length and extent of the data base on these lakes is small and perhaps is insufficient for elucidation of trends.

Table 4. Comparison of current and historical pH, alkalinity, and sulfate data for some Trail Ridge lakes (from Hendry 1983)

	1957-60*			1968-69†			1978-79††		
Lake	pH	Alk.** (ueq/L)	Excess $[SO_4^{2-}]$ (ueq/L)	pH	Alk. (ueq/L)	Excess $[SO_4^{2-}]$ (ueq/L)	pH	Alk. (ueq/L)	Excess $[SO_4^{2-}]$ (ueq/L)
Brooklyn	5.5	30	67	5.5	10	67	5.0	<5	83
Lowery (Sand-Hill)	5.4	30	25	5.3	6	63	5.2	<5	59
Magnolia	5.6	30	35	5.5	8	54	5.1	<5	51
Geneva	5.4	12	89	6.3	24	83	6.1	130	162
Kingsley	6.3	94	74	6.9	114	83	6.7	216	129
Anderson-Cue				4.8	<5	73	4.9	<5	95
Cowpen				5.4	8	94	4.9	<5	259
Galilee				5.6	12	104	5.0	<5	222
McCloud				4.9	<5	73	4.7	<5	127
Santa Rosa				5.2	6	77	5.1	<5	95

* Mean values of semi-annual sampling during 1957-60 (N=6); Clark et al. (1964).

† Mean values of quarterly sampling during 1968-69 [N=3; except Anderson-Cue, McCloud (N=6); Geneva, Kingsley (N=5)]; Shannon (1970).

** Alkalinity values for 1957-60 corrected by subtracting 20 ueq/L. Excess $[SO_4^{2-}]$ determined by difference of total $[SO_4^{2-}]$ minus marine $[SO_4^{2-}]$ (based on Na^+ levels reported in each study).

†† Mean values from this study. Sampling during 1978-79 (N=4).

Importance of Sediment-Water Interactions in Controlling Responses to Acid Loadings in Seepage Lakes

Although the McCloud Lake watershed studies have not been fully interpreted at this time, two important findings have been made. Both results deal with the role of sediment-water interactions in the responses of seepage lakes to acid inputs. First, it is apparent that if aluminun concentrations increase in seepage lakes as a result of acidification, the most likely source of the aluminum is the lake bottom sediments. This statement is based on the fact that aluminum concentrations in rainfall itself are very low (Brezonik et al. 1983) and on the reasonable assumption that most of the water input to seepage lakes comes either directly from the atmosphere or from seepage through the bottom sediments. Because the soils around seepage lakes in the sandhills of Florida are so sandy, there is little surface runoff to these lakes. Preliminary calculations for McCloud Lake indicate that direct precipitation is by far the most important component in the lake's water budget. The data in Table 3 indicate that acidic Florida lakes have higher concentrations of aluminum than do non-acidic ones. However, even in the acidic lakes the concentrations are rather low (maximum of 150 ug/L) compared to values reported for acidic lakes in granitic watersheds of north temperate environments (e.g. 400-800 ug/L; Driscoll 1980). Nonetheless, results from the 20 lake survey show a trend of increasing values with decreasing pH, and it is reasonable to speculate that further decreases in pH will lead to further elevations in Al levels, perhaps to the toxic range (> 200 ug/L). Evidence that this can occur was found in limnobag studies conducted in McCloud Lake in summer of 1981. Concentrations of Al in a littoral enclosure open to the bottom sediments increased to 250-300 ug/L rapidly when the overlying water was acidified to pH 3.7. In contrast, Al levels in a control enclosure (no change in pH) showed no changes in Al concentrations (Brezonik et al. 1982). Laboratory studies with McCloud Lake sediments also produced similar increases in dissolved Al under more acidic conditions. The only source of Al in these experiments was the sediments, and thus it is apparent that the lake sediments act as a reservoir of aluminum for the overlying water. Further studies are needed so that the importance of these findings can be put into a general context for Florida lakes.

A second finding from the McCloud Lake limnobag studies is that bottom sediments are important as a sink for acidity in lakes; i.e. they exert a significant acid neutralizing capacity (ANC). This conclusion arose from studies undertaken to determine why more acid was required to acidify lake water in a limnobag open to the bottom sediments than we calculated based on the buffer capacity of the water itself. Typical results from laboratory experiments indicate that lake sediments have ANC values of several ueq/g over the pH range ambient to 4.4-4.5. Experiments currently are underway to evaluate rates of neutralization in intact cores and to obtain ANC values in terms of sediment surface area.

At least two mechanisms are involved in the observed buffering phenomenon, but present results are insufficient for generalizations about the relative importance of each. The experiment illustrated in Figure 6 was done with sterilized sediments, and the results are interpretable in terms of an ion exchange process, whereby base cations on the sediment are exchanged for hydrogen ions in the water. Other experiments in our lab (Baker, unpublished data) point to the importance of biological sulfate reduction and/or sulfate absorption by lake sediments. Pore water analyses of sediment cores recently taken from McCloud Lake show that sulfate concentrations decrease rapidly with depth (e.g. to about ten percent of the surficial concentration at a depth of 10 cm in the sediment). Clearly, sulfate is not a conservative constituent in lake-sediment systems. These findings may have important implications regarding the sensitivity of seepage lakes to acidification from atmospheric loadings of acid-sulfate, but further studies are needed to develop quantitative generalizations. On the other hand, buffering experiments performed in our laboratory also indicate that once they are acidified, lake sediments buffer the overlying water to a low pH. Consequently, it appears that lakes may remain acidified long after external loadings of acid-sulfate are decreased by institution of more stringent emission controls.

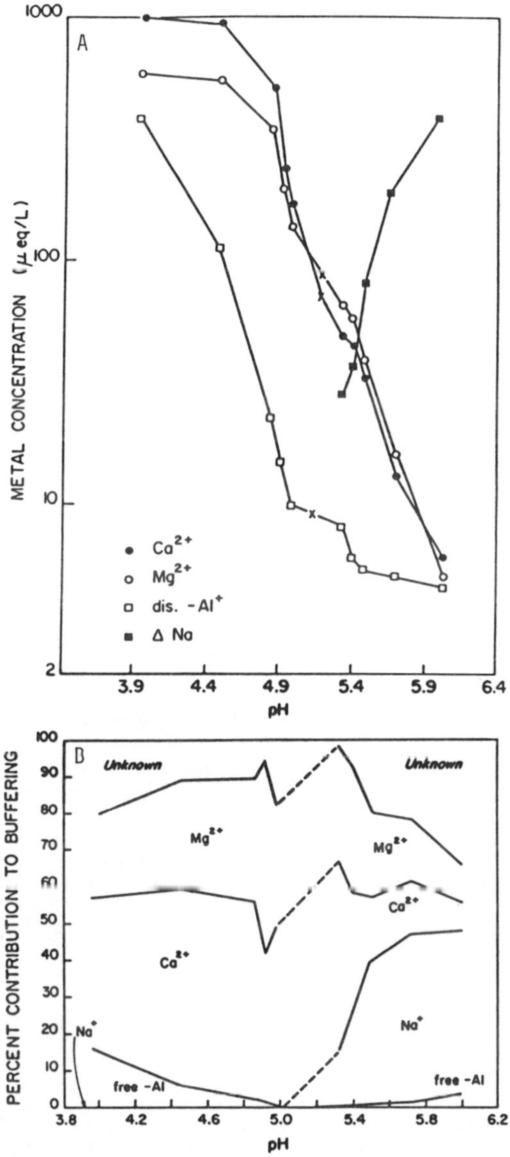

Fig. 6 (A) Changes in cation concentrations in water over McCloud Lake littoral sediments vs. pH and (B) Percent contributions of metal ion-exchange to buffering of the sediments

Summary and Perspective

Results presented in this paper and documented more fully in the cited references allow the following conclusions to be made. Annual average pH values less than 4.7 now occur in rainfall over the northern two-thirds of Florida, and slightly less acidic precipitation occurs over the southern third. Sulfuric acid accounts for about 70% of the observed acidity, and emissions of SO_2 within the state are more than sufficient to account for the non-marine sulfate deposited in rain over the state. Annual deposition of H^+ over northern Florida is about 250-500 eq/ha, and annual deposition of excess sulfate averages about 23 kg/ha over northern Florida and about 15 kg/ha over southern Florida. No long-term temporal trends were observed in the pH of rainfall at Gainesville over a five year period (1976-1981), but more recent data (Hendry and Edgerton, personal communication 1983) indicate lower average pH values for the past two years. A consistent seasonal pattern of lower pH values in summer rains compared to winter rains has been found throughout the state.

Nearly a third of Florida's 7800 lakes can be classified as sensitive to acidification from atmospheric deposition. This conclusion is based on a criterion of 100 ueq/L or less for alkalinity levels in sensitive lakes and a computer search of data on almost 600 lakes in the state. A variety of trends can be seen in the chemistry and biological composition of softwater lakes as a function of pH, but determination of causes for the trends awaits controlled experiments. Observed trends include lower nutrient concentrations and lower population densities of phytoplankton and zooplankton in acidic lakes, but whether the trends are merely correlative or are caused by pH variations is unknown. Acidic lakes also have higher concentrations of aluminum than non-acidic lakes, but concentrations are below levels generally regarded as toxic (> 200 ug/L) even in the most acidic lakes.

Recent studies indicate that sediment-water interactions are a key determinant affecting the way that seepage lakes in Florida respond to inputs of acidity. Sediments play an important (but still poorly quantified) role in neutralizing acid inputs to lakes, and the likely mechanisms for this are cation exchange and biological sulfate reduction. Moreover, sediments release aluminum (and possibly other toxic metals) to overlying water as the system becomes more acidic. Further studies are needed to evaluate the significance of this process.

In brief, our studies over the past seven years have demonstrated the occurrence of acid precipitation throughout the state and have shown that many aquatic systems in the state are sensitive to acidification from the atmosphere. Our surveys have shown the potential for chemical and ecological changes in lakes as they become acidic, and limited historical information indicates that some sensitive lakes in northern Florida have become more acidic over the past 20 years.

However, little is known about the actual damages that may have been suffered by Florida lakes as a result of past deposition, and we are not in a position to predict future consequences of continued acidic inputs. Further research is needed to determine rates of lake acidification as a function of watershed and loading conditions; ideally, relatively simple predictive models should be developed. Controlled experimental studies need to be undertaken to evaluate the ecological consequences of lake acidification. In particular, such studies need to examine direct and indirect impacts of acidification of fish populations and production in softwater lakes, about which very little information currently is available for Florida lakes. Special attention should be paid to interactions among pH, aluminum, and fish. Finally, no information is available on the effects of acidification on mobilization of heavy metals (e.g. mercury and lead) from bottom sediments of Florida lakes. The possible bioaccumulation of such metals by gamefish has not been evaluated, and this should be a high priority for further research.

Acknowledgements

The high quality efforts and cooperation of my graduate students at the University of Florida, who are listed as co-authors of many papers cited in the references, is gratefully acknowledged. Without their ideas, insights, and hard work, the results summarized in this paper would not have been possible. The studies reviewed here were funded by the U. S. EPA, the National Science Foundation, and the Engineering and Industrial Experiment Station of the University of Florida, all of whose support is gratefully acknowledged. Support for preparation of this manuscript was provided by the University of Minnesota.

Literature Cited

Bourne, R. G. 1975. M.S. thesis. Dept. Environ. Eng. Sci., Univ. Fla., Gainesville.

Brezonik, P. L., E. S. Edgerton, and C. D. Hendry. 1980. Science 208: 1027-1029.

Brezonik, P. L., L. A. Baker, R. W. Ogburn, III, E. S. Edgerton, and T. L. Crisman. 1982. Ecological effects of acid precipitation on a sensitive softwater lake in Florida. Rept. to NCSU-EPA Acid Precip. Res. Prog., Dept. Envir. Eng. Sci., Univ. Fla.

Brezonik, P. L., C. D. Hendry, E. S. Edgerton, R. L. Schulze, and T. L. Crisman. 1983. Acidity, nutrients and minerals in atmospheric precipitation over Florida: deposition patterns, mechanisms, and ecological effects. Compl. Rept. Proj. No. 805560, U. S. EPA, Corvallis, Oregon, 170 p.

Brezonik, P. L., R. L. Schulze, and T. L. Crisman. 1983a. Submitted to Canad. J. Fish. Aquat. Sci.

Carlisle, V. W. et al. 1978. Characterization data for selected Florida soils. IFAS, Univ. Fla., Gainesville.

Clark, W. E., R. H. Musgrove, C. G. Menke, and J. W. Cagle, Jr. 1962. Interim report on the water resources of Alachua, Bradford, Clay, and Union Counties, Florida. Fla. Geol. Surv. Inform. Circ. 36. Tallahassee, Fla.

Driscoll, C. T., Jr. 1980. In: Ecological impact of acid precipitation, D. Drabløs and A. Tollan (eds.), pp. 214-215. SNSF Proj., Aas, Oslo, Norway.

Edgerton, E. S. 1981. M.S. thesis. Dept. Environ. Eng. Sci., Univ. Fla, Gainesville, 176 p.

Edgerton, E. S. and P.L. Brezonik. 1981. Acid rainfall in Florida: Results of a monitoring network for 1980. Compl. Rept. to Eng. Ind. Exper. Sta., Univ. Fla., Gainesville, 11 p.

Edgerton, E. S., P. L. Brezonik, and C. D. Hendry. 1981. In: Atmospheric pollutants in natural waters, S. J. Eisenreich, (ed.), pp. 237-258. Ann Arbor Science Publ., Ann Arbor, Mich.

Gmur, N. 1981. General information on acidification chemistry data base. Brookhaven National Lab, Brookhaven, N. Y. (mimeo).

Granat, L. 1972. Tellus 24:550-560.

Hendry, C. D. 1977. M.S. thesis. Dept. Environ. Eng. Sci., Univ. Fla., Gainesville, 141 p.

Hendry, C. D. 1983. Ph.D. thesis. Dept. Envrion. Eng. Sci., Univ. Fla., Gainesville.

Hendry, C. D. and P. L. Brezonik. 1980. Environ. Sci. Technol. 14:843-849.

Hendry, C. D. and P. L. Brezonik. 1983. Submitted to Canad. J. Fish. Aquat. Sci.

Hendry, C. D., P. L. Brezonik, and E. S. Edgerton. 1981. In: Atmospheric pollutants in natural waters, S. J. Eisenreich (ed.), pp. 199-215. Ann Arbor Science Publ., Ann Arbor, Mich.

Huber, W. C., P. L. Brezonik, J. P. Heaney, R. E. Dickinson, S. D. Preston, D. S. Dwornik, and M. A. DeMaio. 1982. A classification of Florida lakes. Rept. ENV-05-82-1 to Fla. DER. Dept. Environ. Eng. Sci., Univ. Fla., Gainesville.

Klopatek, J. M., W. F. Harris, and R. J. Olson. 1980. In: Atmospheric sulfur deposition, D. S. Shriner, C. R. Richmond, and S. E. Lindberg (eds.), pp. 539-553. Ann Arbor Science Publ., Ann Arbor, Mich.

Likens, G. E. 1972. The chemistry of precipitation in the central Finger Lakes region. Cornell Univ. Water Resour. Mar. Sci. Cent. Tech. Rept. 50, Ithaca, N.Y., 62 p.

Likens, G. E. 1976. Chem. Eng. News 54(48):29-44.

Likens, G. E., R. F. Wright, J. N. Galloway, and T. J. Butler. 1979. Sci. Amer. 241:43-51.

McFee, W. W. 1980. *In:* Atmospheric sulfur depositon, D. S. Shriner, C. R. Richmond, and S. E. Lindberg (eds.), pp. 495-505. Ann Arbor Science Publ., Ann Arbor, Mich.

Schulze, R. L. 1980. M. S. thesis. Dept. Environ. Eng. Sci., Univ. Fla., Gainesville, 146 p.

Shannon, E. E. 1970. Ph.D. thesis. Dept. Environ. Eng. Sci., Univ. Fla., Gainesville, 258 p.

ATMOSPHERIC DEPOSITION PATTERNS IN NORTH-CENTRAL FLORIDA

H. Riekerk
School of Forest Resources and Conservation
University of Florida, Gainesville, FL 32611

M. W. Treacy
U. S. Peace Corps, Nepal

B. F. Swindel
U.S. Forest Service, Southeastern Forest Experiment Station
University of Florida, Gainesville, FL 32611

Abstract

Atmospheric deposition was measured for 5 years by a wet/dry fall collector at the weather station of the Bradford Forest Experimental Watersheds about 50 km northeast of Gainesville, FL. The data show gradually increasing acidity of rainwater at the rate of 0.2 pH units per year, mainly associated with increasing sulfate and decreasing potassium levels. The change in rainwater pH had no detectable effect on watershed runoff acidity.

Atmospheric deposition was also measured for one year by multiple collectors in an eight-km radius around the oil-fired Deerhaven I powerplant about 10 km northwest of Gainesville, FL. The data show a standard deviation of 0.5 pH units or a 10 percent coefficient of variation. However, spatial variability around the mean hydrogen-ion concentration per collection period was 71 percent. Some increased rainwater acidity was detected around the power plant during the summer months as well as decreased acidity during the spring months. No correlation with wind direction was detected.

Continuation of rainwater and runoff monitoring is needed to assess future trends. Longterm impacts of changing atmospheric inputs on soil leaching processes and on nutrient balances of forested watersheds need to be evaluated. A possible change in atmospheric deposition around the newly installed coal-fired Deerhaven II powerplant could be quantified using the existing data sets as a baseline.

Introduction

Atmospheric deposition has gained considerable attention because of an observed gradual acidification of rainwater in recent decades (1). Comparisons of acidity data calculated from older atmospheric deposition studies, however, have been brought into question because of possible neutralization by higher dust levels at that time (2).

Rainwater becomes acid when the pH drops below the 5.7 point of carbon dioxide equilibrium (3). Acidification is due to the presence of strong mineral acids, especially sulfuric and nitric acids. Precursors of these acids are released into the atmosphere as a result of the burning of fossil fuels by industry, power companies, and automobile traffic.

Atmospheric deposition of these and other substances represent an input of nutrient elements into forest ecosystems. Nutrient balance studies of experimental watersheds of poorly-drained flatwoods on leached sandy soils in North-Central Florida have shown atmospheric deposition to be a critical source of nutrient elements (4). The studies also showed that harvesting/regeneration practices generated a significant and persistent loss of potassium with runoff proportional to treatment intensity. However, little is known about the effect of chronic exposure of forest ecosystems of Florida to substantial increases of acidic substances (5).

This paper is based on data from two studies: Longterm monitoring of wet/dry fall as part of the National Atmospheric Deposition Program (NADP) since 1978 at the weather station of the Bradford Forest Experimental Watersheds near Starke, FL, and short-term monitoring of bulk deposition during 1979 to assess spatial variability of atmospheric deposition rates around the Deerhaven I power plant near Gainesville, FL.

These studies were funded in part by the Center for Environmental and Natural Resources Programs and the School of Forest Resources and Conservation, IFAS, University of Florida, Gainesville, FL, and in part by the Southeastern Forest Experiment Station, US Forest Service, Atlanta, GA. Access to and use of private lands for these studies is gratefully acknowledged.

Procedures

The site of the long-term NADP monitoring study was at the weather station of the Bradford Forest Experimental Watersheds (6). This location was approximately 8 km northwest of Starke (1980 population 5,300), approximately 40 km north northeast of Gainesville (pop. 81,400), approximately 35 km southeast of Lake City (pop. 9,300), and approximately 60 km southwest of Jacksonville (pop. 540,400). These experimental watersheds comprised about 250 ha total and were near the northeast boundary of an 24,000 ha industrial forest managed for wood fiber production. Other nearby lands, if not also forested, were devoted mainly to cattle grazing and row crop farming. Thus, the monitoring site was located in rural forest and farmland relatively free from nearby industrial or urban air pollution centers.

A standard NADP wet/dry fall collector has been in continuous operation since January 1978. Data recorded at weekly intervals included the amount of precipitation and the field pH. Weekly samples were analyzed by the NADP Central Laboratory at Champaign, IL, for chemical contents.

The site of the short-term atmospheric deposition study was in an area of 8-km radius around the Deerhaven I oil-fired powerplant about 10 km northwest of Gainesville, FL. Thirty-six bulk precipitation collectors were placed on five equidistant concentric rings. Samples were collected alternately from half the number of the collectors each week and analyzed for quantity of rain, pH and chemical contents at the School of Forest Resources and Conservation, University of Florida. A number of collectors were repeatedly vandalized and finally abandoned during the course of the study.

Results and Discussion

The volume-weighted pH data of wetfall-only samples from the long-term NADP study at the Bradford Forest have been plotted in Figure 1. Each observation was placed at the midpoint of its collection period. Superimposing the several years of data on one scatter diagram suggested that pH exhibited a seasonal pattern of the following model:

$$\text{Rain pH} = b_0 + b_1 T + b_2 \sin (360\ T) \quad [1]$$

Repeated runs of the regression model narrowed the best fit for T (in years) with an origin on February 21, 1978. Least-squares fitting of 194 data points to equation [1] yielded the following solution:

$$\text{Rain pH} = 5.4586 - 0.1926\ T + 0.1055 \sin (360\ T) \quad [2]$$

The correlation coefficient R^2 accounted only for 24 percent of the variation, and the standard errors of the regression coefficients were ± 0.0785 for b_0 , ± 0.0274 for b_1 and ± 0.0594 for b_2 . The t-test for the null hypothesis $H_0 : b_1 = 0$ was statistically highly significant at t = 7.03 (probability level <0.001). The t-test for b_2 was just below significance for t = 1.78 at the 0.08 probability level. However, the season (sine) variable was kept in the equation and superimposed on the data of Figure 1. An additional variable of amount of rainfall per collection period was not significant and dropped from the multiple regression equation.

The decline of pH over the 5-year observation period is highly significant at about 0.2 pH units per year from 5.5 in early 1978 to 4.5 in late 1982. A separate multiple regression was run to test the relationship between pH and sulfur (S) content of rainwater over time (T):

$$\text{Rain pH} = b_0 + b_1 T + b_2 S \quad [3]$$

Least-squares regression of 154 data points to equation [3] yielded the following solution:

$$\text{Rain pH} = 5.2610 - 0.1100\ T - 0.1296\ S \quad [4]$$

Standard errors of the regression coefficients were ± 0.1031 for b_0, ± 0.0339 for b_1 and 0.0436 for b_2. The t-tests for the null hypotheses were statistically significant at t = 3.24 for b_1 (probability level = 0.001) and t = 2.97 for b_2 (probability level = 0.004). This information shows that the correlation between acidity of rainwater and its sulfur content is more important than that of acidity and time. The implication is that sulphur is the cause of increased rain acidity.

Stepwise linear regressions of the hydrogen ion concentration of rainwater vs. other chemical constituents were run for the 1979 data set. Besides sulfur as noted above, only potassium was found to be correlated in a highly significant negative relationship (probability level <0.001). The implication is that less potassium is supplied by atmospheric deposition with increasing rain acidity at a time when leaching losses of soil potassium are becoming more serious (4).

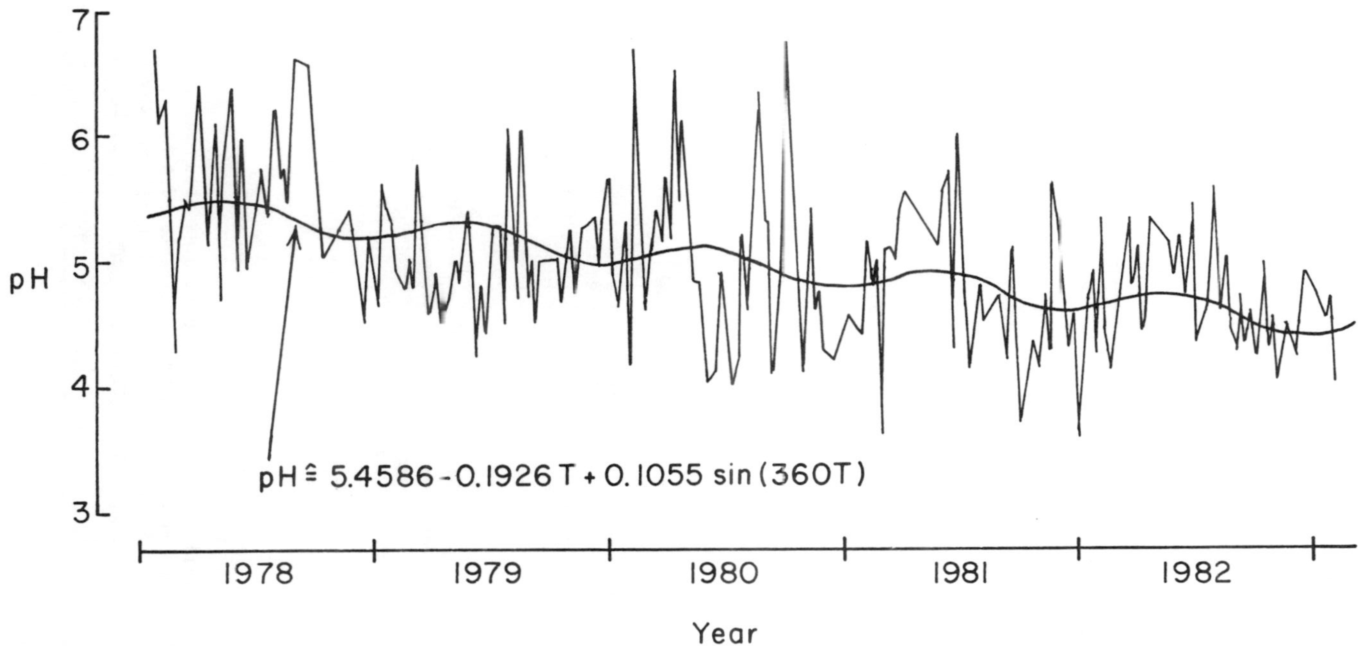

Fig. 1 Observed rainfall pH (broken line) from collections on Bradford Forest Experiment Watersheds in rural north Florida from January 25, 1978, through February 8, 1983, and predicted pH (smooth curve)

It was of interest to assess the impact of rainwater acidification on the pH of runoff from the watersheds of the study site. Data on runoff pH from the undisturbed control watershed (WS 3) were matched with the rain pH data and a multiple regression derived with 115 data points over time (T) as before:

$$\text{Runoff pH} = 4.3507 - 0.0490\ T - 0.0680\ (\text{Rain pH}) \quad [5]$$

Standard errors of the regression coefficients were ± 0.4223 for b_0, ± 0.0286 for b_1 and ± 0.07554 for b_2. The t-tests for the null hypotheses were not significant at t = 1.71 for b_1 (probability level 0.090) and t = 0.90 for b_2 (probability level 0.370). This result suggests that the forested flatwoods watershed had enough absorptive capacity to buffer the increasingly more acid rainwater inputs since the runoff pH remained practically constant over time.

The above results provided information on longterm and seasonal changes of rainwater chemistry. Estimates of areal variability of atmospheric deposition have been made during the short-term study of the Deerhaven site. These data have been plotted in Figure 2 together with the relevant information from the Bradford Forest study.

The standard deviation from about 14 samples per collection period averaged 0.5 pH units or a 10 percent coefficient of variation due to location. This narrow band of spatial variation is deceptive because of the logarithmic nature of pH values. Calculations of mean hydrogen ion concentration values per collection period yielded a spatial coefficient of variation of 71 percent.

Emissions from the Deerhaven I power plant did not make much of a difference when comparing the average rain pH of 4.6 to that of 5.0 at the remote Bradford Forest site (Figure 2). However, the collection of dryfall in the bulk precipitation collectors of the Deerhaven study may have neutralized some of the acidity. A slightly lower pH of rainwater during July and August and a somewhat higher pH during April around the Deerhaven power plant are evident. Increased summertime power demand for air-conditioning has been reported to routinely increase sulfur emissions in Florida (7).

The spatial arrangement of the collectors at the Deerhaven site provided for a test of atmospheric deposition in relation to average wind directions. A multiple regression of rain pH, wind direction and wind speed did not show any significant correlations. This suggests that there was little evidence of a "plume" from the Deerhaven power plant nor of a regional "chute" of mainland air flow (8).

Conclusions and Research Needs in Florida

The combined results of the two studies above suggest that for the 5-year study period acidification of rainwater due to sulfur compounds is significant at 0.2 pH units per year with a spatial variation of about 0.5 pH units or 71 percent of mean hydrogen-ion concentration.

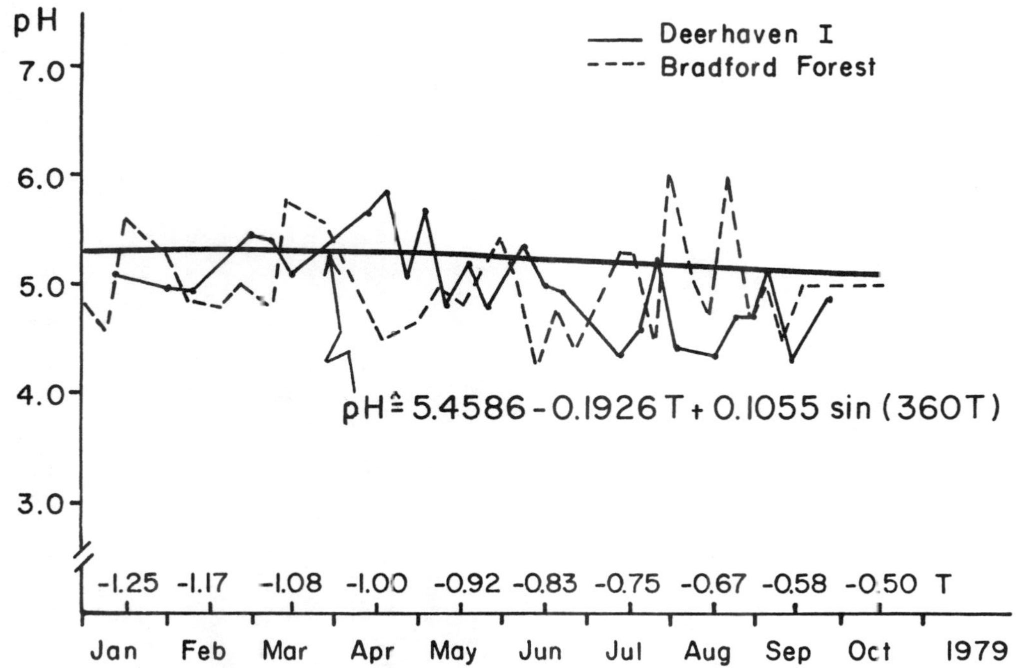

Fig. 2 Observed rainfall pH (solid broken line) from collections from around the Deerhaven power plant near Gainesville and from the Bradford Forest near Starke (dashed broken line) from January to October 1979, and predicted pH (smooth curve)

Concurrently decreasing atmospheric deposition of potassium, and persistently accelerated potassium losses with runoff after forest harvesting suggest a precarious balance for this mobile important nutrient element.

The five-year observation period did not reveal any effects of increasingly acid deposition on the acidity of runoff water from an undisturbed forested flatwoods watershed.

Continued monitoring for trends of atmospheric deposition is an urgent need for forest research in North-Central Florida. Chronic exposure to acidic substances may exhaust the neutralizing capacity of the forested watersheds. Changes of nutrient input rates may affect the mineral cycling and productive capacity of poorly buffered forest ecosystems on Florida's sandy soils.

The recent addition of the coal-fired Deerhaven II power plant makes it necessary to reassess the atmospheric deposition patterns around that site for quantification of environmental impacts. The longterm NADP record of the remote Bradford Forest site could be used by this effort to account for the area-wide general changes in atmospheric deposition due to the operation of more coal-fired power plants in the region.

Literature Cited

1. Likens, G. E., and Butler, T. S. 1981. Recent Acidification of Precipitation in North America, Atmospheric Environment 15:1103-1109.

2. Stensland, G. J., and Semonin, R. G. 1982. Another Interpretation of the pH Trend in the United States, Bull. Am. Meteor. Soc. 63(11): 1277-1284.

3. Barrett, E., and Brodin, G. 1955. The Acidity of Scandinavian Precipitation, Tellus 1:251-257.

4. Riekerk, H. 1983. Impacts of Silviculture on Flatwoods Runoff, Water Quality, and Nutrient Budgets, Water Resources Bulletin 19(1):73-79.

5. Puckett, L. J. 1982. Acid Rain, Air Pollution, and Tree Growth in Southeast New York, J. Environ. Qual. 11(3):376-381.

6. Swindel, B. F., Lassiter, C. J., and Riekerk, H. 1982. Effects of Clearcutting and Site Preparation on Water Yields from Slash Pine Forests, Forest Ecol. Managem. 4:101-113.

7. Florida Electric Power Coordinating Group. 1979. Ten-Year Plan for the State of Florida. Tallahassee, FL.

8. Dohrenwend, R. E. 1978. The Climate of Alachua County, Florida, Institute of Food and Agricultural Sciences, Univ. of Florida, Gainesville, FL., Agricultural Stations Bull. 796. 25 p.

FLORIDA ACID DEPOSITION STUDY - PRELIMINARY MONITORING RESULTS FOR THE PERIOD SEPTEMBER 15, 1981 THROUGH SEPTEMBER 14, 1982

John D. Schert, Kennard F. Kosky,
Eric S. Edgerton and Charles D. Hendry
Environmental Science and Engineering, Inc.
Gainesville, FL 32601

Abstract

A summary of wet-only precipitation monitoring data at 14 locations throughout Florida is presented for the period September 15, 1981 through September 14, 1982. Parameters included field and laboratory measurements for rainfall amount, pH and conductivity, and laboratory measurements for calcium, magnesium, potassium, sodium, ammonium, nitrate, chloride, sulfate, ortho-phosphorus, fluoride, vanadium, aluminum, total/strong acidity, total organic carbon and alkalinity. Field and laboratory data indicate a north-south gradient in precipitation pH (volume-weighted mean values). For annualized field pH volume-weighted means, a low of 4.31 was observed in North Florida (Caryville site), and a high of 4.70 (Tamiami site) was observed in South Florida. With the exception of the Leesburg site, volume-weighted mean laboratory pH values for monitoring stations in the northern and central portions of the state are below 4.60, while those across the southern part of the state are uniformly above 4.70.

Nitrate and excess-sulfate (total sulfate corrected for sea-salt sulfate) concentrations generally decrease from north to south. Nitrate-nitrogen and excess-sulfate concentrations in northern Florida average 0.15 mg/L and 1.20 mg/L, respectively, while the corresponding levels in South Florida average 0.11 mg/L and 0.64 mg/L, respectively.

Concentrations of metal cations (Ca^{+2}, Mg^{+2}, K, Na^{+}) increase from north to south, reflecting higher levels of Ca^{+2} and Mg^{+2} carbonates in South Florida soils, as well as the north-south transition from a continental to a marine environment.

The data for most sites indicate an appreciably higher laboratory pH than field pH. Possible mechanisms accounting for this neutralization include slow dissolution of basic particulate matter, decomposition of organic acids, and loss of hydrogen ion to the wall of the sample bucket.

Introduction

The Florida Acid Deposition Study was approved in November 1980 by the Florida Electric Power Coordinating Group, Inc. (FCG). The study's overall objective is to assess the magnitude, variability, sources, effects, and control options of acid deposition in Florida. Phase I of the 4-phase study was initiated in January 1981 with the design, installation, and operation of a 14-station acid deposition monitoring network for an 11-week period (June 30, 1981, through September 15, 1981). In addition, literature searches were performed in source attribution and long-range transport and ecological and material effects. Phase I results were published in January 1981 (1,2,3).

Phase II of the Florida Acid Deposition Study was initiated on September 15, 1981, and continued through September 14, 1982. Approximately 1,140 wetfall sample buckets and 110 dryfall sample buckets were collected from the 14-station acid deposition monitoring network during the 52 weeks comprising Phase II. A summary of these data was published in January 1983 (4).

The purpose of the paper is to present a summary of Phase II data set with respect to the following questions:

1. What are the spatial patterns in concentrations and depositions of acidity and related parameters across Florida?
2. What is the seasonal variability of precipitation concentrations and deposition?
3. What chemical species account for precipitation acidity?
4. What procedures and equipment factors influence data variability?

Methodology and Network Operation

Each of the 14 stations (Figure 1) was equipped with an Aerochem Metrics Model 301 wet/dry precipitation collector. Samples were collected and analyzed every Tuesday between the hours of 7 a.m. and 9 a.m. At four sites, two samplers were collocated to determine potential variability in the methodology, equipment, and site location. At two of those four sites, one sample was collected on a daily basis and one sample was collected on a weekly basis for comparing daily-versus-weekly data; at the other two collocated sites, both samples were collected weekly to develop precision data.

Rainfall, pH, and conductivity were measured under temperature-controlled conditions at each field laboratory by a trained and qualified field technician. Prior to analysis of the rain sample, the conductivity meter was standardized and the pH meter was calibrated. In addition, a low ionic strength solution was analyzed to assess response to a dilute acid solution. Samples were then shipped to the central laboratory for further analyses.

Laboratory analyses were performed on wet samples for the parameters below and the summary data for these are presented in Tables 1 and 2.

pH (on unfiltered and filtered samples)	conductivity
sulfate (SO_4^{2-})	chloride (Cl^-)
nitrate (NO_3^-)	phosphate (PO_4^{3-})
ammonium (NH^{4+})	sodium (Na^+)
potassium (K^+)	magnesium (Mg^{2+})
calcium (Ca^{2+})	aluminum
vanadium	fluoride
total/strong acidity	total organic carbon
alkalinity	

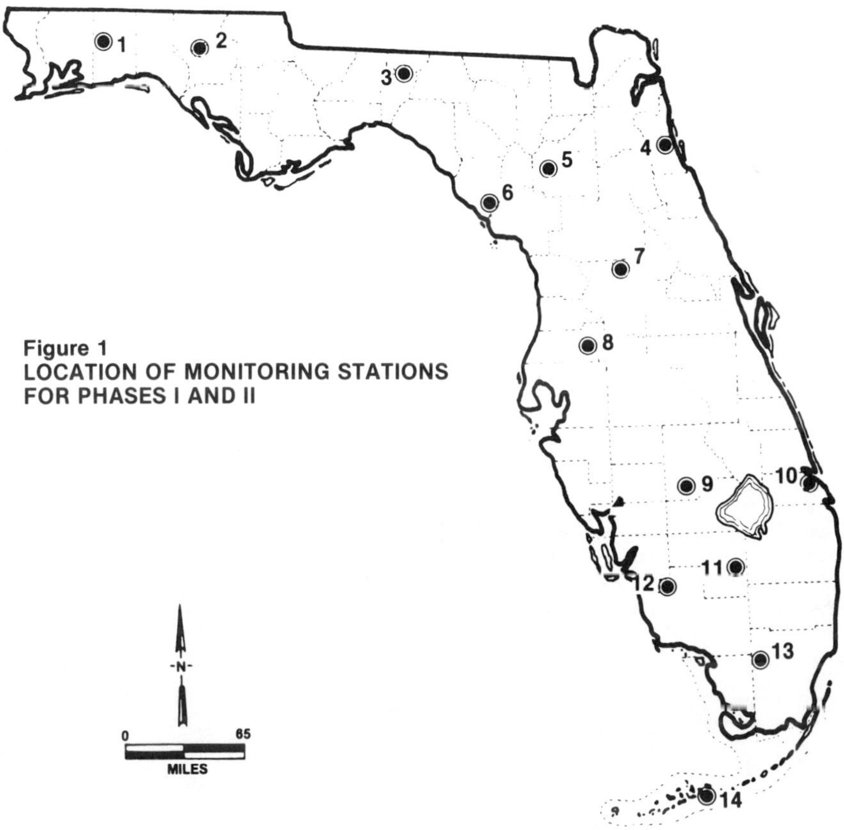

Figure 1
LOCATION OF MONITORING STATIONS
FOR PHASES I AND II

Table 1. Concentrations of Phase II Category I Parameters for the Sampling Period September 15, 1981 to September 14, 1982 (Volume Weighted Means)*

Site Number†	Field pH	Laboratory pH	DpH**	Ca	Mg	K	Na	NH_4-N	NO_3-N	Cl	SO_4	Cation/ Anion Ratio	Total Rainfall Amount (rain gage)
1	4.38	4.47	4.48	0.080	0.037	0.025	0.229	0.07	0.155	0.56	1.41	1.01	66.96
2	4.33	4.42	4.44	0.083	0.043	0.025	0.275	0.09	0.183	0.60	1.50	1.06	49.74
2C	4.31	4.41	4.44	0.085	0.040	0.023	0.274	0.08	0.181	0.55	1.48	1.11	49.74
3	4.36	4.49	4.53	0.089	0.041	0.039	0.227	0.06	0.132	0.70	1.09	1.03	37.33
4	4.41	4.58	4.59	0.093	0.097	0.035	0.805	0.03	0.114	1.55	1.07	0.94	64.25
5	4.51	4.56	4.58	0.086	0.033	0.022	0.187	0.05	0.134	0.41	1.16	1.05	63.66
5D	4.49	4.55	4.59	0.088	0.039	0.030	0.225	0.07	0.135	0.48	1.15	1.08	62.91
6	4.42	4.53	4.56	0.125	0.048	0.023	0.296	0.06	0.130	0.78	1.24	0.99	53.71
7	4.57	4.69	4.69	0.128	0.057	0.061	0.277	0.10	0.151	0.58	1.23	0.98	68.68
8	4.51	4.59	4.60	0.143	0.052	0.040	0.342	0.14	0.173	0.69	1.59	0.99	48.26
8C	4.45	4.50	4.49	0.149	0.052	0.026	0.335	0.10	0.176	0.78	1.59	0.97	48.26
9	4.60	4.77	4.79	0.106	0.055	0.021	0.349	0.06	0.120	0.78	0.79	0.97	52.52
9D	4.66	4.88	4.87	0.120	0.067	0.024	0.337	0.07	0.123	0.67	0.81	0.99	53.09
10	4.61	4.71	4.70	0.152	0.121	0.045	1.008	0.06	0.132	1.91	1.02	1.00	57.21
11	4.61	4.78	4.77	0.156	0.059	0.023	0.355	0.07	0.125	0.86	0.81	0.99	56.90
12	4.63	4.77	4.79	0.127	0.077	0.021	0.386	0.04	0.107	0.99	0.65	0.98	51.22
13	4.70	4.84	4.86	0.211	0.071	0.031	0.500	0.06	0.133	1.00	0.79	1.00	58.30
14	4.63	4.89	4.87	0.130	0.136	0.040	1.058	0.02	0.070	2.02	0.71	0.99	46.76

* All concentrations as mg/L, except pH and cation/anion ratio (which are unitless) and rainfall amount (measured as inches)

† C = collocated
D = daily

** DpH = pH of sample after being filtered in the laboratory

Source: ESE, 1983

Table 2. Concentrations of Phase II Category II Parameters for the Sampling Period of September 15, 1981 to September 14, 1982 (Volume Weighted Means)*

Site Number†	PO_4-P**	F^-	Al^{3+}	V^{4+}**	DOC	Total Acidity	Bound Acidity	Alkalinity
1	<9	4	3.2	<0.6	1.32	52.2	24.0	<2
2	<9	4	3.5	<0.6	2.02	57.1	25.9	<2
2C	<9	4	3.6	<0.6	2.94	63.2	32.1	<2
3	<9	<4	3.4	<0.6	1.97	51.2	25.2	<2
4	<9	<4	2.8	<0.6	1.29	47.8	26.7	<2
5	<9	4	3.1	<0.6	1.67	48.6	25.8	<2
5D	<9	4	2.6	<0.6	0.66	42.5	21.9	<2
6	<9	4	3.3	<0.6	1.57	51.4	26.1	<2
7	11	6	3.0	<0.6	1.10	41.3	21.5	<2
8	14	14	4.2	0.6	1.10	47.0	29.9	<2
8C	<9	15	4.0	0.6	1.37	51.6	21.1	<2
9	<9	4	2.0	<0.6	1.23	34.1	19.6	<2
9D	11	4	2.0	<0.6	0.79	31.6	20.6	<2
10	<9	<4	2.7	<0.6	1.07	38.6	22.0	<2
11	<9	<4	<2.0	<0.6	1.02	34.5	20.2	<2
12	<9	<4	2.4	<0.6	1.05	32.7	18.1	<2
13	<9	<4	2.6	<0.6	1.19	33.5	21.5	<2
14	<9	<4	3.5	<0.6	1.56	32.0	20.3	<2

* All concentrations as ug/L, except DOC (mg/L) and total acidity, bound acidity, and alkalinity (ueg/L)

† C = collocated
D = daily

** Greater than 90 percent of analyses for each site were below the detection limit

Source: ESE, 1983

Results of Data Analysis

Phase II results indicate that significant patterns in precipitation chemistry exist across Florida. In general, the annual rainfall acidity decreases from north to south in Florida by about a factor of 4. North Florida monitoring sites observed annual average laboratory pH values in the range of 4.4 to 4.6, while South Florida sites observed averages in the range of 4.7 to 4.9. Nitrate and non-marine sulfate exhibited similar statewide patterns for laboratory-pH. Sodium and chloride concentrations increased markedly near the coastal monitoring sites (4, 10, and 14). Statistical tests demonstrated that the observed spatial variation in acidity and various other parameters was significant, provided sites were treated in groups rather than individually. Results of this grouping procedure show that the Panhandle group (Sites 1 through 3) has significantly higher concentrations of H+, XS-SO_4^{2-}, and NO_3- than the northern peninsula group (Sites 4 through 8) which in turn had higher concentrations than the southern peninsula group (Sites 9 through 14).

Comparison of precipitation data for summer and winter demonstrated that concentrations of many species are higher in the summer for North Florida sites but higher in the winter for South Florida sites. This apparent anomaly may be related to seasonal patterns of rainfall in Florida. Atmospheric deposition of acidity and related parameters across the state was found to be roughly 2 to 2.5 times higher in summer than winter. Exceptions to this finding were the sea-salt species (sodium, magnesium, and chloride) and the soil-derived species (calcium). Presumably, higher wintertime deposition of these species is due to enhanced airborne concentrations caused by high wind speeds and lower quantities of precipitation.

Multiple regressions analyses show that nitrate, ammonium, and non-marine sulfate, chloride, and calcium account largely for precipitation acidity. At most sites during both seasons, sulfate appears to contribute the largest share to acidity, followed by nitrate and chloride. Ammonium and calcium apparently serve to neutralize a portion of the acidity donated by sulfate, nitrate, and chloride.

Comparison of data from weekly-weekly and weekly-daily collocated samplers shows that concentrations in nearby collectors frequently differ by greater than 10 percent. The minor cations NH_4+-N and K^+ differed substantially at three of the four collocated sites, while laboratory-pH, sulfate, and nitrate were typically very similar in collocated samplers. Weekly-daily comparisons showed greater variability than weekly-weekly comparisons, perhaps due to the inability to analyze very small samples from daily monitors.

Comparison of field-pH and laboratory-pH shows that the two differ, to a high degree of certainty, at all sites. Results also show that sulfate, nitrate, and chloride are not present in sufficient amounts to account for observed field-pH values. This finding suggests that unknown acids, such as naturally derived organic acids, may contribute to the acidity of Florida precipitation. Ongoing studies during Phase III will address this question.

Historical Data

Since 1974, four monitoring networks have collected wet-only precipitation in various parts of the state: the University of Florida (UF), the National Aeronautics and Space Administration (NASA), the National Atmospheric Deposition Program (NADP), and the Florida Acid Deposition Study. Bulk precipitation data from the UF network is not included in this comparison due to the large differences in the methodology of sample collection between wet-only sampling and bulk sampling. Results of these investigations (Table 3) suggest that precipitation pH as measured in the laboratory has remained stable during the last 5 years and that the various monitoring efforts produce data which are generally consistent. At Tallahassee, the location with the longest time interval between measurements, the annual pH was 4.53 in 1974 and 4.59 in 1980. At Gainesville, the site with the longest continuous period of analysis (1976 to 1981), annual pH values ranged from 4.51 to 4.66. Florid Acid Deposition Study Phase II laboratory-pH results for Monticello (30 km east of Tallahassee) and Gainesville (15 km west of the UF site) were 4.49 and 4.56, respectively.

Table 3. Summary of Historical Volume-Weighted Mean pH Data for Florida Wet-Only Precipitation (Laboratory Measurements Only)

Sampling Location	Time Period							
	1974	1976-77	1977-78	1978	1978-79	1979	1980	1981
Jay						4.63(5)	4.51(6)	4.52(7)
Tallahassee	4.53(8)†						4.59(6)	
Bradford Forest							4.69(9)	4.65(9)
Austin-Carey Forest							4.66(9)	4.76(9)
Gainesville		4.53(10)		4.65(5)		4.53(5)	4.54(6)	4.51(7)
Cedar Key						4.51(5)	4.53(6)	
Apopka				4.57(5)				
Tampa							4.53(6)	
Kennedy Space Center			4.47(11)		4.61(11)			
Belle Glade*				5.24(5)		5.49(5)		
Corkscrew Swamp							4.82(6)	
Tamiami Trail							4.83(6)	
Homestead							4.95(9)	5.34(9)

* Examination of these data reveal high concentrations of calcium (Ca^{2+}) and magnesium (Mg^{2+}). This monitor was located east of Belle Glade in the Everglades Agricultural District at the University of Florida, Institute of Food and Agricultural Sciences (IFAS) Research Station. This area is characterized by muck deposits, calcareous soils, and intense agricultural activity. These pH data are artificially high due to localized entrainment of alkaline soil-derived materials and particulates from sugar cane-burning activities in the wet-only samples.

† (5) = Brezonik et al., 1982
(6) = Edgerton and Brezonik, 1981
(7) = Edgerton, 1981
(8) = Likens, 1976
(9) = Peden, 1983
(10) = Hendry and Brezonik, 1980
(11) = Madsen, 1980

Source: ESE, 1983

Previous data also support the spatial variation of pH observed during Phase II. Results of the current network, as well as the UF and NADP networks, indicate roughly a 0.3 pH unit increase (i.e., a halving of free acidity) along a north-south transect of the state. Relatively high pH values for Belle Glade and Homestead (1981) may reflect site-specific rather than area-wide conditions.

In summary, the available data for wet-only precipitation pH in Florida indicate neither a major increase or decrease in free acidity since about 1974. Examination of pH data for Gainesville since about 1976 shows values that are consistent over time and consistent with Florida Acid Deposition Study Phase II results for nearby Site 5. Comparison of data from three statewide networks indicates a persistent north-south gradient in pH, despite relatively high pH values reported at the NADP sites.

Literature Cited

1. Environmental Science and Engineering, Inc., 1982. Florida Acid Depositon Study: Monitoring Program, Phase I Summary Report, Gainesville, Florida, January 1982.

2. Environmental Science and Engineering, Inc., 1982. Florida Acid Deposition Study: A Literature Review of the Ecological and Material Effects of Acid Deposition. Gainesville, Florida, January 1982.

3. Environmental Science and Engineering, Inc., 1982. Source Attribution Evaluation: Phase I Summary Report.

4. Environmental Science and Engineering, Inc., 1983. Florida Acid Deposition Study: Phase II Report, Gainesville, Florida.

5. Brezonik, P.L., Hendry, C.D., Edgerton, E.S., Schulze, R.L. and Crisman, T.L., 1982. Acidity, Nutrients and Minerals in Atmospheric Precipitation Over Florida: Deposition Patterns, Mechanisms, and Ecological Effects. EPA Grant No. 805560, University of Florida, Gainesville, Florida.

6. Edgerton, E.S. and Brezonik, P.L., 1981. Acid Rainfall in Florida: Results of a Monitoring Network for 1980. Completion Report for Project Funded by the Engineering and Industrial Experiment Station, Special Water Research Fund, College of Engineering, University of Florida, Gainesville, Florida.

7. Edgerton. E.S., 1981. Personal Communication, Department of Environmental Engineering Sciences, University of Florida, Gainesville, Florida.

8. Likens, G.E., 1976. Acid Precipitation, Chemical and Engineering News, 54(48): 29-44.

9. Peden, M., 1983. Personal Communication, Illinois State Water Survey, Champaign/Urbana, Illinois.

10. Hendry, C.D. and Brezonik, P.L., 1980. Chemistry of Precipitation at Gainesville, Florida, Environmental Science and Technology, 14:843-849.

11. Madsen, B.C., 1980. Acid Rain at Kennedy Space Center, Florida: Recent Observations, Atmospheric Environment, 15(5): 853-862.

ACID PRECIPITATION IN FLORIDA: A BIBLIOGRAPHIC ESSAY

Ann King, Dolores Jenkins and Barry Hartigan
Energy Committee, University of Florida Libraries
Gainesville, Florida 32611

Abstract

A combination of on-line information retrieval and utilization of specialized reference sources can provide strong support for research on environmental topics. This paper will discuss the development of appropriate terminology--both controlled and free-text--in the context of combined on-line/manual literature retrieval dealing with acid rain in Florida. It will indicate the rationale for choosing among a number of available on-line databases in terms of coverage of specialized or unique literature, overlap in the publications indexed, degree and depth of indexing, cost vs. quality and quantity retrieved. The paper also presents problems which influence the development of an appropriate search terminology strategy.

Introduction

The bibliography which accompanies this discussion is the result of the combined efforts of several members of the University of Florida Libraries Energy Committee. This committee was formed by the Library Administration during 1980 in response to the dramatic price increases for energy resources. Its initial mission was to review and evaluate energy-related materials held by University of Florida Libraries. Subsequent activities of the group have included similar bibliographic searches for other research projects.[1]

Search Strategy and Technique

Nature of the Databases

The selection of the most useful and appropriate databases is crucial to the performance of a successful on-line search. Databases, like their print counterparts, vary in scope and in their coverage of government publications, report and journal literature, conference proceedings, research-in-progress summaries, and the patent literature. They also differ in their emphasis on specific subject specialties, overlapping to some degree on the publications indexed, while varying in the level of indexing terminology used. In addition, databases may display unique features such as the Registry Number capability developed by Chemical

---

[1]A. E. S. Green, editor, 1981. An Alternative to Oil: Burning Coal With Gas. University Presses of Florida, Gainesville.

A. E. S. Green, editor, 1980. Coal Burning Issues. University Presses of Florida, Gainesville.

Abstracts, or the Classification Code developed by INSPEC (Information Services in Physics, Electrotechnology, Computers and Control). Such tools enable the information specialist to perform precise and cost-effective searches.

When the Energy Committee was first approached for a search of the literature concerning acid rain, an initial appraisal of the available databases (now over 170) revealed 18 databases which should contain appropriate citations. The databases considered may be grouped into three broad categories:

BIOLOGICAL AND CHEMICAL SCIENCES

- Agricola
- Aquatic Science Abstracts
- BIOSIS
- CA Search
- Environmental Bibliography
- Enviroline
- Pollution Abstracts

PHYSICAL SCIENCES AND ENGINEERING

- DOE Energy
- Electric Power Database
- INSPEC
- Compendex
- Georef
- Meteorological and Geoastrophysical Abstracts
- Water Resources Abstracts
- APTIC (Air Pollution Technical Information Center)

GOVERNMENTAL AND REPORT LITERATURE

- GPO Monthly Catalog
- NTIS

INTERDISCIPLINARY

- Conference Papers

Terminology and Search Strategy

The importance of terminology cannot be overemphasized. The more variation incorporated into the search, the more comprehensive and accurate will be the results. Indeed, adding variety and subtlety to the search strategy can do much to neutralize indexing problems such as limited fields open for searching, lack of abstracts, lack of depth in indexing terminology used. Truncation of terms with the question mark allows the searcher to add variation to the strategy without repetition. Use of the Boolean logical operator "or" will increase the variation of terms while allowing the system to eliminate duplicate records. The command "and" provides a means of combining groups of words to specify the detailed aspects of a broader topic or the combination of topics.

The initial round of discussions about strategy focused on acid rain/precipitation in general, with no geographic restrictions. In August, ICAAS asked that our preliminary search activities be presented at a luncheon meeting. At this meeting the broad search strategy was presented, along with a comparison of the various databases with their print versions. A number of databases were reviewed through a Dialindex search survey using the broad terminology, "acid?(w)precip?" and "acid?(w)rain?." Over 5000 citations were posted.

Following the luncheon meeting, additional terms were suggested for addition to the search strategy. These terms included "sulfur dioxide," and "nitrous oxide," to be used in conjunction with "coal burning," and "power plants." Later discussions brought out the terms "auto emissions," "pulp and paper mills," and "acid deposition." Once subject terminology had been developed, it was apparent that some regional limitations should be included. At this point, geographical terms were incorporated into the terminology.

After further discussion, the following search strategy was used:

acid?(w)rain	Florida
acid?(w)precipitation	Alabama
sulfur(w)dioxide	Georgia
sulphur(w)dioxide	gulf(3w)Mexico
so2	gulf(3w)states
nox	
nitrous(w)oxide	
acids?(w)deposition	

The terms in the first column were grouped together using the operator "or" to allow selection of citations having any one of these terms, as is the case with the second column. The two groups of citations were then paired using the command "and" to produce citations having the regional limitation.

The final task was the choice of databases to be searched based on the terminology. Eight databases were selected for their ability to pinpoint the environmental literature efficiently. DOE Energy was chosen for its comprehensiveness; it covers governmental publications, report literature, and journal articles in the broad area of energy. Meteorological and Geoastrophysical Abstracts, Electric Power Database, and Water Resources Abstracts were chosen for their ability to cover, with a degree of overlap, the physical sciences and engineering aspects. Water Resources Abstracts is particularly important for its emphasis on state and regional publications. Enviroline was chosen for its thorough coverage of the conference and report literature, and for its document support through the Envirofiche collection, available in the Agriculture Library. Pollution Abstracts was needed for its coverage of the technical literature. Agricola offers unique coverage of the impact of acid rain on agriculture. Paperchem was chosen as an additional database because it covers publications dealing with kraft mill pollution.

By focusing on a group of databases which are predominantly environmental, the search relies on the depth of indexing within the databases. Specificity is improved and it becomes unnecessary to search much broader databases such as CA Search (Chemical Abstracts).

Other Sources

Although on-line searching does allow the information specialist to access the literature in unique and imaginative ways, a literature review would not be complete without a manual search of the printed sources. Serendipity has led to a number of articles concerning Florida's acid rain problems. An EPRI (Electric Power Research Institute) bibliography has provided a citation to a Florida Power and Light study on acid rain produced by Environmental Sciences and Engineering, Inc., a consulting firm located in Gainesville. This study contains a substantial general bibliography on acid rain. Thus, by utilizing both on-line and manual sources, this comprehensive bibliography on acid rain has been developed.

Bibliography

Acid Air and Lung Cancer: The Numbers Add Up in Florida, 1982. University of Florida Engineering and Industrial Experiment Station Bulletin Series, 7(3):14-15.

Beaver, J.R. and Crisman, T.L., 1980. Acid Precipitation and the Response of Ciliated Protozoans in Florida Lakes, Proceedings: International Association of Theoretical and Applied Limnology, 21(1): 353-358.

Benzing, D., 1982. Everglades Park Bromeliads Face Possible Danger, Journal of the Bromeliad Society 32(2):75-78.

Brezonik, P.L., 1972. Nitrogen: Sources and Transformation, Nutrients in Natural Waters, H.E. Allen and J.R. Kramer (Eds.) Wiley, p. 1-50.

Brezonik, P.L., 1976. Nutrients and Other Biologically Active Substances in Atmospheric Precipitation, First Specialty Symposium on Atmospheric Contribution to the Chemistry of Lake Waters, International Association for Great Lakes Research, p. 166-186.

Brezonik, P.L., Edgerton, E.S. and Hendry, C.D., 1980. Acid Precipitation and Sulfate Deposition in Florida, Science 208(4447):1027-1029.

Brezonik, P.L., Hendry, C.D., Edgerton, E.S., Schultz, R.L. and Crisman, T.L., 1981. Nutrient and Acid Deposition by Atmospheric Precipitation in Florida: Spatial Patterns and Ecological Significance, EPA Project Completion Report.

Brezonik, P.L., Hendry, C.D. and Edgerton, E.S. Loading Rates and Ecological Significance of Nutrients and Acidity in Wet and Dry Atmospheric Precipitation, EPA Project Report.

Brezonik, P.L. and Messer, J.J., 1977. Analysis of Trophic Conditions and Eutrophication Factors in Lake Weir, Florida, North American Project-- A Study of U.S. Water Bodies, L. Seyb and K. Randolph (Eds.), Environmental Protection Agency EPA-600/3-77-086 p. 1-24.

Brezonik, P.L., Morgan, W.H., Shannon, E.E. and Putnam, H.D., 1969. Eutrophication Factors in North Central Florida Lakes, University of Florida Engineering and Industrial Experiment Station Bulletin Series 133:101.

Brezonik, P.L. et al., 1979. Factors Affecting Primary Productivity in Lake Okeechobee, Florida, University of Florida Department of Environmental Engineering Sciences Report No. Env. 05-79-01.

Conservation Consultants, Inc., 1979. Estech General Chemicals Corporation, Duette Mine, Manatee County, Florida, Resource Documents: Air Quality, Draft Environmental Impact Statement, NTIS PB 110091, 215 p.

Craig, J.M., Bell, B.A., Lipuma, T.A. and Allison, J.K., 1974. Lime/Limestone Scrubbing in a Pilot Dustraxtor-Key West, NTIS Report PB-237 622.

Crisman, T.L., Schultz, R.L., Brezonik, P.L., Bloom, S.A., Drabloes, D. and Tollan, A., 1980. Acid Precipitation: The Biotic Response in Florida Lakes, Ecological Impact of Acid Precipitation. International Conference on the Ecological Impact of Acid Precipitation, Sandefjord, Norway, NTIS CONF-800365, p. 296-297.

Department of Agriculture, 1979. Draft Environmental Impact Statement, Seminole Plant Units 1 and 2 and Associated Transmission Facilities, Information Resources Press.

Dry Scrubbing of Utility Emissions, 1975. Environmental Science and Technology 9(8):712.

Duxbury, J.M., Bouldin, D.R., Terry, R.E., and Tate, R.L. III, 1982. Emission of Nitrous Oxide from Soils, Nature 298(5873):462-464.

Edgerton, E.S., Brezonik, P.L. and Hendry, C.D., 1981. Atmospheric Deposition of Acidity and Sulfur in Florida, Atmospheric Pollutants in Natural Waters, S.J. Eisenreich (Ed.), Ann Arbor Science Press, p. 237-258.

Environmental Protection Agency, 1974. Implementation Plan Review for Florida as Required by the Energy Supply and Environmental Coordination Act, Environmental Protection Agency, EPA 450/3-74-078, 23 p.

Environmental Protection Agency, 1981. Jacksonville Electric Authority, St. Johns River Power Park. FDER State Analysis Report: Technical Reference Document, Environmental Impact Statement, 919 p.

Ewel, K.C., Gamble, J.F. and Lugo, A.E., 1976. Aspects of Mineral-Nutrient Cycling in a Southern Mixed Hardwoods Forest in North-Central Florida, Mineral Cycling in Southeastern Ecosystems, F.G. Howell, J.B. Gentry and M.H. Smith (Eds.), ERDA Symposium Series, Conference 740513.

Florida Acid Deposition Study, Gainesville, 1981. Environmental Sciences and Engineering, Inc., v.1-

Florida Sulfur Oxides Study, Inc., 1978. Florida Sulfur Oxides Study. Orlando, FL. 10 vols.
Winchester, J.W. and Leslie, A.C.D., January 1978. Florida Sulfur Aerosol Characterization.
SRI International, January 1978. Florida Utility Pollution Control Options and Economic Analysis. 2 vols.

Interdisciplinaty Center for Aeronomy and (Other) Atmospheric Science (ICAAS), February 1978. An Interdisciplinary Study of the Health, Social and Environmental Economics of Sulfur Oxide Pollution in Florida.

Dames and Moore, February 1978. Dispersion, Meteorology and Climatology. 3 vols.

Environmental Research and Technology, Inc., February 1978. Florida Sulfur Oxide Study: Environmental Effects.

State of Florida Department of Environmental Regulation and PEDCO Environmental, Inc., April 1978. Air Quality Monitoring and Data Assessment.

Florida Sulfur Oxides Study, Inc., May 1978. Final Report of Florida Sulfur Oxides Study.

Galloway, J.N. and Whelpdale, D.M., 1980. An Atmospheric Sulfur Budget for Eastern North America, Atmospheric Environment 14:409-417.

Green, A.E.S., Schwartz, J.M., Singhal, R.P. and Bolch, W.E., 1982. Wind Roses for Florida, Journal of the Air Pollution Control Association 32(8):822-825.

Hansen, D.A. and Hidy, G.M., 1982. Review of Questions Regarding Rain Acidity Data, Atmospheric Environment 16(9):2107-2126.

Helms, G.T., 1976. Overview of Regional Fuel Combustion Regulations: Particulate Matter and Sulfur Oxides, Proceedings of the Stationary Source Combustion, Atlanta, GA, Vol. 1. Fundamental Research, p.429-441.

Hendry, C.D., 1977. The Chemistry of Precipitation in North-Central Florida, M.S. Thesis, University of Florida, College of Engineering.

Hendry, C.D., Brezonik, P.L. and Edgerton, E.S., 1981. Atmospheric Deposition of Nitrogen and Phosphorus in Florida, Atmospheric Pollutants in Natural Waters, S.J. Eisenreich (Ed.), Ann Arbor Science Press, p. 199-215.

Hendry, C.D. and Brezonik, P.L., 1980. Chemistry of Precipitation at Gainesville, Florida, Environmental Science and Technology 14(7): 843-849.

Hendry, C.D., Brezonik, P.L., Edgerton, E.S., Drabloes, D. and Tollan, A., 1980. Acid Precipitation in Florida (USA): Results of a Statewide Monitoring Network, International Conference on the Ecological Impact of Acid Precipitation, Sandefjord, Norway, NTIS CONF 800365, p. 118-119.

Henmi, T. and Reiter, E.R., 1978. Regional Residence Time of Sulfur Dioxide Over the Eastern United States, Atmospheric Environment 12: 1489-1495.

Hiers, C.W., Libey, J.A. and Wisely, F.E., 1980. Lakeland, Florida: A Simple Approach to Burning Refuse as a Power Plant Fuel, Proceedings of 1980 National Waste Processing Conference.

Jenkins, S.D. and Lyall, A.E., 1981. Technical/Economic Issues and Reliability Requirements for Tampa Electric Company's Big Bend Unit Number 4 FGD System, American Society for Mechanical Engineers Paper 81-JPGC-Pwr-35.

Joyner, B.F., 1974. Chemical and Biological Conditions of Lake Okeechobee, Florida, 1969-72, U.S. Geological Survey Report of Investigation no. 71.

Kappelmann, R.L. and Dyer, R.T., 1978. Utilization of Power Production and Air Quality Models to Develop a Cost Effective Sulfur Dioxide Emission Regulation: Jacksonville, Florida...A Case Study, Air Pollution Control Association 71st Annual Meeting and Exhibition: Abstracts of Papers.

Madsen, B.C., 1981. Acid Rain at Kennedy Space Center, Florida: Recent Observations, Atmospheric Environment 15(5):853-862.

Malchon, J., 1979. Health and Property Effects of Air Quality, Clearing the Air: Policy and Technique, Illinois-Indiana Bi-State Commission Air Quality Conference, Chicago, IL, USA, p. 7-12.

Messer, J., 1975. An Analysis of Nutrient Loading and Trophic Conditions in Lake Weir, Florida, M.S. Thesis, University of Florida, College of Engineering.

Messer, J., 1978. Nitrogen Mass Balances in Florida Ecosystems, Ph.D. Thesis, University of Florida, Department of Environmental Engineering Sciences.

Messer, J. and Brezonik, P.L., 1981. Importance of Atmospheric Fluxes to the Nitrogen Balance of Lakes in the Florida Peninsula, Atmospheric Pollutants in Natural Waters, S.J. Eisenreich (Ed.), Ann Arbor Science Press, p. 217-236.

Messer, J. and Brezonik, P.L. Regional Nitrogen Mass Balance Model for the Florida Peninsula, Proceedings of the Fourth International Symposium on Environmental Biogeochemistry, Canberra, Australia.

Meyer, C.D. and Pechan, E.H., 1979. Analysis of Environmental Factors Affecting Technology Deployment: A Case Study, Department of Energy, Office of Resource Applications, 69 p.

Moseley, C.L., 1980. Health Hazard Evaluation Determination Report: HE 79-124-682: Orlando International Airport, Orlando, Florida, National Institute for Occupational Safety and Health, 10 p.

Munger, J.W., and Eisenreich, S.J., 1983. Continental-Scale Variations in Precipitation Chemistry, Environmental Science and Technology 17 (1):32A-42A.

Ogren, J.A., Blumenthal, D.L., Anderson, J.A. and White, W.H., 1978. Pollutant Measurements in Plumes from Power Plants and Cities, Summer 1975, February 1976, and February 1977, A Project MISTT Report. Final report, 18 June 1975--18 October 1977, Meteorology Research, Inc., 63 p.

Roddy, C.P., 1975. Maintaining Clean Heat Exchangers, Power 119(3):56.

Rural Electrification Administration, 1979. Final Environmental Impact Statement. Seminole Plant Units 1 and 2 and Associated Transmission Facilities, Information Resources Press, 163 p.

Saulmon, W.D., 1977. Tougher Air-Quality Standards Face Sulfur-Recovery Plants, Oil and Gas Journal 75(21):53.

Sheih, C.M., Wesely, M.L. and Hicks, B.B., 1979. Estimated Dry Deposition Velocities of Sulfur Over the Eastern United States and Surrounding Waters, Atmospheric Environment 13:1361-1368.

Spaite, P., Esposito, M.P., Szabo, M.F. and Devitt, T.W., 1980. Acid Rain: The Impact of Local Sources, PEDCO-Environmental, Inc. NTIS DOE/METC/14787-103, 28 p.

Tanaka, S., Darzi, M. and Winchester, J.W., 1980. Short Term Effect of Rainfall on Elemental Composition and Size Distribution of Aerosols in North Florida, Atmospheric Environment 14:1421-1426.

Tanaka, S., Darzi, M. and Winchester, J.W., 1980. Sulfur and Associated Elements and Acidity in Continental and Marine Rain from North Florida, Journal of Geophysical Research 85(C8):4519-4526.

Terry, R.E., Tate, R.L. and Duxbury, J.M., 1981. The Effect of Flooding on Nitrous Oxide Emissions from an Organic Soil, Soil Science 132(3):228-232.

Terry, R.E., Tate, R.L. III and Duxbury, J.M., 1981. Nitrous Oxide Emissions from Drained, Cultivated Organic Soils of South Florida, Journal of the Air Pollution Control Association 31(11):1173-1176.

Valencia, J.A., Peirson, J.F. Jr. and Ramans, G.J., 1981. Evaluation of the Limestone Dual Alkali Prototype System at Plant Scholz: Final Report, Report for September 1978-March 1981, 113 p.

Watkins, J.J., 1978. St. Regis Study Finds Hydropyrolysis More Beneficial than Tomlinson Furnace, Southern Pulp and Paper Manufacturer 41(4):20, 22, 24-25; 41(6):26, 28-30, 32-34.

Winchester, J.W. and Vogt, J.R., 1977. Atmospheric Aerosol Chemistry of Sulfur: Nuclear Accelerator Methods in Energy-related Research, Nuclear Methods in Environmental and Energy Research, International Conference on Nuclear Methods in Environmental and Energy Research, Columbia, MO, 10 October 1977, p. 1-10.

Wisniewski, J. and Cotton, W.R., 1977. Chemical Analysis of South Florida's Environment, Precipitation Scavenging Conference, Champaign, IL, 1974. Energy Research and Development Administration.

THE SCIENTIFIC NEEDS OF ENVIRONMENTAL ECONOMISTS

Thomas A. Lynch
Florida Department of Environmental Regulation
Tallahassee, Florida 32304

Abstract

The economist must have physical and biological science conclusions on environmental effects that not only indicate the nature of the problem but the correct functional form of that causal relationship to enable meaningful behavioral inference into the extent of direct and indirect economic costs and benefits. The economist must help bridge the gap between science models and economic models. As in the sciences, these models will often reflect the levels of our ignorance as well as the levels of our knowledge. The economist, therefore, starts socioeconomic analysis near one half the distance of the chasm. These models once specified, with appropriate limits clarified, become the foundation of estimating direct effects. Such direct effects are more closely understood even if the precision of direct measures are still under considerable debate. Secondary economic effects, while more subtle and complex, are still less quantifiable than direct effects. Their complex nature may, however, render them more important than (often limited) direct effects. The task of the environmental economist, then, is to inquire into direct and indirect economic effects as deeply as the physical and biological literature allows, with large uncertainty and critical thinking as his constant companions.

Introduction

The role of the environmental economist in estimating the costs and benefits of acid rain controls is similar to the role an economist plays in any other evaluation procedure. Since economics is the study of the distribution and consumption of scarce resources among competing uses, environmental economics' charge is to track this theoretical foundation in evaluating which scarce resources are or will be consumed at varying levels of acidic deposition and what level of benefits, or satisfaction, is associated with this consumption?

Economists share the same large measures of uncertainty facing physical and biological scientists examining acid rain related effects. The unknowns are due to the limits in precision of scientific understanding of interacting biosphere mechanisms and the lack of adequate historic data. Furthermore, only recent emphasis has been given to acid rain research; therefore, most research must, by its very nature, be considered preliminary. These limits facing the physical scientist pose a major analytical constraint on the economist as well. It does not, however, preclude the environmental economic analyst from taking the first steps of attempting to merge physical and biological sciences research into a meaningful theoretical model of economic costs and benefits, and begin to bound this universe of uncertainty.

Constraints of the unknown face all economists, and there are limits on every analytical inquiry. We do not know enough in most areas due to the large number of interacting factors and the uncertainty of cause and effect relationships among them. An example of this is our inability to estimate, with any degree of certainty, the results of a given economic policy such as fiscal-monetary controls, price regulations, tax policy, supply side economics or any other of a myriad of economic policies. These and a host of other measures are enacted daily at the national and state levels with little concern for the extent of that uncertainty. We make policy and reap harvest without the wrangling or wringing of hands over the uncertainty equal to or greater than that of dealing in environmental standards and estimations of costs and benefits associated therein. The economic costs of erring by a fraction of a percent in such measures frequently dwarf the costs of erring in great magnitudes on the environmental issues. While we do not know enough to suggest that there are no substantial and costly economic damages from acidic deposition, at this point we clearly do not know enough to throw caution to the wind and go about with our business-as-usual attitude with no concern for these unknowns and their potentially staggering adverse economic costs.

The environmental economist must have physical and biological science conclusions on environmental effects that indicate both the nature and form of these causal relationships to enable meaningful inferences into the extent of direct and indirect costs and benefits. The economist must help bridge that gap between physical science and economic models, and as in the physical and biological sciences, these models will often reflect the level of our ignorance as well as the level of our knowledge.

The following gross outline is a preliminary overview of the areas of investigation of the environmental economist in translating physical and biological science findings into estimates of Costs and Benefits (1-5).

A. Direct Effects: 1) Agriculture and Soil, Losses or Gains, 2) Forestry and Soil, Losses or Gains, 3) Aquatic Effects, 4) Material Effects, 5) Health Effects, 6) Other.

B. Indirect Effects: 1) Change in Employment Mix, 2) Change in Quality of Life, 3) Change in Energy, Manufacturing Costs and Inflation, 4) Other.

C. Economic Cost of Compliance for Regulatory Measures: 1) Transportation, 2) Industrial, 3) Governmental Institutional Costs.

Traditional Dose-Response Model--Economic Impact Analysis

The functional form of the biological response of a population of interest can take many forms. For some species it may be linear; others may react in an all-or-nothing step function where below a clearly delineated threshold the species simply cannot survive. Frequently biological populations respond to pollution loadings in a nonlinear pattern. One familiar form for a biological dose-response function in pollution related analysis is the well known logistic "S" curve. The populations of interest respond very slowly at first to changes in the independent (pollution) variable but tend to react more swiftly to changes near some midrange value. The midpoint or point of inflection is where the rate of change slows before the curve approaches the limit. This particular form reflects a probability distribution where the value of B controls the rate of response throughout the range.

This functional form has seen extensive use in the ecological economic literature in examining human population morbidity and mortality responses to air pollution (6,7) and aquatic species response to changes in pollutant loadings (2). A more complex and theoretically interesting dose-response methodology for use in benefit-cost analysis was developed by Loehman, Green, et al. (5). They developed a composite index of several air pollutants and estimated a dose-response model for the population at risk from international literature sources. This approach is consistent with the theory that synergistic reactions such as sulfur oxides and particulates may be far more detrimental to human health than either single pollutant at most levels. This dose-response methodology may hold great promise for the future.

The major problem and limitation to such analysis is the requirement for extensive and detailed biological dose-response information and economic information of the population at risk. This experimental evidence is difficult and costly enough to collect for a select few species and parameters, but when one begins to consider the wide number of biological communities that may be involved, the complexity and diversity of the ecosystems under analysis and the synergistic nature of pollutant factors at work, the enormity of the task slows even the ambitious.

Costs of Acid Rain Regulatory Measures

A full environmental benefit/cost analysis which calls upon all the biological and physical scientific inputs undoubtedly represents the future method for developing acid rain regulatory measures. However, at this time estimating the range of cost alternatives of regulatory measures which would restrict the creation of precursors to acid rain is a much simpler and more feasible task. The wide range of potential cost increases for sulfur oxide and nitrous oxide control will include the net increase cost of fuel, capital expenditures and operation and maintenance costs incurred over and above the otherwise unregulated condition. For nitric oxides this may include such things as low excess air burners and pollution abatement equipment for automobiles.

The major sources of precursors for sulfur oxides are electrical generation and industrial manufacturing. The cleanup costs for these types of sources include such expenditure categories as fuel switches, fuel preparation such as coal washing and chemical treatment, in-plant operational modifications, increased operation and maintenance costs, and ultimately, if regulatory requirements are sufficiently stringent, the potentially very large costs of capital investment for flue gas desulfurization or other removal technology.

Finally, the often overlooked cost of public administration and enforcement of any regulatory scheme is important to consider. These costs may be considerable, especially where corrective actions such as ecosystem repair (such as liming lakes) or grants for abatement research technology and equipment purchases are included.

Obviously each regulatory option involves a wide number of potential cost options. The analysis provided in Figure 1 of Tschinkel (this volume) examines four potential statewide sulfur oxide emission scenarios. The costs of complying with each scenario involve fuel switching only and are estimated for the state of Florida electrical generation facilities alone since they emit 80 percent of the state's total SO_2 loadings. The scenarios presented are:

1) Total statewide SO_2 emissions of plants coming on line between now and 1990 are uncontrolled (clearly an unrealistic scenario since New Source Performance Standards for all new coal fired units are mandated by the state and federal standards).

2) Implementation of New Source Performance Standards for new sources and existing sources are allowed to continue to emit at their current 1983 levels. This scenario estimates the most realistic future based on the state electrical industry forecast as set forth in the 1982 "Florida Ten Year Site Plan". No net increase in costs above what is <u>currently</u> forecast to transpire result.

3) Incorporates scenario 2 plus the reduced volume of emissions associated with restricting all existing oil fired units to 1% sulfur oil.

4) Incorporates scenario 2 and 3 plus the increment of reduced emissions associated with restricting all existing coal fired units to 1% sulfur coal.

The respective costs for scenarios 3 and 4 are presented in Table 1. These costs are based on an historic "stable" differential of $1.84 that exists between 1% and 2.5% sulfur fuel oil. Anyone familiar with the vagaries of the oil market realizes that the price differential has been as high as $4.50/barrel and as low as $0.73/barrel over the past five years. Our analysis for the Gulf Coast Spot Market indicates the median differential over the prices of September 1977 through early 1983 is $1.80/barrel. The regression equation presented at the bottom of Table 1 for oil and coal serves as the basis of the "proposed" cost of compliance 1% sulfur oil and coal. This proposed cost increase is then compared with the 1983 cost of the sulfur content that each power plant actually did burn in 1980.

The total cost for scenario 3 therefore becomes $59,605,000 per year. This results in a net reduction of 99,011 tons of sulfur oxide annually in 1983 or 12 percent of the state 1983 total sulfur emissions. Scenario 4 results in an increase in cost of $34,182,000 for purchase of higher price 1% sulfur coal across the state with the resulting reduction in emissions by 252,364 tons per year or over 24 percent of annual 1983 levels. These reductions could improve sulfur oxide air quality in affected areas by as much as 60 $\mu g/m^3$ from existing 24 hour averages, and concurrently reduce particulate levels by four to five $\mu g/m^3$ from their current 24 hour average. The change in pH is not possible to estimate at this time.

Table 1. Acid Rain Scenarios

PLANT	BBLS OIL (1000'S) (1980)	AVERAGE % SULFUR	GAS (MCF)	COST OF FUEL PRESENT PER BBL	PRESENT TOTAL (1000$)	PROPOSED PER BBL	PROPOSED TOTAL (1000$)	ESTIMATED EMISSION PRESENT (TONS)	PROPOSED (TONS)	PROPORT. #6
FLA POWER										
ANCLOTE	7610.00	2.24	0	23.68	180205	25.38	193104	53711	24383	
AVON PARK	332.00	2.21	968	23.70	7868	25.38	8425	2312	1064	
CRYSTAL RIVER	54.00	.21	0	36.89	1992	36.89	1992	40	40	
BAYBORO	114.00	.28	0	33.25	3790	33.25	3790	110	110	
DEBARY	306.00	.67	0	26.88	8226	26.88	8226	667	667	
TURNER	1018.00	1.78	998	24.03	24466	25.38	25832	5730	3262	
HIGGINS	1024.00	2.24	751	23.68	24248	25.38	25984	7227	3281	
INTERCESSION	614.00	.32	0	31.88	19575	31.88	19575	670	670	
BARTOW	3889.00	1.97	106	23.87	92820	25.38	98683	24185	12461	
SUWANNEE	354.00	1.75	6687	24.06	8518	25.38	8983	1959	1134	
TOTAL	15315		9510.00		371709		394594	96611	47071	0.89
COST DIFF.							22885			
FLA POWER & LIGHT										
CAPE CANAVERAL	4208.00	2.20	17299	23.70	99750	25.38	106778	31747	13483	
CUTLER	0.00	0.00	0							
FORT MYERS	5254.00	2.11	0	23.76	124857	25.38	133320	38021	16834	
LAUDERDALE	1094.00	.88	11804	25.79	28217	25.79	28217	3314	3314	
MANATEE	9227.00	.96	0	25.50	235312	25.50	235312	30476	30476	
PORT EVERGLADES	6775.00	0.90	21706	25.72	174221	25.72	174221	20987	20987	
PALATKA	0.00	0	0					0	0	
MARTIN	1131.00	.65	0	27.02	30564	27.02	30564	2536	2536	
PUTNAM	1376.00	.61	0	27.33	37610	27.33	37610	2897	2897	
RIVERIA	1883.00	.92	20539	25.64	48282	25.64	48282	5962	5962	
SANFORD	4725.00	2.13	6416	23.75	112221	25.38	119897	34516	15139	
TURKEY POINT	3728.00	.93	13548	25.61	95457	25.61	95457	11931	11931	
TOTAL	39401		91313		986492		1009658	182387	123559	0.98
COST DIFF.							23167			
FT. PIERCE	0.70	.25	3136	34.56	24	34.56	24	1	1	0
COST DIFF.							0			
GAINESVILLE										
J R KELLY	22.20	1.74	5183	24.07	534	25.38	563	122	71	
DEERHAVEN	247.50	1.49	2833	24.37	6031	25.38	6280	1170	793	
TOTAL	269.70		8015		6566		6844	1292	864	0.99
COST DIFF.							278			
GULF POWER - SMITH	16.80	.45	0	29.12	489	29.12	489	14	14	0
COST DIFF.							0			
JACKSONVILLE E. A.										
KENNEDY	1877.60	1.00	0	25.38	47644	25.38	47644	6490	6490	
NORTHSIDE	5061.00	1.67	0	24.15	122206	25.38	128423	29165	16216	
SOUTHSIDE	2101.50	1.00	0	25.38	53326	25.38	53326	7263	7263	
TOTAL	9040.10		0		223175		229393	42918	29969	0.98
COST DIFF.							6217			
LAKE WORTH	66.80	2.20	3066	23.70	1583	25.38	1695	514	214	1
COST DIFF.							112			
LAKELAND										
LARSEN	142.10	2.32	6054	23.63	3358	25.38	3606	1124	455	
PLANT #3	1045.70	1.24	2271	24.78	25915	25.38	26535	4432	3350	
TOTAL	1187.80		8326		29273		30140	5556	3806	0.97
COST DIFF.							867			
ORLANDO										
INDIAN RIVER	2200.90	1.64	12487	24.18	53218	25.38	55848	12580	7052	
HIGHLAND	32.80	.73	3026	26.51	869	26.51	869	84	84	
TOTAL	2233.70		15513		54087		56717	12664	7135	1.00
COST DIFF.							2630			

Table 1. (Continued)

TALLAHASSEE										
PURDOM	323.60	2.19	2138	23.71	7673	25.38	8211	2479	1037	
HOPKINS	627.20	2.32	8700	23.63	14823	25.38	15915	5091	2010	
TOTAL	950.80		10838		22495		24127	7570	3046	1.00
COST DIFF.							1631			
TECO										
GANNON	3551.10	.86	0	25.87	91879	25.87	91879	10292	10292	
HOOKERS POINT	796.10	.93	0	25.61	20385	25.61	20385	2492	2492	
BIG BEND	201.10	0.30	0	32.52	6540	32.52	6540	209	209	
TOTAL	4548.30		0		118803		118803	12993	12993	0.95
COST DIFF.							0			
KEY WEST										
MONROE	291.40	1.88	0	23.94	6977	25.38	7394	1867	934	
STOCK ISLAND	451.40	2.37	0	23.60	10655	25.38	11454	3644	1446	
TOTAL	742.80		0		17632		18849	5511	5511	0.97
COST DIFF.							1217			
VERO BEACH	332.60	2.44	3184	23.57	7839	25.38	8440	2839	1164	1.00
COST DIFF.							601			
TOTAL	74106.1		152900		1840168	$	1899773	370871	235347	
							59605			
COAL PLANTS										
CRYSTAL RIVER	1920.00	1.94		48.27	92670	51.41	98709	72634	37440	
SANFORD	51.00	0.70		52.41	2673	52.41	2673	696	696	
DEERHAVEN	26.10	0.62		52.68	1375	52.68	1375	316	316	
BIG BEND	2719.70	2.85		45.22	122987	51.41	139822	151147	53034	
GANNON	1127.20	1.28		50.47	56894	51.41	57950	28135	21980	
CHRIST	1712.60	2.52		46.33	79336	51.41	88046	84157	33396	
SCHOLTZ	170.60	3.14		44.25	7549	51.41	8771	10446	3327	
SMITH	956.80	1.10		51.08	48870	51.41	49190	20523	18658	
TOTAL	8684.00				412356	$	446538	368054	168846	
							34182			
TOTAL					2252524	$	2346311	738925	404194	
							93787		-334732	

ASSUMPTIONS			
% SULFUR OIL	1.00		
PRICE/BBL OIL	25.38		
% SULFUR COAL	1.00		
PRICE/TON COAL	51.41	OIL	COAL
(PRICE REGRESSION MODEL)		(INVERSE)	(LINEAR)
		22.313	54.757
		3.062	-3.346

While the direction of the analysis and data presented by this report is only preliminary, it reflects the kind of economic valuations of costs and benefits that are possible in the real world and essential for society to make informed judgements on what levels of costs and benefits are associated with various regulatory control levels. As the physical and biological science models are refined, the economist will be able to more accurately measure the costs and benefits of acid deposition upon society and better gauge the results of our policy and regulatory decisions.

Literature Cited

1. Crocker, T. D. and W. Shultz. Methods Development for Assessing Air Pollution Control Benefits, Vols. 1-7. EPA Contract No. 600/5-79-001 through 600/5-79-007, EPA Grant R805059010, University of Wyoming, Laramie, Wyoming. February 1979.

2. Association of Environmental and Resource Economists (AERE) Newsletter, Vol. 3 no. 1. Essay: Environmental Benefit-Cost Analysis Activities at Mathtech. January 1983.

3. Bell, Frederick. Benefits from Water Pollution Abatement--Coastal Waters, Vols. I and II, PB 252172-02, National Commission on Water Quality, April 1976.

4. Loehman, E.J., S.V. Berg, A.A. Arroya, R.A. Hedinger, J.M. Schwartz, M.E. Shaw, R.W. Fahien, V.H. De, R.P. Fishe, D.E. Rio, W.F. Rossley and A.E.S. Green, Distributional Analysis of Regional Benefits and Cost of Air Quality Control, Journal of Environmental Economics and Management, 6, September 1979.

5. Manual, E.H., Jr., R.L. Horst, Jr., K.M. Brennan, W.N. Lanen, M.C. Duff and J.K. Tapiero. Benefits Analysis of Alternative Secondary National Ambient Air Quality Standards for Sulfur Dioxide and Total Suspended Particulates. Draft Final Report for EPA Contract No. 68-02-3392, Mathtech, Inc., Princeton, N.J.

6. Liu, B. and E.S. Yu. Physical and Economic Damage Functions for Air Pollution by Receptors. Environmental Protection Agency, Corvallis. 1976.

7. Hagerhall, B. Estimation of Economic Damages to Aquatic Ecosystems in Sweden in Relation to Different Emissions Control Cases for 1985. A paper prepared for OECD Workshop on Acid Precipitation, Paris, France (April 24-26, 1979).

SOURCES OF ACIDIC SUBSTANCES IN FLORIDA'S ATMOSPHERE

Hamilton S. Oven, Jr.
Florida Department of Environmental Regulation
Tallahassee, Florida 32304

Abstract

There is a need to know how much intra-state sources of acidic substances contribute to acid deposition in Florida. Utilities are the largest source of sulfur oxides. Vehicles are the largest source of nitrogen oxides. In the recent past, ambient levels of sulfur oxides and nitrogen oxides have been the same. We need to know how this correlates to acid deposition trends and how emissions of acidic precursors can be reduced.

Sources of Acidic Substances

There are a number of issues that need to be addressed in this workshop on acid deposition. The title of this session is "Sources, Atmospheric Transformations, and Transport of Acid Substances in Florida". In reviewing the papers scheduled for presentation here today it appeared appropriate to lay out a framework of questions that need to be answered for us to understand acid deposition.

We need to know what are the sources of acidic substances in the atmosphere, natural and man-made. We need to know how much Florida sources of acidic material contribute to the current situation. We need to know if there is a trend in acid deposition levels, what sources of acidic materials are controllable and which sources we can control feasibly.

Although there is no clearly understood knowledge of what causes acid deposition, correlations can be made between anthropogenic levels of sulfates and nitrates and the pH of acidic depositions. In the United States today it is estimated that some 30 million tons per year of sulfur oxides are discharged to the atmosphere. Twenty million tons per year are emitted east of the Mississippi River. Approximately 16 million tons per year of that come from electric power plants. Currently in Florida over a million tons per year of SO_x are emitted. In 1975 the "Final Report of Florida Sulfur Oxides Study" estimated that two-thirds of the sulfur oxides in Florida come from electric utilities (1). The same study estimated that by 1985 Florida utilities would emit 78% of the sulfur oxides (Table 1). The SO_x study also predicted sources of NO_x (Table 2).

The records of the Department of Environmental Regulation indicate that sulfur oxide emissions were essentially the same from 1975 through 1980 (2). Ambient monitoring data shows no trend in ambient levels of sulfur dioxide or nitrogen dioxide during the same period. It would be interesting to know what acid deposition levels were doing in the state during that same time period.

Of the anthropogenic sources of acid deposition it might be assumed that sulfur oxides comprise two-thirds and nitrogen oxides one-third.

Table 1. Estimated (1975) and Projected (1985) Sulfur Oxides Emissions in Florida

Category	1975 Tons/Year	%	1985 Tons/Year	%
Utilities	654,300	67	939,000	78
Industry*	255,100	26	173,300	14
Mobile Sources	70,300	7	97,200	8
Other	700	<1	700	<1
Total	983,400		1,210,200	

* Includes fuel combustion and sulfuric acid production
Source: "Final Report of Florida Sulfur Oxides Study", Florida Sulfur Oxides Study, Inc., May 1978.

Table 2. Estimated (1975) and Projected (1985) Nitrogen Oxides Emissions in Florida

Category	1975 Tons/Year	%	1985 Tons/Year	%
Mobile Sources	629,300	57	688,100	53
Utilities	272,500	24	415,400	32
Industry	202,300	18	181,800	14
Open Burning	4,900	<1	4,400	<1
Other	300	<1	2,400	<1
Total	1,109,300		1,292,100	

Source: "Final Report of Florida Sulfur Oxides Study", Florida Sulfur Oxides Study, Inc., May 1978.

Hydrochloric acid from refuse incineration and industrial acid sources contribute trace amounts. A summary of 1980 emissions compiled by Environmental Science and Engineering for the Florida Electric Power Coordinating group "Florida Acid Deposition Study" is illustrated by Table 3 (3).

The location of electric power plants in Florida is illustrated by Figure 1. The figure shows existing power plants as solid dots and proposed plants or plants under construction with circles. Table 4 shows a listing of the ten highest utility SO_x sources and the ten highest industrial sources (3). It is interesting to note that the combined total of those sources produces 2/3 of the 1980 total. Table 5 shows the top ten utility and industrial sources of NO_x (3). In this instance those sources produce less than half of the statewide NO_x. Transportation and other area sources produce the majority of the NO_x.

The areas where maximum SO_x and NO_x emissions occur are the Tampa Bay Region, the Jacksonville Area, the Pensacola Area, Crystal River, the Broward County/Dade County Area and the Brevard County Area. Emissions in the Jacksonville, Crystal River and Tampa Bay areas will increase when new power plant units are completed.

With the construction of new coal-fired power plants and conversion of existing oil-fired units, emissions of SO_x and NO_x may increase. The new coal-fired plants will have lower SO_2 emission rates than some existing oil-fired units, 1.2 lbs/MMBtu vs. 2.75 lbs/MMBtu. New power plants will control NO_x emissions to the extent of good combustion controls. The Department is considering a policy that would limit the SO_2 emissions from units converted to coal to no more than the current emission limit from those units. However, after conversion to coal the NO_x emissions may more than double. In order to try to maintain the status quo we can try to insure that Best Available Technology is used on new and converted plants to try to keep total emissions from increasing significantly. Some existing oil- and coal-fired units can reduce emissions by burning cleaner fuels. Many utilities are already using cleaner fuels than necessary to meet the existing emission limitations. Maintaining Florida's total emissions and current levels will be made even more difficult as our population increases bringing more vehicles to emit NO_x as well as increasing demands for electricity.

As knowledge is gained on sources, transformations, and deposition patterns, we may be able to develop a rational policy of control and prevention. I am looking forward to hearing the papers being presented today as steps along the path to understanding what acid deposition is.

Table 3. Summary of the Estimated 1980 Emissions for Florida

	SO_2		NO_x	
	Emissions (tons/yr)	Percent of Total	Emissions (tons/yr)	Percent of Total
Point Sources				
Utility*	738,500	72.8	331,593	42.9
Utility--other†	3,851	0.4	4,292	0.6
Nonutility	189,717	18.7	51,424	6.7
TOTAL FOR POINT SOURCES	932,068	91.8	387,309	50.1
Area Sources				
Fuel Consumption	39,533	3.9	25,333	3.3
Solid Waste	260	0.03	1,618	0.2
Transportation	42,603	4.2	347,191	44.9
Miscellaneous	360	0.04	11,551	1.5
TOTAL FOR AREA SOURCES	82,756	8.2	385,693	49.9
TOTAL FOR POINT AND AREA SOURCES	1,014,824	100.0	773,002	100.0

* Does not include gas turbines. † Includes gas turbines.
Sources: FCG, 1982; DER, 1982; EPA NEDS, 1982.

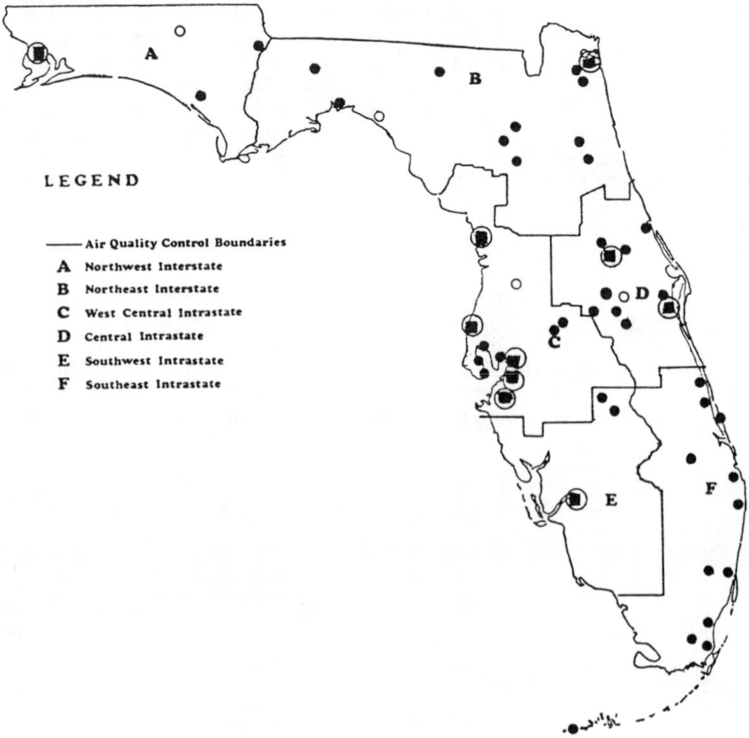

Fig. 1 Power Sources in Florida

- ◉ Ten largest SO_2 emitters
- • Other existing plants
- o Plants certified or proposed

Sources: US EPA; Florida DER

<u>Literature Cited</u>

1. "Final Report of Florida Sulfur Oxides Study", Florida Sulfur Oxides Study, Inc., May 1978.

2. "Exceedances of the Ambient Air Quality Standards in Florida 1981", Florida Department of Environmental Regulation, October 1982.

3. "Florida Acid Deposition Study, Phase II Report", Environmental Science and Engineering, Inc., E.S.E. No. 81-631-600, Doc. No 009F, January 1983.

Table 4. Listing of 10 Highest SO_2 Emissions from Utility and Nonutility Point Sources

	Utility Sources			Nonutility Sources		
Rank	Name	County	Emissions (tons/year)	Name	County	Emissions (tons/year)
1	TECO-Big Bend	Hillsborough	159,000	Union Camp	Duval	17,056
2	GPC-Crist	Escambia	74,720	Container Corp. of America	Nassau	15,842
3	FPC-Crystal River	Citrus	69,400	Int. Minerals Chem.	Polk	11,845
4	FPC-Anclote	Pasco	56,800	Exxon USA Jay Chemical	Santa Rosa	11,431
5	TECO-Gannon	Hillsborough	39,800	Occidental	Hamilton	10,964
6	FPL-Fort Myers	Lee	38,580	Southwest Forest Industries	Bay	9,969
7	FPL-Sanford	Volusia	34,460	Buckeye Cellulose	Taylor	8,220
8	FPL-Cape Canaveral	Brevard	33,190	St. Regis Paper	Duval	7,689
9	FPL-Manatee	Manatee	31,410	Georgia Pacific	Putnam	7,301
10	JEA-Northside	Duval	29,870	Alton Box Board	Duval	5,810
	TOTAL		567,230			106,127

Sources: FCG, 1982.
DER, 1982.

Table 5. Listing of 10 Highest NO_x Emissions from Utility and Nonutility Point Sources

	Utility Sources			Nonutility Sources		
Rank	Name	County	Emissions (tons/year)	Name	County	Emissions (tons/year)
1	TECO-Big Bend	Hillsborough	42,400	Container Corp. of America	Nassau	6,251
2	GPC-Crist	Escambia	31,500	Alpha Chemicals	Polk	5,188
3	FPC-Crystal River	Citrus	29,500	Alton Box Board	Duval	4,675
4	TECO-Cannon	Hillsborough	27,100	Pratt and Whitney	Palm Beach	4,137
5	FPL-Manatee	Manatee	20,430	U.S. Gypsum	Duval	3,320
6	FPL-Port Everglades	Broward	17,160	Central Phosphate	Hillsborough	3,258
7	GPC-Lansing Smith	Bay	13,900	St. Regis Paper	Duval	2,497
8	FPL-Sanford	Volusia	13,510	ITT Rayonier	Nassau	2,284
9	Lakeland-McIntosh	Polk	13,400	Monsanto Textile	Escambia	1,891
10	FPL-Cape Canaveral	Brevard	12,960	Air Products Chemicals	Santa Rosa	1,462
	TOTAL		221,860			34,963

Sources: FCG, 1982.
DER, 1982.

THE POTENTIAL IMPACT OF HIGH SULFUR OIL UTILIZATION IN FLORIDA ON ACID PRECIPITATION

Charles L. Proctor,II
Department of Mechanical Engineering
University of Florida, Gainesville, FL 32611

Abstract

This paper evaluates the impact of SO_2 emissions limiting standards established by the Florida Department of Environmental Regulation (DER) for liquid fueled stationary sources in Florida on the potential for acid precipitation. Currently, liquid fueled stationary sources use relatively clean oil; however, emissions standards, as they now exist, permit the use of high sulfur oil. The emissions of existing liquid fueled sources, as permitted by law, can in certain regions exceed SO_2 emissions of new solid fuel sources by as much as a factor of two. Investigation of the potential absolute emissions of SO_2 from such sources is made. An examination of the trace species that will exist in these dirtier oils, and the influence such species might have on reactions within the boiler or furnace as related to acid forming species are also examined. The emission of these trace species complicates atmospheric chemistry considerations for acid precipitation; therefore, trace species emissions resulting from the use of high sulfur oil also are presented.

Discussion

Acid deposition results from a myriad of complex chemical events occurring in the atmosphere. Its initiation is the emission of various compounds typical of combustion processes, such as sulfur dioxide, the oxides of nitrogen, and to a lesser extent carbon dioxide and chlorine, from natural, stationary and mobile sources. Stationary coal fired sources have long been considered the principal contributor to the "acid rain" problem, and strict controls are being imposed on sulfur dioxide emissions from such sources; however, in Florida, a serious impact on sulfur dioxide emissions could be realized from the burning of high sulfur oil.

The Florida Emissions Limiting Standards established by the Department of Environmental Regulation now limits sulfur dioxide emissions from new source liquid and solid fueled steam generators to 0.8 and 1.2 lb sulfur dioxide/million BTU heat input - maximum two hour average, respectively. Standards for existing liquid fueled sources, however, range from 1.1 to 2.6 lb/million BTU, with most of the state in the greater than 2.0 lb/million BTU category (1). Anticipating an eventual phasing out of oil fired units, the net impact of these standards would be a lowering of total sulfur dioxide emissions (assuming high sulfur oil were being used), but considering the unpredictable nature of oil prices (and their consequent effect on the desirability of oil as a fuel), realizing a reduction of sulfur dioxide emissions through this path is also unpredictable. The impact of high sulfur oil, however, can have a very severe and direct impact on Florida's total sulfur dioxide emissions simply through the continued operation of existing units if and/or when economical sources of high sulfur fuel oils (typically No. 6 grade fuel oil) are found.

Fuel oils vary significantly in their sulfur content. No. 1 grade fuel oils have a range of 0.01 - 0.05 percent sulfur by weight, whereas No. 6 grade fuel oils contain 0.7 - 3.5 percent sulfur by weight (2). No. 6 grade fuel oils are the ones of significance in this analysis.

The gross calculated heating value of No. 6 type fuel oils varies from 17,410 to 18,990 BTU/lb oil. Considering the maximum allowable emissions of sulfur dioxide as set by the DER for existing liquid fueled stationary units, this translates to a maximum weight percent sulfur content of approximately 2%. This value is well within the attainable sulfur levels in No. 6 fuel oils and is equal to the typical maximum sulfur content in No. 4 fuel oils (0.2 - 2.0 percent by weight sulfur). Consequently, any user shift to the higher content sulfur fuel oils could have an immediate impact on Florida's atmospheric sulfur dioxide loadings.

Assuming an existing plant currently using 0.7% sulfur No. 6 fuel oil begins using (less expensive or more readily available) 2.0% sulfur fuel, its net impact on Florida's atmospheric sulfur dioxide loading is equivalent to the installation of a new coal fired stationary source of 10% greater size without the benefit of added power production. Current scenarios base evaluation of SO_2 emissions in Florida on 1.5% sulfur, and usually the lowering of SO_2 emissions is evaluated by considering replacement of these units by new source coal-fired units and/or using cleaner oil. However, since regulations permit higher sulfur content oils, the downside scenario must be considered, and its immediate impact recognized. Furthermore, a shift to higher sulfur oils will not only affect sulfur dioxide emissions, it will also impact other trace species emissions from liquid fossil fueled sources.

In the dirtier residual fuel oils, many elements appear in conjunction with sulfur. These include: vanadium, iron, sodium, nickel, calcium, zinc, aluminum, and copper(2). Of these, vanadium, nickel and sodium are of most serious concern to the boiler designer due to their severe corrosion effects on boiler components. Consequently, boiler operators will attempt to control excessive occurrence of these materials in the fuel oils. Finite amounts of these metals, nevertheless, do appear in stack gases, and are usually in an oxidized form.

For the subject of sulfur acid deposition (less any impact these materials might have on atmospheric chemistry), oxidized vanadium compounds and sodium can impact sulfur oxides emissions through reactions within the combustion device. Sodium reacts with SO_2 to form Na_2SO_4. This then reacts with V_2O_5 to form SO_3. Formation of SO_3 can also occur through catalytic reactions on vanadium-rich ash deposits on boiler components. Consequently, SO_3 as well as SO_2 must be anticipated as sulfur emissions when examining high sulfur fuel oils.

In long range transport and cloud chemistry the effect of trace species may, due to atmospheric dilution, be negligible. But considering the importance of local sources to Florida's acid rain problem, these species may influence the nature of acid fog and plume chemistry. Consideration should not only be given to long range transport and cloud chemistry in Florida, but significant attention should be provided plume chemistry and near field deposition and precipitation.

As a final note, this brief analysis considered conventional liquid fuels; future considerations must, however, include fuels such as solvent refined coal (e.g., SRCII), which are "liquid" fuels, but contain many of the contaminants of the original coal. Indeed, techniques are being tested to reduce the amount of sulfur in these fuels, but because they may someday be introduced into existing liquid fuels stationary sources that are permitted to produce sulfur dioxide emissions at the upper end of the regulations, little sulfur cleaning may be done. Wishful thinking that SO_2 emissions will be reduced or even maintained at the current level is environmentally dangerous while the current emission standards are in place. Utilities and industries will, understandably, do what is most economically advantageous (hopefully within regulations), and this must be the realistic scenario.

Literature Cited

1. Green, A. E. S. (1981). An Alternative to Oil: Burning Coal with Gas, University Presses at Florida.

2. Babcock and Wilson (1972). Steam, Babcock and Wilcox.

OIL BACKOUT AND ACID RAIN

Alex E. S. Green and Krishna M. Pamidimukkala
Department of Mechanical Engineering
University of Florida, Gainesville, Florida 32611

Abstract

Utility and industrial oil backout for the purposes of reducing the United States' balance of payments problem and our reliance upon foreign oil is a national necessity. Oil to coal conversions as demonstrated in Australia and Denmark provide the most immediate path for relief. We first list the various coal conversion options including (1) COM (coal oil mixture), (2) DCO-NFGD (dual coal and oil fired boiler with no flue gas desulfurizer), (3) DCO-WFGD (with FGD), (4) NCFBWFGD (new coal-fired boiler with FGD), (5) CWS (coal water slurry), (6) C/G (coal with gas augmentation), and (7) CWS/G (coal water slurry with gas augmentation). The problem of SO_x and NO_x suppression in oil-coal conversions to minimize acid rain is next addressed. Dry sorbent injection into multistaged burners (LIMB) and dry FGD and fabric filtration are given the primary attention since retrofitting an existing oil boiler to accommodate a wet FGD would be extremely demanding in space and capital funds.

The major scientific purpose of this paper is to compare the combustion kinetics of several blends. First we simplify and improve the MIT coal devolatilization model for use in our kinetic calculations. Our present model uses a non-Arrhenius rate constant for the devolatilization of coal to an intermediate state which subsequently decays to volatiles and char with branching ratios which depend explicitly upon the temperature via an Arrhenius type factor.

Gas augmentation with coal or coal/water is found to speed up the time course of char burnout and CO and H_2 evolution. It also appears to favor the reduction of NO_x and SO_x emissions by providing fuel rich conditions in the initial burning zone and by increasing the calorific fuel content without adding sulfur. We also describe a number of collateral studies which favor the concept of burning coal cleanly and efficiently with the aid of gas augmentation. Finally we describe an R & D program needed to bring this concept to reality in Florida.

Introduction and Background

Oil backout endeavors at the University of Florida have their roots in ICAAS environment-energy studies on Florida problems dating back to 1970 (1-9). In connection with recent studies (9) we compiled a "spaghetti" chart of energy use in Florida which is illustrated in Figure 1a. When compared to the corresponding energy use chart for the United States as a whole (Fig. 1b) it should be obvious that percentage wise Florida has been using oil in great excess in electricity generation (50% compared to a national average of about 15%). Thus Florida has contributed and still is contributing disproportionately to our national balance of payments problem. To correct this ICAAS reached an early consensus that Florida should rely more fully on conservation measures and upon coal <u>burned</u> <u>cleanly</u> (2, 7, 9).

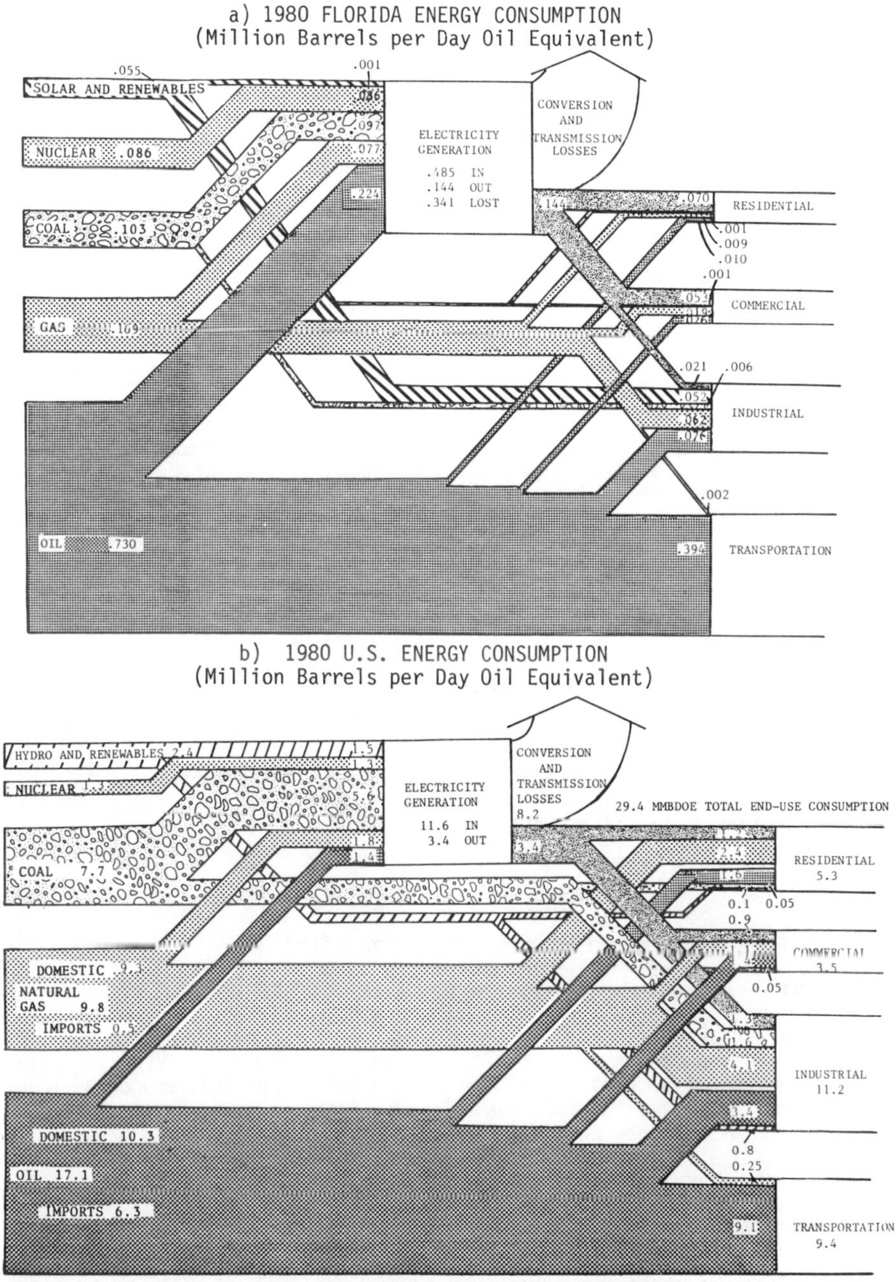

Fig. 1 Spaghetti chart of energy use: a) Florida, b) USA

In 1974 ICAAS (2) performed a review of the proposed Gainesville Regional Utilities Board's (RUB) application for site certification of Deerhaven Unit No. 2 at the request of the Florida Department of Pollution Control (now the Department of Environmental Regulation). At that time the nominal 250 MW generator was designed to burn a low sulfur-low ash oil. Among ICAAS's recommendations, in view of the uncertainties of oil supplies and prices in 1974, was that consideration be given to alternative fuels particularly coal. A year later the United States Department of Energy denied RUB a permit to use oil as a primary fuel. Since the oil boiler had been ordered in the interim the question of retrofitting an oil designed boiler received attention in public hearings in 1975-1976 when "conventional wisdom" viewed the retrofitting of an oil boiler to burn coal as impractical. The cancellation of the oil boiler contract and the purchase of a boiler designed for coal firing was associated with substantial cost penalties.

Recent successful conversions of oil designed boilers to coal or dual coal or oil firing in Australia (10, 11) and in Denmark (12) have resolved the question as to the practicality of oil to coal conversions and have stimulated a number of comparative studies of coal conversion alternatives in the United States (13-17). Table 1 gives a summary listing of estimates made in 1980 (9) of total conversion costs for most of the options now under various degrees of consideration. Table 2 gives more detailed economic comparisons of these alternatives for a 200 MW unit. More recent detailed engineering-technical analyses on many of these options have been made by Kimel (16) and Kimel and Kurtzrock (17). It would be expected that these studies of oil to coal conversion options will become subjects of intensive public discussions in the coming months, particularly after the forthcoming completion and release of DOE sponsored studies (18, 19) on the conversion of Florida utility oil boilers to coal.

From the viewpoint of the acid rain issue it would be well to scope the status of emissions and control technology particularly as it pertains to imminent oil backout issues.

Emissions and Control Technology (20)

Pollutants emitted from fossil-fuel-fired plants include sulfur oxides (SO_2 and SO_3), nitrogen oxides (NO and NO_2, generally referred to as NO_x), and solid material carried with the gas stream (ash from the fuel, unburned carbon, and other non-gaseous particles--all generally referred to as "particulate matter"). Most of these come from the fuel itself, by reaction of sulfur and nitrogen compounds with oxygen supplied by the combustion air, and by burning out the combustible compounds leaving the ash as small solid particles. In addition, some NO_x is formed by reaction of nitrogen and oxygen in the combustion air.

The amounts of such pollutants vary with type of fuel, design and size of boiler, and capacity factor. Typical data are given in Table 3 for an uncontrolled 500 MW unit. Although the tonnages listed are high, the concentrations in the flue gas are quite low because of the very large flue gas volume, which is composed mainly of nitrogen, carbon dioxide and water vapor. For example, the 500 MW boiler in Table 3 would

Table 1. Total Conversion Costs for Various Options

	Percent Coal Energy	Total Conversion Cost $/KW	Total Annual Cost 10^6\$/yr	Differential Annual Cost 10^6\$/yr	Payback Period
Oil	0	--	46.0	Base	--
COM	30	84	43.9	- 2.1	3.3
Dual NFGD	80	144	29.5	-16.5	1.5
Coal Oil FGD	80	216	34.4	-11.6	2.4
NCFBWFGD+	100	376	36.0	-10.0	3.2
Gas/Coal	70	127	25.8	-20.2	1.2
Coal/Water	100	132	31.3*	-14.7	1.5
Gas-Coal/Water	70	132	29.4*	-16.6	1.4

+ denotes new coal fired boiler with flue gas desulfurization.

* includes extra energy cost of water evaporation and fine coal grind but not cost of peaking unit to replace lost capacity.

Table 2. Economic Comparisons of Alternatives (200 MW Unit)**

Economic Considerations	Oil	COM.	Dual NFGD	Coal/ Oil FGD	New Coal FGD	Gas/ Coal (B)	Coal Water*	Gas/ Coal/ Water
Annual Coal Consumption, % by Energy	0	30	80	80	100	70	100	70
Conversion Cost								
Direct Cost, $/KW	-	84	144	216	376	127	132	132
Direct Cost, 10^6\$	-	16.8	28.8	43.2	75.2	25.4	26.4	26.4
Direct Cost+I DC, 10^6\$	-	20.3	34.8	52.3	91.0	30.5	31.7	31.7
Annual Cost								
Fuel Costs	46.0	39.4	22.0	22.9	16.6	19.2	23.8	21.9
O & M Costs	Base	0.4	0.5	1.0	1.2	0.5	1.2	1.2
Total Operating Costs	46.0	39.8	22.5	23.9	17.8	19.7	25.0	23.1
Differential Operating Costs	Base	-6.2	-23.5	-22.1	-28.2	-26.3	-21.0	-22.9
Fixed Charges	-	4.06	6.96	10.46	18.2	6.1	6.3	6.3
Total Annual Costs	46.0	43.86	29.46	34.36	36.0	25.8	31.3	29.4
Differential Annual Cost	Base	-2.14	-16.54	-11.64	-10.0	-20.2	-14.7	-16.6
Payback Period	-	3.3	1.5	2.4	3.2	1.2	1.5	1.4

NOTES: Annual Capacity Factor, 60%; Fixed Charges, 20%/yr; Interest During Construction, 10%/yr; Capital Costs Annualized for 15yr period; *Costs for boiling water off and finer coal grind included but not replacement costs for lost power; **All costs in million dollars per yr. 1980 Fuel Costs: Fuel Oil, 4.80 $/10^6$ BTU (30$/BBL); Coal, 1.67 $/10^6$ BTU (40$/Ton); COM, 4.11 $/10^6$ BTU; Coal with water, 47$/Tons + 8%; Gas, 2.60/$10^6$CF.

produce about 60,000 tons of flue gas per day, at full power; at 60% capacity factor, this is equivalent to 13 million tons of flue gas per year. Also shown in Table 3 are the emissions expressed in lbs/10^6 BTU for the various cases. For SO_2 these can be computed using the nomogram in Figure 2. The accepted best available control technology (BACT) are 1.2 lbs $SO_2/10^6$ BTU, 0.2 lbs $NO_x/10^6$ BTU and 0.1 lb particulate/10^6 BTU.

Table 3. Typical Uncontrolled Emissions of Pollutants[a]

Fuels	Sulfur oxides		Nitrogen oxides		Particulates Matter	
	b	c	b	c	b	c
Natural gas	--	--	4	0.28	--	--
Oil (1.0% Sulfur)	15	1.1	4.8	0.34	1.1	0.08
(3.0% Sulfur)	45	3.3	4.8	0.34	∿2	∿0.2
Coal						
Low-sulfur (0.7%)	15	1.1	8.2	0.57	110	7.7
High-sulfur (4.0%)	88	6.2	8.2	0.57	110	7.7

[a] 500 MW boiler, 60% load factor. Coal burned about 1.1 million tons/yr.

b in 1000 tons per year, c in lbs/10^6 BTU

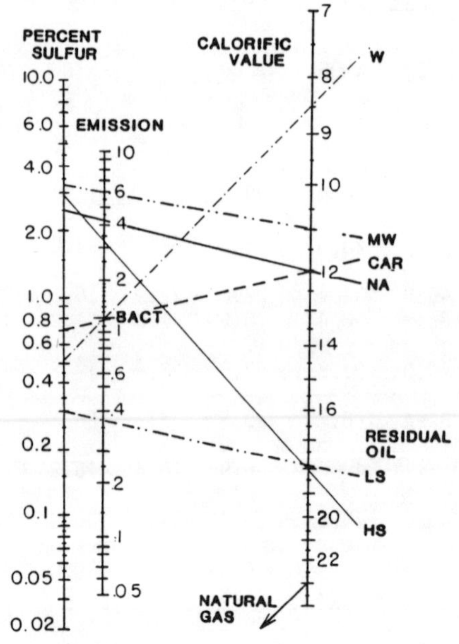

Fig. 2 Nomogram for Calculating Emission Levels (adapted from EPA-600/9-77-017). Emission levels are in σ = 1 lb $SO_2/10^6$ BTU. Calorific values are in 10^3 BTU/lb. W (western) MW (mid-western) CAR (central Appalachian and Rockies) NA (Northern Appalachian) LS (low sulfur and HS (high sulfur)

The technologies currently in use or available for SO_2 reduction include: (1) fuel substitution, (2) physical coal cleaning, (3) blending with low-sulfur fuel, (4) oil desulfurization, (5) wet FGD throwaway processes (lime/limestone, dual alkali, alkaline fly ash), (6) dry FGD throwaway process (spray dryer FGD), (7) regenerable FGD processes (Wellman-Lord, Mag Ox), and (8) lime/limestone scrubbing with chemical additives e.g., adipic acid.

For NO_x reduction, technologies in use or available include: (1) flue gas recirculation (FGR) (gas-fired utilities), (2) low excess air, (3) staged combustion, (4) low-NO_x burners, and (5) flue gas treatment (FGT).

For SO_2 reduction, emerging technologies include: (1) atmospheric fluidized bed combustion, (2) pressurized fluidized-bed combustion, (3) fuel gasification, (4) gasification with combined cycle operation, (5) coal liquefaction, direct (SRC-I and SRC-II) and indirect (e.g., SASOL), (6) limestone injection with multi-stage burner (LIMB process), and (7) baghouse particulate-SO_2 technology.

For NO_x reduction, emerging technologies include: (1) flue gas treatment (FGT) and (2) advanced low-NO_x burners.

Advanced low-NO_x coal burners, projected to be capable of NO_x emission of 0.2-0.3 lb per million Btu, have begun commercial demonstration. It is projected that these advanced low-NO_x burners may be commercially available in the 1983-85 period.

Among the existing technologies the wet FGD (item 5 above) seems prohibitively costly in capital and space requirements for application to retrofitting of oil boilers in Florida.

One of the newer FGD developments is injection of a lime slurry into a spray-dryer concurrently with the flue gas. The lime reacts with the SO_2 to form a dry, solid product which is subsequently collected in an electrostatic precipitator (ESP) or fabric filter. The main advantages are relative simplicity of equipment, production of a dry waste, lower energy requirement, no reheat requirement, and possibly lower maintenance, and greater reliability. The drawbacks are the greater cost for lime (compared to limestone in wet FGD) and a relatively high stoichiometry to achieve a sufficient degree of sulfur removal from high sulfur coal.

Operating cost advantage depends mainly on operating labor and maintenance, the amount of lime required, and the price margin over limestone and the energy used. A TVA study estimate shows lower direct costs (including absorbent) for wet scrubbing but overhead and capital charges may be less for dry processes.

The possibility of using a baghouse to remove the particulates including the dry solid products from a spray-dryer SO_2 removal system has much appeal from the viewpoint of simplicity, reduced materials handling problems associated with the wet sludge and reduction of water consumption. Most recently EPRI (the Electric Power Research Institute) (21) is evaluating the technical and economic aspects of injecting dry sorbents for SO_2 removal into the flue gas just ahead of a fabric filter (baghouse). Fly ash and the SO_2 by-product are then both collected in the baghouse. Compounds of interest as dry sorbents are narcolite and trona naturally occurring sodium bicarbonate and sodium carbonate compounds. For low sulfur western coal levels of SO_2 removal efficiencies of approximately 70% were achieved at very reasonable capital costs (~\$60/kW) and levelized busbar costs (2.5 mills/kWh).

Burner Scrubbing (22)

Among the three approaches to pollution reduction, precombustion coal cleaning, flue gas cleaning or modifying the combustion process to prevent the formation of pollutant species, the last is most likely to have a minimal impact on power generation costs. This approach has been intensely investigated by the Energy and Environmental Research Corporation in a recent study (22). This section is extracted directly from their report with the kind permission of the authors.

Coal sulfur does not necessarily convert quantitatively to sulfur oxides in the combustion products. Coals with high alkali metal oxide contents retain significant quantities of sulfur in the ash as sulfates. Thus, one method of preventing the emission of sulfur oxides is to effectively increase the alkali metal concentration in the system. This may be accomplished by adding alkali to the coal (23) or by injecting a sorbent which will capture sulfur into the combustion chamber. The injected sorbent (generally calcium carbonate) flash-calcines and reacts with gaseous sulfur species forming solid calcium sulfite and/or calcium sulfate. The solid sulfur compounds are then collected with the fly ash at the boiler exit. Considerable effort was expended in the late 1960's and early 1970's on development and demonstration projects (24) based on this process. Although pilot plant studies showed promise, the results could not be duplicated in full-scale systems. The lack of success was attributed to deadburning and maldistribution of the sorbent. Recent pilot scale studies (25, 26) with low-NO_x coal burners suggest that sorbent injection could be more effective under conditions which minimize NO_x formation.

The minimization of NO_x formation in pulverized-coal flames requires the generation of extensive fuel-rich zones and the reduction of peak flame temperatures. Thus, limestone sorbent injection into multistage low-NO_x burners (LIMB) may provide the opportunity to control emissions of both sulfur and NO_x by modifying the combustion process to prevent their formation. Sulfur capture by sorbent injection involves three processes, namely:

- o Sorbent activation - the sorbent particles are heated and calcined. Particle reactivity depends upon initial properties and peak particle temperature.
- o Capture - sulfur species (H_2S, COS, or SO_2) react with the sorbent producing either sulfate or sulfide.
- o Regeneration - the spent sorbent may decompose, regenerating gas-phase sulfur species.

The general reaction describing sulfur capture under oxidizing conditions:

$$CaO + SO_2 + \frac{1}{2} O_2 \rightarrow CaSO_4 \qquad [1]$$

has been studied extensively in thin-bed and dispersed-flow reactors by several workers (27, 28). However, none of these studies duplicated the time-temperature conditions that prevail in pulverized-coal flames. Borgwardt (29) has suggested that reactions such as:

$$CaCO_3 + H_2S \rightarrow CaS + H_2O + CO_2 \qquad [2]$$

and

$$CaO + H_2S \rightarrow CaS + H_2O \qquad [3]$$

could become significant under fuel-rich conditions. Extrapolation of rate data for such reactions (29) indicates that reaction [2] is fast enough to allow significant sulfur capture in coal flames.

Consequently, there are two possible modes by which calcium-based sorbents can capture sulfur species in a pulverized-coal-fired combustor operating under low-NO_x conditions. In fuel-lean regions reaction [1] may proceed because the lower temperature will reduce deadburning. In the fuel-rich region, reaction [2] may be significant, and then retention of sulfur in the solid phase under fuel-lean conditions becomes an important factor in the overall process. Equilibrium calculations indicate that conversion of calcium to the sulfate or sulfide is very high (> 80 percent) under either fuel-rich or fuel-lean conditions but that the sulfate decomposes at much lower temperatures than the sulfide. At stoichiometric conditions the sulfate is relatively unstable suggesting that prolonged residence times at stoichiometric conditions could reduce overall sulfur capture even though high captures occur in the fuel-rich zone.

The Energy and Environmental Research Corporation (22) have conducted experimental studies at both pilot and bench scale. Pilot scale tests were conducted in two research furnaces firing at up to 20.5 MW ($\sim$70 x 10^6 Btu/hr), with a prototype distributed mixing burner. With limestone injection, sulfur dioxide captures of 30 to 40 percent were obtained for Ca/S molar ratios of 2. Test data show that burner design, coal type, and mode of sorbent injection all influence the sulfur capture process. Thermal environment in the combustor has a significant impact on sorbent utilization, depending upon the location of sorbent injection.

Under conditions of low NO_x operation two modes of sulfur capture are possible: capture under fuel-rich and fuel-lean conditions. A series of bench scale ($\sim$1 x 10^6 Btu/hr firing rate) studies were carried out to determine the parameters controlling capture in these two modes. Under fuel-lean conditions, the thermal history of the sorbent dictates sulfur capture. Under fuel-rich conditions, the process is more complex. The bench scale studies indicate that sulfur is retained in the coal in the fuel-rich region and that regeneration of gaseous sulfur during second-stage burnout is possible. Rich-zone stoichiometry had an impact on sorbent utilization in the bench scale studies; utilization was low under highly staged conditions (SR_1 < 0.7). The effect of sorbent and fuel type was investigated. Five sorbents were used: two calcites, one marl, one dolomite, and one hydrated line. Under all of the conditions investigated, the dolomite had the highest utilization (a maximum of 30 percent). Three fuels were used: a bituminous coal, a lignite coal, and propane doped with SO_2. It was found that fuel type has an impact on sulfur capture and that this effect was not entirely attributable to thermal effects.

Kinetic Simulation of Coal Combustion

The major scientific purpose of the present paper is to compare the combustion kinetics of several fuel blends from the viewpoint of retrofitting oil burners. In order to do so we must first utilize or develop a coal devolatilization and char burn out model to enable us to treat coal combustion in parallel with gas phase combustion. In a previous work (30) we used the empirical devolatilization model of Kobayashi, Howard and Sarofim (31) which was based upon experimental MIT data. In the present work we utilize a somewhat simpler model based upon the mathematics of nuclear (or atomic) decay (32) to fit the MIT data while maintaining more realistic high temperature limits.

When coal is heated in an inert atmosphere it produces volatiles (gases) and char which is mostly carbon and ash. The volatile composition is highly temperature dependent but the principle components are mostly CH_4, CO and H_2, as illustrated in Fig. 3a and small amounts of higher hydrocarbons. Kobayashi et al. (31) report higher yields of volatiles under rapid heating conditions, than estimated from proximate analysis. Their empirical model assumes two channels of decay and is illustrated in Fig. 3b and by Eqs. 1-4 in Table 4. Parameters used with the Kobayashi model are also given. During the course of our investigations we examined somewhat different models in an effort to avoid some unphysical results which kinetic simulations gave at higher temperatures. We first used a single decay model with α dependent upon T via an Arrhenius type equation. However we could fit the MIT data better if we used the intermediate state model illustrated in Fig. 3c. This intermediate state or Nuclear Decay Model (see Eqs. 5-8) with $k_1 = k_2 = k$ gives an early time behavior which goes as $\alpha\ k^2t^2$ and the later time behavior which saturates at α. Thus it has a slight sigmoid type behavior. Without the intermediate state the function would start out proportional to t and rise as a straight line which then saturates at α.

Our intermediate state model utilizes a branching ratio given by Eq. 7 in Table 4 and an empirical analytic representation for the reaction rate given by Eq. 6 rather than the usual Arrhenius rate expression. The volatile yield versus time and temperature is then represented by the three parameter (T_0, T_1 and M) analytic function given by Eq. 8. Figure 4 illustrates the fits of Eqs. 6-8 to the MIT data and the fits given by Kobayashi et al. One can reduce the standard deviations even further with our model by letting $k_1 \neq k_2$ but the scatter in devolatilization data does not warrant such refinement at this time. Nevertheless further generalizations of this decay scheme utilizing one or more intermediate states and temperature dependent branching ratios might provide highly realistic phenomenological simulations of the devolatilization process. For example, we could reflect the influence of coal particle size, ash

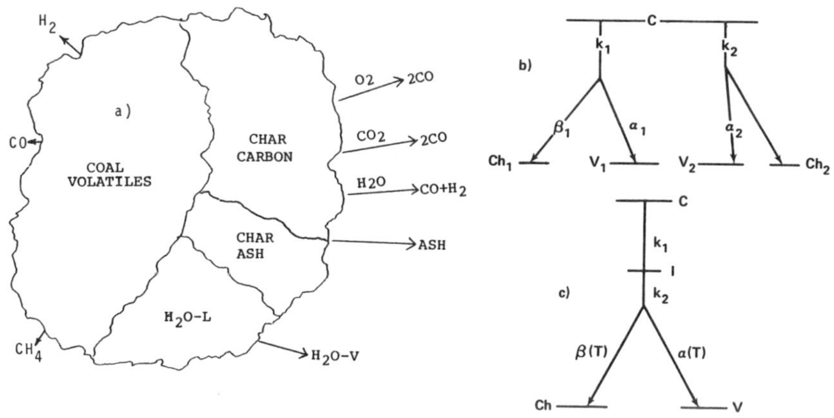

Fig. 3 (a) Coal Particle Gasification Model (b) MIT (c) U.F.

Table 4. Coal pyrolysis - Equations and parameters

Kobayashi et al Model (parameters α_1, A_1, T_1, α_2, A_2, T_2)

$$dV/dt = (k_1 \alpha_1 + k_2\alpha_2)\ C \tag{1}$$

$$C = C_0 \exp - (k_1 + k_2)t \tag{2}$$

$$k_1 = A_1 \exp(-T_1/T),\ k_2 = A_2 \exp(-T_2/T) \tag{3}$$

$$V = \frac{k_1\alpha_1 + k_2\alpha_2}{k_1 + k_2}\ [\,1 - \exp - (k_1 + k_2)t] \tag{4}$$

$\alpha_1 = 0.3$ $\quad A_1 = 2.0\text{E}05\ S^{-1}$ $\quad T_1 = 12580$ K

$\alpha_2 = 0.7$ $\quad A_2 = 1.3\text{E}07\ S^{-1}$ $\quad T_2 = 20130$ K

Nuclear Decay Model (parameters T_0, M, T_1)

$$dV/dt = kI,\ dI/dt = kC \tag{5}$$

$$k(T) = M \exp\ (T/T_1)^2 \tag{6}$$

$$\alpha(T) = \exp(-T_0/T) \tag{7}$$

$$V = \alpha[1 - (1 + kt)\exp\ (-kt)] \tag{8}$$

Bituminous

$$T_0 = 1045\text{K},\ T_1 = 906\text{K},\ M = 1.65\text{s}^{-1}$$

content, the hydrogen to carbon ratio of the volatiles at various temperatures and the various hydrocarbon and pollutant emissions by multiple channels.

A rigorous calculation of the combustion of coal/natural gas blends in a vertical cyclone-type reactor such as used in our development program (33) which realistically considers the turbulent, chemically reacting and radiation flow field, is beyond the current state-of-the-art. Efforts to describe idealized linear coal/air systems under various simplifying assumptions have been made (34-36) and we hope before long to have such a model working. At this time, we have available a sophisticated code drawn from the type of kinetics used in atmospheric studies (37-38). We have adapted this code to give the time evolution of various species concentrations in a parcel of combustible mixture threading its way through a vertical cyclone reactor under fixed temperature conditions. The usefulness of the code was demonstrated in our recent work (30).

In the present work fifty-six gas phase reactions for CH_4-air system are included using reaction rates given in the literature (39-41). We are not including the NO producing reactions although we allow for 77% by weight of the nitrogen to serve in 3 body reactions.

To place the combustion of pulverized coal on the same kinetic basis as that for gaseous fuels we assumed hypothetical coal consisting of 75% carbon, 5% hydrogen, 10% oxygen, and 10% ash with a molecular weight of 100. We also assumed in view of our spectroscopic studies on burning of natural gas with coal additives (42,43) that coal is devolatilized to an intermediate state which produces the volatiles CH_4, CO, H_2, and C_2 and a "charash". Thus, the net concentration of char depends on the production reaction from charash and the slow oxidation reactions of char with O_2, CO_2, and H_2O as shown in Fig. 3a.

Figure 5 presents the comparative production and burnout of char for A) coal burning at 1500°K, B) the gas/coal blend described in Figure 6 burning at 1600°K, C) A coal/water blend 63/34 by weight burning at 1400°K and D) a gas/coal/water blend in 13/58/29 ratios burning at 1450°K. These temperatures are estimates based upon our cyclone reactor experience.

Note here that coal burnout occurs more rapidly in the gas/coal case than in the coal case. While the increased temperature is influential in the speedup the overall reaction process in the gas/coal case is also facilitated by the intermediate radicals OH, H, CH_3, etc. ... which are made available more quickly by the rapid combustion processes associated with methane (30). The speed up of char burnout is also facilitated by the fact that CO_2 and H_2O formed at earlier times (10^{-3} sec.) by the combustion of methane start reacting with char to produce CO and H_2 at intermediate times (10^{-2} sec.).

For the coal/water case to incorporate the energy costs (∿10 kcal/mole) of water vaporization we have slightly lowered the combustion temperature to 1400° K. Note here that the burnout time of the char has been extended despite the fact that this system leads to greater concentrations of H_2O, as a consequence of the liquid water vaporization, and the H_2O attack on the char. Water might also promote the combustion of coal as a consequence of the greater availability of H and OH radicals in a coal/water flame. However, the break up of the H_2O molecule into H and OH costs about 5eV (∿100 kcal/mole) as indicated in an energy level and process diagram for H_2O (44). While this energy is probably recovered in the boiler, its loss in the immediate reaction zone might be expected to slow down the combustion processes with respect to the processes in gas/coal combustion.

Figure 5 also presents the char versus time for the combustion of a gas/coal/water mixture at 1450° K. As might be expected, the results are intermediate between the gas/coal and coal/water cases and suggest that some of the combustion problems created by the use of water to facilitate pulverized coal transport might be overcome by the addition of gas in the combustion zone.

Figure 6 presents a kinetic simulation of gas/coal combustion in a 30/70 ratio by energy release (17/83 by weight) in a swirl type reactor. The curves represent the (a) stable and (b) free radicals species concentrations as they depend upon time. In our gasification model, the coal produces volatiles CH_4, CO, H_2 and C_2 which reach maxima around 5 x 10^{-2} seconds. The gaseous products begin reacting quickly in the gas phase, whereas char is slowly oxidized to species which remain steady at higher concentrations in intermediate times but start to decrease at a later time scale. The CO and H_2 concentrations first go up somewhat with the decrease of char depending upon the initial O_2 in complex fashions. The final combustion products CO_2 and H_2O increase steadily.

More detailed kinetic simulations of gas/coal combustion are underway (45). Of particular interest are the processes leading to NO_x and SO_x production. In both cases the use of gas/coal blends appear to favor their reduction which would be helpful from the viewpoint of mitigating acid rain.

Augmentation with Gas

In its general work "Coal Burning Issues" (8) and in its specific work on the "Impact of Increased Coal Use in Florida (IICUF)" (7) ICAAS reach an internal consensus that Florida should shift the burden of electricity generation from oil and gas to coal burned cleanly.

When evidence emerged that North American production of natural gas could be maintained at the 20 TCF level for 30 to 50 years we proposed that oil boilers existing in Florida and elsewhere be retrofitted to simultaneous burn gas and coal (9). Our group also initiated a development and scientific program on blended fuel combustion intended to provide a basis for such retrofits. The work continues to support the idea that the following four direct measures would enable us to replace oil with coal/gas in an environmentally acceptable way without serious derating problems and without the use of wet flue gas desulphurization. At this time we would also consider the fifth measure in the list below:

1. choose a good grade of coal
2. clean the coal to reduce further the ash and sulfur
3. augment the coal with natural gas in the combustion chamber
4. incorporate combustion chamber scrubbing
5. incorporate dry sorbent FGD.

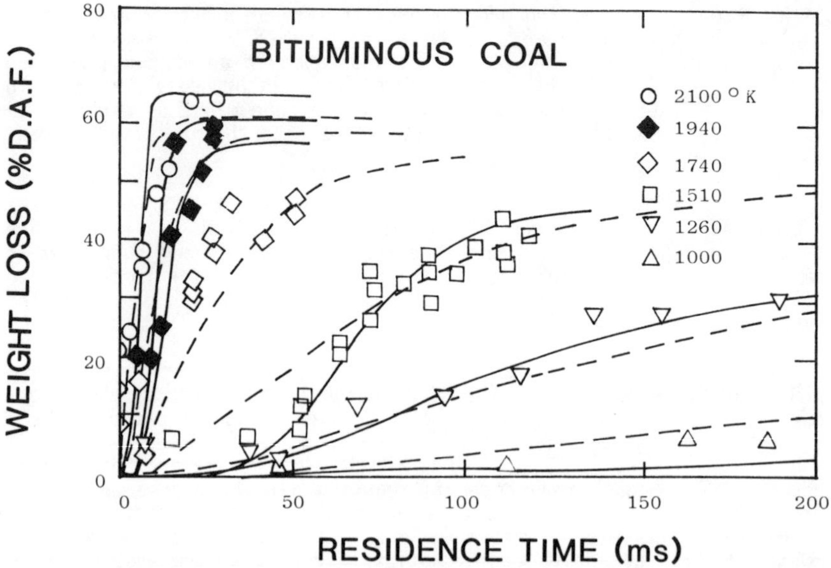

Fig. 4 Solid curves represent MIT model (2 reactions, 6 parameters); dashed curves represent intermediate state-temperature dependent branching ratio (UF) model (1 reaction, 3 parameters). Note: the behavior at early times of the 1510° data may be experimental artifacts (A. Sarofim, private communication).

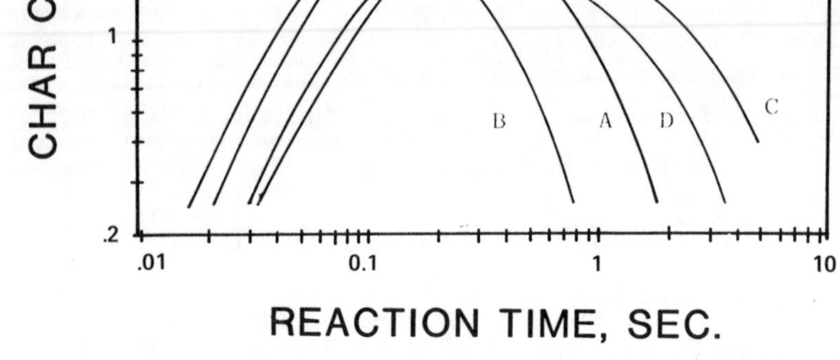

Fig. 5 Char concentration versus time for A) coal, B) gas/coal, C) coal/ water and D) gas/coal/water.

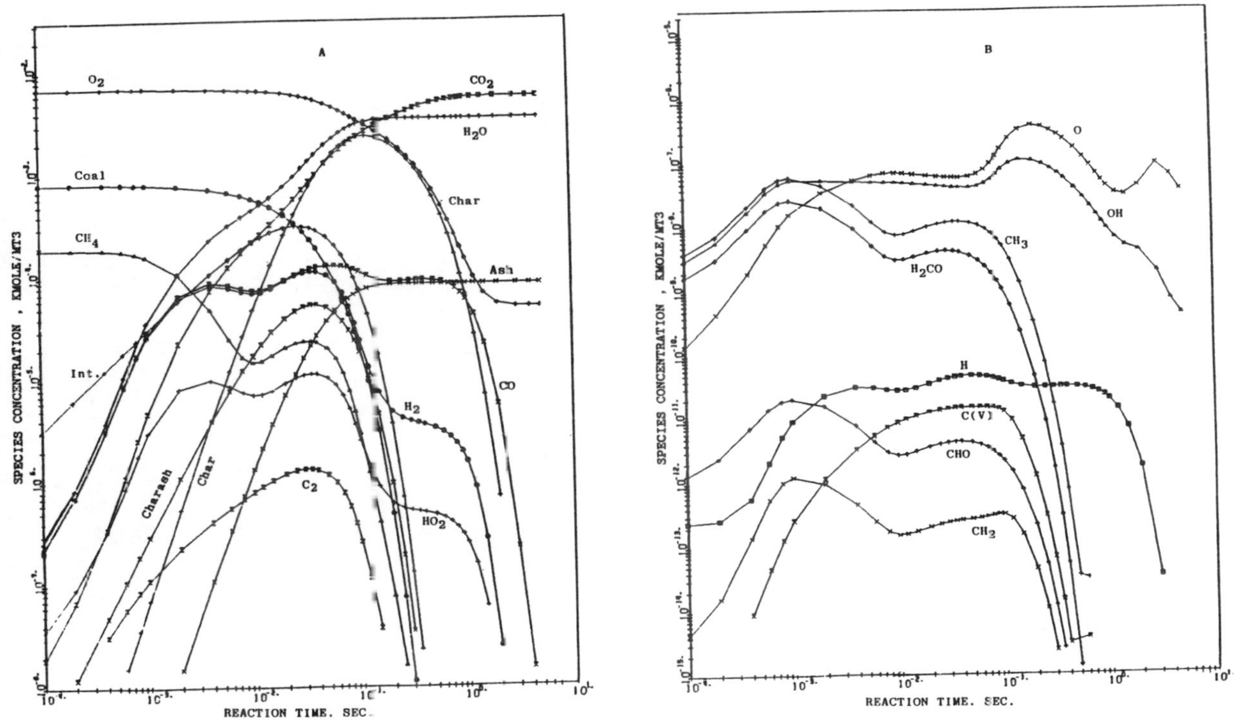

Fig. 6 Species Concentration vs Time for Coal/Gas System at 1600°K (A) Stable species (B) Intermediate species

All of these measures would mitigate acid rain problems by reducing the sulfur output of Florida oil boilers. Measure 3, augmentation with gas, is useful in a benign way by taking a share of the energy load without contributing sulfur dioxide. Simultaneous combustion of coal with gas also helps in achieving rapid burnout of coal particles and improves flame stability at low load levels. At high load levels the proportion of gas energy used would be increased to about 40% to help achieve the original boiler capacity. In Florida higher demand usually goes with the heavy air conditioning load in the summer. Fortuitously this occurs when natural gas in the United States is in abundant supply because of the reduced demands for gas heating in the north. Since the probability for atmospheric transformations of sulfur dioxide to acid rain is also high in the summer, shifting more of the energy load to gas should ameliorate the problem.

One of the major obstacles to the direct retrofitting of an oil boiler to coal burning is the high cost of the modifications required to accommodate slagging and fouling properties of ash. A promising type of oil boiler retrofit would be to use a first stage coal/natural gas system whose function is to de-ash the coal and extract hot producer gas augmented with natural gas for afterburning in the original boiler (see reference 9, pg. 135). One might argue that, "Why not just use a simple coal gasifier? Why add gas?" Here, the advantages are the much higher BTU content of natural gas than producer gas and the fact that CH_4 produces the free radicals H, OH, CH_3...which speed up the combustion kinetics.

We have built and tested two vertical cyclone reactors designed to produce a rich gas for afterburning. The first reactor assembled and fired at the Dragon Fire and Clay Company in Micanopy (for want of a suitable University combustion laboratory) operates at the 1,000,000 BTU/hr level and can burn gas/coal in all proportions. The second reactor at the 100,000 BTU/hr level which is fired on campus, can only burn coal with some gas augmentation. The afterburning characteristic and the char burnout are very encouraging. The objective of the vertical cyclone is to separate out the ash particles from the hot fuel rich gases so that most of the ash is collected in the first stage system before the fuel rich gases are afterburned in the oil boiler (see reference 9, Fig. 11.4). The program has produced promising qualitative results in terms of carbon burn out and ash separation. Both vortex combustors have also been fired using coal/water slurries with natural gas assistance.

Collateral Studies

There are a number of collateral endeavors which give credibility to the gas/coal-oil backout concept. The first is the practical experience with simultaneous gas/coal burning at the James River Power Plant in Springfield, Missouri (46). Their boilers were originally designed to burn natural gas as the primary fuel and pulverized coal as the secondary fuel. However, during the middle 70's the boilers were retrofitted to burn coal as the primary fuel with natural gas as a standby. In the 70's as they were forced more and more to depend upon coal as the primary fuel, they burned a combination coal/gas fuel in a 60/40 ratio until they overcame some problems in the design of their coal systems and their coal characteristics. Thus they acquired practical experience in gas/coal combustion not in relation to an oil backout concept but rather in relation to problems in plant design and coal characteristics. The practical experience in gas/coal combustion at James River nevertheless provides valuable physical information for use in retrofitting an oil boiler to coal with gas augmentation.

A recent paper on "The Conversion of Oil Designed Boilers to Dual Coal/Natural Gas Firing" (16) is another important collateral study since Kimel, the author, is a senior supervising engineer of Burns & Roe, who have been world leaders in the retrofitting of oil boilers. Indeed, Burns & Roe served as consultants on the Kwinana utility plant conversion in Perth, Australia and have been leaders in the general analysis of oil to coal conversions even when "conventional wisdom" viewed the retrofitting of an oil boiler to burn coal as impractical.

A third recent development is the use of staged injection of air to keep the burning zone in a fuel-rich state so as to minimize the production of NO_x (47). This technique has advanced very considerably in recent years. The addition of natural gas is another controllable way of achieving fuel rich conditions in a burning zone.

A similar program of ash separation has been initiated in Sweden based on a similar vertical vortex ash separation technique (48). In addition, lectures by the senior author at several leading Institutes of Technology in India on the concept of vertical cyclone ash separators and gas augmentation have led to the initiation of plans for collaborative endeavors (49). India has very high (30%) ash content coal in its Bihar province.

Thompson, Ramo and Woolridge (TRW) (50) has recently proposed a modification of the horizontal cyclone burner as a first stage combustor-ash separator. They go to high, ash melting, temperatures (2700°F) and while they do not mention it, their performance would probably be favored by natural gas augmentation in a retrofit.

Another collateral endeavor counteracts the problems of coal/water as a fuel for oil boilers by augmentation with methanol (51). The physical and economical arguments, for the use of methanol as an additive to obtain combustion characteristics are probably even more applicable to augmentation by methane. Indeed, Dooher et al. (52) has used gas augmentation of a coal/water slurry in a 15,000,000 BTU/hr boiler to overcome the marginal flame characteristics of the coal/water slurry in an industrial boiler, being converted from oil to coal, and to minimize derating.

A study of wood pyrolysis by scientists at Brookhaven (53) indicated that the addition of methane in the combustion promotes the production of higher hydrocarbons. While their objective is to produce synthetic fuel, their work supports the gas augmentation concept.

Finally we might note that two economic analyses on gas/coal use have recently been initiated, a Florida assessment sponsored by the Department of Energy (18) and a National Assessment by Environmental Research Technology sponsored by the American Gas Association and the Gas Research Institute (54).

Research Need

The research needed to promote the clean burning of coal, this country's most abundant fossil fuel, are for the most part, reasonably well provided for by federal and quasi-federal agencies concerned with fossil fuels. Although several scientific papers have emerged from our efforts (30, 42 and 43), thus far our gas/augmentation concept has not been viewed favorably as compared to the coal-oil slurry or coal/water slurry approaches to oil backout (55-57). Yet the James river experience, the coal/water/methanol experience, the Dooher et al. experience, the Brookhaven experiment, as well as our own field level experiments all point to the value of gas augmentation as a means for upgrading a marginal fuel to an acceptable fuel. An analysis by Payne et al. (58) suggests that coal/water slurrys would have difficulty at low load levels and here augmentation with gas should be particularly helpful. Thus, there is a need for an R & D program on gas augmentation of any lower grade fuel than oil, e.g., coal, coal/water, lignite, peat, wood chips, and urban waste to develop the general concept of gas augmentation as a means of upgrading a lower quality fuel. While one can design a new boiler to accommodate almost any fuel, to develop the technology of retrofitting there is a need for understanding fuel blends. This type of work can be carried out to some extent at a laboratory scale level and in part by theoretical models which numerically simulate the chemistry, flow field and radiation in laboratory and industrial flames.

There is also a need for studies in a 1-10 million BTU/hr level boiler which cannot presently be carried out in a campus laboratory. In this connection there is a great need for the development in Florida of a combustion laboratory to pursue fuel blend studies including coal/water fuels with gas augmentation.

In light of the industrial experience at the James River Plant (46) and elsewhere supporting the concept of gas augmentation, there is a need for detailed feasibility studies by a commercial architect/engineering firm, in collaboration with a university R & D program on the feasibility of retrofitting an existing industrial boiler and utility boiler. For example, a selection of one of the "A" classification plants given in reference 9 (pg 56) would greatly advance the concept of coal burning with gas augmentation.

The possibility of a demonstration of a retrofit of a small industrial boiler to coal with gas augmentation also needs support. The coal/water slurry concept seems to have the greatest momentum in DOE and EPRI circles at this time. Gas augmentation might well provide the margin needed to make this concept practical over wide load levels. The pursuit of the concept in the direct burning version and the two stage system both with and without gas augmentation needs further study. Provision for a demonstration plant of an industrial boiler in the gas augmented retrofit mode would be highly desirable.

Acknowledgements

We would like to thank Professor A. F. Sarofim and Mr. Michael Maxwell for helpful suggestions as to sources of information on environmental-combustion issues. We would also like to thank Dr. Roy Payne of Energy and Environmental Research Corporation for permission to use directly material from one of their reports on limestone sorbent injection into multistage burners (LIMB).

Literature Cited

1. ICAAS, "Report of the Coordinated Program for the Remote Sensing of Atmospheric Aerosols-Clean Air Preservation and Enhancement Research (Caper), 32 authors, published in part in "Light Scattering and the Size-Altitude Distributions of Atmospheric Aerosols", J. of Colloid and Interface Science, A.E.S. Green et al., 39, 520-535, 1972. The complete report includes material from ICAAS proposals, "An Air Quality Index Project" and "Preservation and Enhancement of Air Quality in Florida and the Southeast (PEAQ)" (1971).

2. ICAAS, "Review of Application for Site Certification Deerhaven Generator Unit No. 2", for Florida Department of Pollution Control, by ICAAS Power Plant Siting Group, 30 authors, University of Florida, February (1975).

3. ICAAS, "Response to Energy Problems Which Protects Air Quality (REPPAQ)", proposal which led to ICAAS designation by Florida Department of Pollution Control, Directive 74-12, to review Florida Air Implementation Plan and provide professional and scientific expertise on air pollution problems (1974).

4. ICAAS, An Interdisciplinary Study of the Health, Social and Envrionmental Economics of Sulfur Oxide Pollution in Florida, prepared for the Florida Sulfur Oxide Study, February 10 (1978) published in part in Journal of Environmental Economics and Management, "Distributional Analysis of Regional Benefits and Cost of Air Quality Control, E. J. Loehman et al., 6 222-243 (1979).

5. A.E.S. Green et al., "Pollution Dispersal Modeling and Regional Public Policy Implications", Proceedings of Symposium held April 17-19, 1979 on Environmental and Climatic Impact on Coal Utilization, Williamsburg, VA., Academic Press, 479-540, (1980).

6. ICAAS, "Interdisciplinary Study of Public Policy Decision Methodologies for Sulfur Oxide Control in the Southeast (SESOS)" and "Mitigation of the Impact of Increased Coal Use in the Southeast (MIICUSE)", ICAAS interdisciplinary proposals (1980) published in part in Atmospheric Environment, A.E.S. Green et al. 14, 327-338 (1980).

7. ICAAS, Impact of Increased Coal Use in Florida (IICUF)", Interdisciplinary report for Florida Governor's Energy Office, November (1980).

8. A.E.S. Green (Ed.), Coal Burning Issues, University Presses of Florida, Gainesville, Florida (1980) with 32 coauthors.

9. A.E.S. Green (Ed.), An Alternative to Oil: Burning Coal with Gas, University Presses of Florida with contributions by J.R. Jones, M.J. Ellerbrock, J.M. Schwartz, S.J. Kuntz and B. Zeiler, (1981).

10. J.B. Kirkwood, D.H. Kregg and H.D. Schoenwetter, "Conversion to Coal of Gas/Oil-Fired Boilers", Power Engineering, p. 66, July (1978).

11. J.B. Kirkwood, "Commissioner's Corner", Energy Commission News, Western Australia, March (1979).

12. Staff Report, Denmark Converts to Coal, The Energy Daily, Vol. 11 No. 81, pg. 2, Thursday, April 28, 1983.

13. J. Philipp and D. Kregg, "Conversion from Oil to Coal-Firing", Burns and Roe, Inc. Report, City of Gainesville, Department of Utilities, June 17, (1980).

14. J. Philipp, "Oil to Coal Conversion Options", Burns and Roe, Inc., Paper Presented at EPRI/Florida Power and Light Seminar on Use of Coal in Oil-Design Utility Boilers, Lake Buena Vista, Florida December 2-4 (1980).

15. S. Erhlech, S. Drenker and R. Manfred, Coal Use in Boilers Designed for Oil and Gas Firing, American Power Conference, Chicago, Illinois, April 21-23 (1980).

16. E.J. Kimel, "Conversion of Oil Designed Boilers to Dual Coal-Natural Gas Firing", Proceedings of International Conference on Combined Combustion of Coal and Gas, Cleveland, Ohio, October 12-13 (1982).

17. E. Kimel and R. Kurtzrock, "Comparison of Coal Conversion Alternatives", Proceedings of Fifth International Symposium on Coal Slurry Combustion and Technology, Tampa, Florida, April 25-27 (1983).

18. Office of Fuels Programs, "Conversion of Florida Electric Powerplants From Oil to Coal Burning", Economic Regulatory Administration, U.S. Department of Energy, April 1983.

19. Science Applications, Inc., Oak Ridge, Tennessee, Draft Implementation Plan for Florida Regional EIS, Oak Ridge, Tennessee (1982).

20. M.E. Rivers and K.W. Riegal, "Emissions, Costs and Engineering Assignment", U.S.-Canada. Memorandum of Intent on Transboundary Air Pollution, Work Group 3B, Final Report June (1982).

21. K. Yeager, "SO_2 Control by Dry Sorbent Injection", EPRI Journal, p. 36 March (1983).

22. P.L. Case, M.P. Heap, R. Payne and D.W. Pershing, "LIMB Testing: The Use of Dry Sorbents to Reduce Sulfur Oxide Emissions from Pulverized-Coal Flames Under Low-NO_x Conditions", Proceedings of the EPA/EPRI Joint Symposium on Stationary Combustion of NO_x Control, Dallas, Texas, November 1-4 (1982).

23. Freund, H., and R.K. Lyon, "The Sulfur Retention of Coal Containing Calcium During Fuel Rich Combustion", Paper presented at the Western States Section, The Combustion Institute, Los Angeles, CA, October 1980.

24. Gartrel, F.E. "Full Scale Desulfurization of Stack Gas by Dry Limestone Injection", EPA-650/2-73-019a,-c (NTIS PB228447, 230384, 230385), August 1973.

25. Zallen, D.M. et al. "The Generalization of Low Emission Coal Burner Technology", Proceedings of the Third Stationary Source Combustion Symposium, Vol. II, EPA-600/7-79-050b (NTIS PB292539), February 1979. pp. 73-109.

26. Flament, G. "The Simultaneous Reduction of NO_x and SO_2 in Coal Flames by Direct Injection of Sorbents in a Staged Mixing Burner", IFRF Doc. No. G 19/a/10, 1981.

27. Coutant, R.W. et al. "Investigation of the Reactivity of Limestone and Dolomite for Capturing SO_2 from Flue Gas", EPA Report No. APTD-0802 (NTIS PB204385), October 1971.

28. Borgwardt, R.H. "Kinetics of the Reaction of SO_2 with Calcined Limestone", Env. Sci. and Tech. 4:1 (1970).

29. Borgwardt, R.H., G.R. Gillis, and N.F. Roache. "IERL-RTP Research on Sulfur Capture in LIMB", Paper presented at Joint Symposium on Stationary Combustion NO_x Control, Dallas, Texas, November 1982.

30. A.E.S. Green and K.M. Pamidimukkala, "Kinetic Simulation of the Combustion of Gas/Coal and Coal/Water Mixtures", Proceedings of First (Intnl) Conference on Combined Combustion of Coal and Gas, Cleveland, Ohio (1982).

31. K. Kobayashi, J.B. Howard and A.F. Sarofim, "Coal devolatilization at high temperatures", Sixteenth Symposium (Intnl) on Combustion, p. 411 (1976).

32. A.E.S. Green, Nuclear Physics, International Series of Physics, McGraw-Hill, New York, New York (1955) p. 169.

33. A.E.S. Green, B.A.S. Green, K.M. Pamidimukkala, K. Zawoy and J. Samuels, "Development of Practical Gas/Coal Combustor", (1981).

34. P.J. Smith, "Theoretical Modeling of Coal and Gas Fired Turbulent Combustion and Gasification Processes", Ph.D. Thesis, Brigham Young University, Provo, Utah (1979).

35. P.E. George and N.M. Laurendeau, "Gasification in Pulverized Coal Flames", Investigation of Pulverized Coal Combustors and Gasifiers: Multiple Sub-reactor Model and Experimental Results, Part II, DOE-FE-2029-10, April (1982).

36. L.D. Smoot, "Pulverized coal diffusion flames: A perspective through modeling", Eighteenth Symposium (Intnl) on Combustion (1978).

37. W.L. Chameides and D.D. Davis, "Chemistry in the Troposphere", Chemical and Engineering News, Special Report, October 4 (1982).

38. R.M. Baum, "Stratospheric Science Undergoing Change", Chemical and Engineering News, September 13 (1982).

39. F. Westley, "Table of recommended rate constants for chemical reactions occurring in combustion", NSRDS-NBS 67, NBS, Department of Commerce, Washington, D.C. 20234, April (1980).

40. L.D. Smoot and D.T. Pratt, Pulverized-coal Combustion and Gasification, Plenum Press, New York (1979).

41. C.K. Westbrook, "Hydrogen Oxidation Kinetics in Gaseous Detonations", Combustion Science and Technology, Rev. I (1981).

42. D.B. Vaidya, J.J. Horvath and A.E.S. Green, "Remote Temperature Measurements in Gas and Gas-Coal Flames Using the OH (0,0) Middle-UV Band", Appl. Optics, 21, 3357, (1982).

43. J.J. Horvath, K.M. Pamidimukkala, W.B. Person and A.E.S. Green, "Spectroscopic observations of methane-pulverized coal flames", Western States Section Symposium/The Combustion Institute, JPL, CA (1983).

44. A.E.S. Green, J.J. Olivero and R.W. Stagat, "Biophysical Aspects of Radiation Quality", International Atomic Energy Agency, Vienna, p. 79-97 (1971).

45. K.M. Pamidimukkala and A.E.S. Green, in preparation.

46. R.J. Drummond, "Combustion of Fuel Burning at James River Plant", Proceedings of International Conference on Combined Combustion of Coal and Gas, Cleveland, Ohio, October 12-13, (1982).

47. J.T. Kelly, R.A. Brown, E.K. Chu, J.B. Wightman, R.L. Pam, E.L. Swenson, E.B. Merrick and C.F. Busch, "Pilot-Scale Development of Low-NO_x Coal-Fired Tangential System", U.S. EPA Report No. EPA-600/S7-81, October (1981).

48. E.R. Lindgren, Proposal Dated July (1982). (private communication with A.E.S. Green).

49. D.S. De, Chemical Engineering Department Indian Institute of Technology Kharagpur, India, private communication.

50. R. Myers, "TRW makes its bid for the boiler business", The Energy Daily, p. 2. December 21, (1982).

51. R.P. Ruskan, "Coal/Oil, Coal/Water Slurries...Now its Coal/Methanol", Power p. 137, October (1982).

52. J. Dooher, T. Kanabrocki and D. Wright, "Co-Firing of CWS with natural gas", title of intended paper Fifth International Symposium on Coal Slurry Combustion and Technology, April 25-27, Tampa, Florida (1983) (private communication).

53. Brookhaven, Wood Pyrolysis with Methane. Quads, Forest Service USDA No. 18, December 1982.

54. Environmental Research and Technology, (private communication with Mathew Reilly) September 1982.

55. R.C. Furman, "Coal-Water Fuel Technology, A User's Perspective", Proceedings of the Coal-Water Fuel Technology Workshop (1981).

56. C.B. Foster, "The U.S. Coal-Water Mixture Combustion Program", Proceedings of 4th International Symposium on Coal Slurry Combustion, Orlando, Florida (1982).

57. S.W. Chun et al., Proceedings of the Fifth International Symposium on Coal Slurry Combustion and Technology, Tampa, Florida April 25-27 1983.

58. R. Payne, S.L. Chen, W. Richter, and J. Wilborn, "A Procedure for the Evaluation of the Combustion Performance of Coal-Water Slurries", Proceedings of the 5th International Symposium on Coal Slurry Combustion and Technology, Tampa, Florida, April 25-27 (1983).

BIOMASS GASIFICATION AS A SOURCE OF ACIDIC SUBSTANCES

Kay M. Eoff
Department of Physics, 467 Little Hall
University of Florida, Gainesville, FL 32611

Lawrance N. Shaw
Agricultural Engineering Department, IFAS
University of Florida, Gainesville, FL 32611

Don M. Post
School of Forest Resources and Conservation, IFAS
University of Florida, Gainesville, FL 32611

Abstract

A brief general description of thermochemical conversion technology for biomass is presented along with those applications and fuels which are likely to become of importance in Florida. The possible air pollutants generated by small gasifier systems including fuel related pollutants, producer gas related pollutants, atmospheric nitrogen conversion in the gasifier and in the user device, as well as some stack gas measurements, are discussed.

Introduction

The rapid increase of oil prices which began about 1974 led to an expanded study of renewable energy sources. Important among these has been biomass. My colleagues and I have been working in the field of thermochemical conversion of biomass to fuel gas for several years. We cannot present to you estimates of the increase or decrease of acid rain attendant to this or that scenario of use of biomass for energy production since that is outside our field of expertise. This conference offers us an opportunity to present information about our work in thermochemical conversion so that a small part of the environmental impact of utilization of biomass for energy production can be discussed. To this end we will discuss thermochemical conversion with emphasis on those aspects of the technology which may lead to production of acid forming gases. First, we will discuss thermochemical conversion in general and those techniques which are likely to be used in Florida; then the chemical composition of typical fuels available in Florida will be given. Application of the technology will be discussed and finally the few measurements of emissions from biomass fueled systems with which we are familiar will be given.

Thermochemical Conversion and Air Gasifiers

The complicated chemical structure of wood and other plant material (generally called biomass) is broken down by heat into smaller molecules (1). Some of the resulting molecules are gases at normal conditions. These include the combustible gases H_2, CO, and CH_4. The other major gaseous component is non-combustible CO_2. This mixture of gases is often called producer gas and is a useful fuel which can replace natural gas and liquid fossil fuels in many applications.

Among the liquid components produced are acetic acid and methanol, both of which have been commercially produced from biomass. Turpentine and other valuable liquid products are produced from destructive distillation of pine. Another liquid formed is "tar" or "creosote," the same material that can condense in the chimney of a wood stove or fireplace and cause destructive chimney fires. The solid component remaining after heating the biomass is called char, or charcoal.

The relative amounts of the various components produced by the heating of biomass depend not only on the type of biomass but also on the rate and temperature at which the heat is supplied, the size of the pieces of biomass and the design of the apparatus. A device which is designed to maximize the production of combustible gas is called a gasifier and many designs are possible. The simplest and most commonly employed class of gasifiers is the air gasifier in which air is introduced into the vessel to burn some of the biomass thus producing the heat for gasification. Most air gasifier designs fall into 4 categories: updraft, in which the air and gases flow upward through the biomass, down draft and cross draft with the obvious meanings, and fluidized bed.

The updraft gasifier was the first type utilized and is illustrated in Figure 1a. Air enters near the bottom of the gasifier and is distributed to the fuel through the grate which also holds up the fuel bed but allows ash to drop through as it is produced. Gas is removed near the top above the fuel bed. Fuel is fed into the gasifier from the top and ash removed from the bottom.

The most commonly used gasifier is the cross draft type illustrated in Figure 1b. Air entering the gasifier flows through the traveling grate and gasifies the fuel. The gases are burned immediately by the addition of combustion air. The flame then enters the furnace, boiler or kiln.

A sketch of a down draft gasifier is shown in Figure 1c. In the down draft gasifier, the hottest zone is not at the grate as in the other designs, but is at the level of the air inlets. The heat generated in this zone converts the fuel above into char. Thus, the air enters a bed of char continuously produced from the biomass above. The gases (and liquids in vapor form) must pass through this bed of hot char to exit the gasifier. The hot, chemically active bed of char breaks down almost all the tar into the gases CO, CO_2, CH_4 and water vapor. The hot char can also "reduce" the water and CO_2 according to the reactions

$$H_2O + C \rightarrow H_2 + CO$$

$$CO_2 + C \rightarrow 2\ CO$$

thus producing more of the useful fuel gases, CO and H_2, while at the same time greatly reducing the tar content of the gas.

The fluidized bed gasifier is of more recent invention. A sketch of such a design is shown in Figure 2. The name derives from the motion of

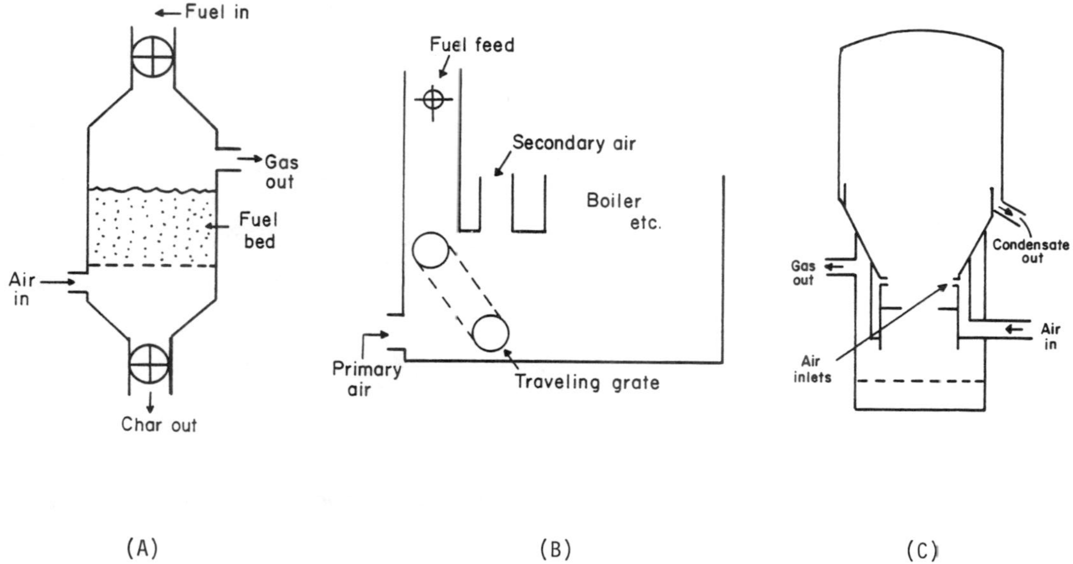

Fig 1. Commonly Used Gasifiers: (A) updraft (B) cross draft (C) down draft

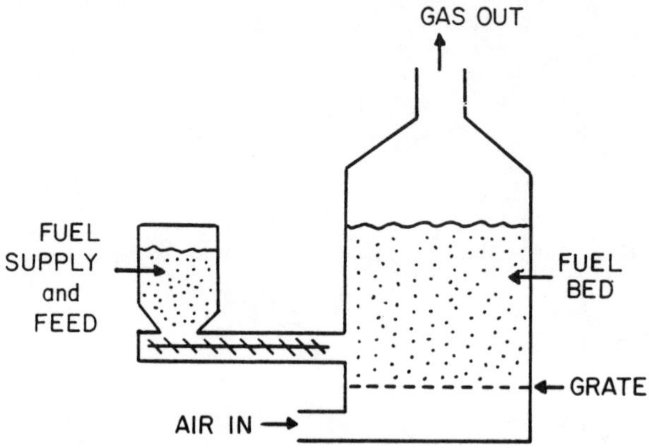

Fig. 2 Fluidized Bed Gasifier

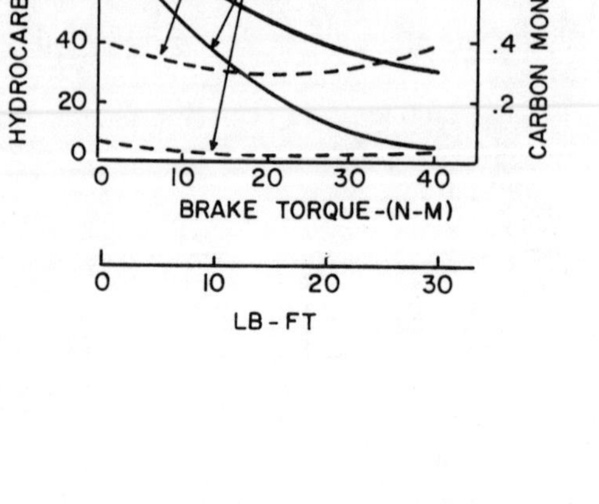

Fig. 3 Emissions from Diesel Fueled with Producer Gas

the fuel bed which is like that of boiling water. The bed is composed of small particles and the air flow is maintained high enough to cause constant agitation. Because of the constant agitation, the temperature and composition are uniform throughout the bed. If the air flow rate is less than that required to produce fluidization, the gasifier would operate in the updraft mode in which the heat for gasification is produced near the grate by burning some of the fuel. In fluidized bed operation the heat for gasification is produced throughout the bed by burning some of the fuel gas produced. The gasifier will operate with air flows several times the minimum value required for fluidization; however, too high a flow rate would blow the fuel bed out of the gasifier. The temperature in the bed is determined by both the air flow rate and the fuel flow rate as also is the total gas production.

Air Pollution Potential of Gasifiers

There are two substances contributing to acid rain which can be generated by gasifiers, volatile hydrocarbons and NO_x from fixation of atmospheric nitrogen. With regard to the latter, temperatures above 2400°F are required for significant nitrogen oxide formation. Temperatures of at least 2200°F are found near the air inlets of a down draft gasifier so some NO_x production may occur. No direct measurement of NO_x concentration in the gas stream from a down draft gasifier is available; however, the condensate which forms upon cooling the gas is slightly alkaline due primarily to NH_4 which suggests that little NO_x would remain in the gas. Fluid bed and updraft gasifiers generally operate at lower temperatures so NO_x production from such types should be minimal. Cross draft gasifiers are a special case since the gas is burned immediately as it is produced. NO_x production would depend on burner design and temperature requirements in the particular application.

The volatile hydrocarbons have their source in the pyrolysis liquid and vapor produced which are toxic and corrosive and contain polycyclic aromatic hydrocarbons (PAH) including known and suspected carcinogens (2). Fugitive emissions of gas or liquid from the gasifier represents an environmental hazard which must be minimized. These dangerous substances are however mostly eliminated from the stack gases when the fuel gas is burned in the user device; for instance no increase in PAH levels were measured in grain dried with stack gas from an updraft gasifier (3).

The fluidized bed gasifier is particularly well suited for utilizing high sulfur fuels. Addition of ground limestone to the fuel bed will keep most of the sulfur in the ash and greatly lower the sulfur content of the gas.

It is seen that the use of gasifiers should not lead to significant production of acid forming gases. To the extent that an increased use of gasifiers would lead to a reduction of open burning of forest wastes by providing a market for these currently under-utilized materials, air pollution from biomass may actually be lowered. Estimates of 1 to 54 g/Kg THC (total hydrocarbons) and 2 g/Kg NO_x are given for emission from burning forest wastes and values as high as 120 g/Kg THC are reported from wood stoves (4). In Canada it is estimated that nearly 4% of the total volatile hydrocarbons emitted into the atmosphere is attributable to forest fires (5).

Gasifier Fuels

About 2% of Florida's energy budget is currently provided by combustion of wood wastes in the forest products industry. This amount could be substantially increased without diverting wood from current markets or changing land use patterns (6). Wood is a very good gasifier fuel and is considered the major biomass feedstock for thermochemical conversion. Some agricultural wastes are likely to be considered for gasifier fuel because they are collected as a by-product during harvest. Among these are corncobs, peanut shells and bagasse. Rice farming may be on the increase in Florida so rise husks may also be used. Some properties of these fuels and, for comparative purposes, coal, are given in Table 1. In order to compare the fuels on an equal energy basis the numbers in parentheses are calculated by multiplying the percent N and S by the ratio of the heating value of unwashed coal to the heating value of the alternative fuel. With the exception of nitrogen in peanut shells, biomass fuels should produce less acid gas emissions than low sulfur coal.

Table 1. Comparative Properties of Various Coal and Biomass Fuels

Fuel*	Heat Value Btu/LB	Weight % Dry Basis N	S
COAL:			
Penn. (washed)[10]	14000	1.6 (1.5)	1.2 (1.1)
Penn.[10]	13000	1.5	2.1
Western[10]	9400	1.0 (1.4)	0.4 (0.6)
WOOD:			
Citrus[11]	7900	-----------	0.03 (0.05)
Slash Pine[10]	8700	0.09 (0.13)	0.04 (0.06)
Oak Bark[10]	8400	0.2 (0.3)	0.1 (0.15)
CROP WASTES:			
Corncobs[12]	7500	0.47 (0.81)	0.1 (0.17)
Peanut Shells[12]	8700	1.1 (1.64)	-----------
Bagasse[13]	8200	0.3 (0.6)	0.16 (0.25)
Rice Husks[9]	7000	-----------	0.2 (0.37)

* Numbers after fuels refer to literature cited.

Gasifier Applications

Producer gas can be used to fuel spark ignition and dual fueled diesel engines and gas turbines. Down draft or fluidized bed gasifiers are preferred for these applications because the gas produced is low in "tar" and more easily cleaned for engine use. Internal combustion engines generally produce some NO_x and hydrocarbon emissions. It is not clear how operation on producer gas would affect these. Some factors concerning producer gas which may relate to this are:

1. The gas burns more slowly than fossil fuels and may not burn completely in modern high speed engines.

2. Since the gasifier produces the gas as it is used, the response time to load changes is less for a gasifier-engine system than for a fossil fueled engine. Carburetion problems are experienced during changes in load.

3. The energy density of a stoichiometric mixture of producer gas and air is about 75% that of a gasoline-air mixture. To achieve rated output of an engine requires operation at higher RPM, increasing the compression ratio or supercharging. Any of these can change the pollution potential of the engine.

Updraft and cross draft gasifiers are used to produce heat for boilers, kilns, dryers, greenhouses and even home heating. The particulates and "tar" in the gas can generally be burned without problems in these applications. The NO_x production would depend on the temperature required of the particular application. Again, it is not clear how the use of producer gas would affect the results. If gasifiers replace the use of direct fired devices in these applications, particularly for home heating, some reduction in air pollution can be expected. Not only are the gasifiers generally more efficient than direct fired devices but also the increased control over the combustion leads to less volatile hydrocarbon emissions.

Flue Gas Measurements on Gasifier Systems

We have had the exhaust gases from a spark ignition engine powered with producer gas analyzed for hydrocarbon and carbon monoxide emissions. The carburetor on this gasifier engine system was very primitive and yet the emissions - 15 ppm for hydrocarbons, and 0.15% for CO - were much lower than for a similar engine run on gasoline and compare well with results of using the best emission control devices.

The exhaust gas from a diesel dual fueled with producer gas has been studied by Ogunlowo et al. (7). The results are given in Figure 3. Both the hydrocarbon and the CO emissions are generally higher when using producer gas than when using diesel alone. The valve timing of the engine was not designed to operate in a dual fueled mode with combustible gas in the intake manifold. Most small diesel engines which are available for conversion to producer gas operation will suffer from this problem. Table 2 contains an analysis of the stack gases from a burner provided with gas from a fluidized bed gasifier (8, 9). Rice husks and wood chips were used as fuel. The higher sulfur emission when using rice husks is expected from fuel compositions given in Table 1. The high NO_x emission when using rice husks is not explained except by a brief statement that the burner was operated at a high temperature.

Table 2. Analysis of Stack Gases from a Fluidized Bed Gasifier Burner (from refs. 8, 9)

FUEL	SO_2 PPM	NO_x PPM	CO PPM	O_2 %
Rice Husks	8	250	35	8.5
Wood	Trace	Trace	100	4.7

Summary

In summary it would appear that increased use of thermochemical gasification of biomass for energy production should contribute very little to acid rain formation. To the extent that the technology would lead to a decreased use of high sulfur fuels or even a decreased use of direct fired biomass devices, some decrease of acid producing emissions may be expected.

Literature Cited

1. SERI/TR-33-239, 1979. A Survey of Biomass Gasification. Solar Energy Research Institute.

2. Lee, M., Prado, G., Howard, J., and Hites, R., 1977. Source Identification of Urban Airborne Polycyclic Aromatic Hydrocarbons. Biomedical Mass Spectrometry 4(3):182-186.

3. Payne, F., Ross, I., Walker, J., and Brasher, R., 1980. Gasification-Combustion of Corncobs and Analysis of Exhaust. Paper No. 80-3025, ASAE 1980 Summer Meeting.

4. Source Assessment: Prescribed Burning, State of the Art. EPA-600/2-79-019h, Nov. 1979.

5. Memorandum on Intent on Transboundary Air Pollution. Emission, Costs and Engineering Assessment Work Group 3B. Final Report. June 1982.

6. IFAS Fuelwood Commodity Committee, 1983. Fuelwood in Florida. Institute of Food and Agricultural Sciences, University of Florida, Gainesville.

7. Ogunlowo, A., Chancellor, W., and Goss, J., 1981. Dual-Fueling a Small Diesel Engine with Producer Gas. Transactions of the ASAE 24(1):48-51.

8. DOE/RA/50303, 1981. Feasibility Study for Alternative Fuels Production: Fluidized-Bed Gasification of Wood. U.S. Department of Commerce.

9. DOE/RA/50378, 1980. Feasibility Study for Alternative Fuels Production: Fluidized-Bed Gasification of Rice Hulls. U.S. Department of Commerce.

10. Pringrey, D., 1976. Forest Products Energy Overview. Forest Products Research Society. Proceeding No. P-76-14:1-16.

11. Analysis by Thornton Laboratories, Inc. Laboratory No. 578976.

12. Summer, H., Summer, P., Hammond, C., and Monroe, G., 1981. Energy Available from Biomass for Grain Drying. Paper No. 81-3014 ASAE 1981 Annual Meeting.

13. IGT, 1982. Thermochemical Gasification Characteristics of Biomass. Institute of Gas Technology.

COASTAL AND INLAND NATURAL H_2S SOURCES

Dian R. Hitchcock
Hitchcock Associates
1702 S.W. 35th Place
Gainesville, Florida 32608

and

Edward S. Deevey, Jr.
Florida State Museum
University of Florida
Gainesville, Florida 32611

Abstract

We report the most recent of a series of field experiments exploring the question of the importance of natural S emissions to the atmosphere, and the possibility that S introduced into the environment by human activities is recycled to the atmosphere by natural processes. Comparisons of the ambient H_2S concentrations at a productive coastal salt marsh (Wallops Island, VA) and a marshlike freshwater habitat (near Rochester, NY) show that emissions are spatially extremely variable in both kinds of habitats, and that the release rates are similar. Cautious extrapolation of the limited data indicates that summertime biogenic S emissions in a part of the Lake Ontario Lake plain ($\sim$8000 km^2) are of the same order as the local anthropogenic emissions. A survey of the biogeochemical characteristics of Florida lakes and wetlands suggests that (a) many are probably significant sources of atmospheric S, and (b) human activities may promote biological S releases to the atmosphere.

Introduction

The quantitatively most important sulfur cycle in the biosphere is the oxidation of organic matter at the expense of sulfate in solution. This process produced H_2S as a metabolic byproduct, and occurs wherever metabolizable organic matter and sulfate are present in a habitat devoid of oxygen. The reaction is mediated by a variety of bacteria known as the sulfate reducers. Sulfate reduction is the dominant anoxic respiration in marine sediments, because seawater contains abundant sulfate (2700 ppm). Very high rates of sulfate reduction have been recorded on organic-rich coastal marine sediments (1,2), but sulfate reduction also occurs in freshwater habitats, where its importance varies with the trophic state of the system and the sulfate concentration.

About 10 years ago it occurred to us and others (3) that many human activities tend to promote conditions in which sulfate reducers thrive. Organic and nutrient pollution directly or indirectly increase the availability of metabolizable organic matter, and lead to anoxia in sediments and/or the water column, and typically increase the rate of sulfate reduction in harbors, estuaries, and other habitats. This may also occur in freshwater systems, which may also be sensitive to the addition of sulfate, either directly in the form of industrial releases, acid mine drainage, or fertilizer runoff, or indirectly, via the deposition pollutant sulfur dioxide or precipitation sulfate.

It is obvious that if there is a natural cycle of sulfur from coastal or inland aquatic habitats into the atmosphere, then these human activities may increase the atmospheric release of biogenic S. Evaluating this potential requires first determining whether there is a significant natural cycle of biogenic S into the atmosphere, and then identifying the physical and biogeochemical correlates of the "sensitive" systems where human activities may increase emissions.

One of us (DRH) therefore began a program of studies of the natural atmospheric sulfur cycle, focusing initially on coastal salt marsh habitats where bacterial sulfate reduction rates are naturally high. We reasoned that if sulfide emissions are negligible in these areas, then human activities are unlikely to have an important impact on "natural" sulfur emissions anywhere. These studies have entailed measuring the atmospheric abundance of S compounds in the summer at coastal sites and employing a number of methods to identify the biogenic component of the SO_x. The most successful studies have used evidence of $^{34}S/^{32}S$ ratios to distinguish SO_x delivered to our site by air mass transport from that locally produced from biogenic sulfide precursors. These latter studies were conducted at Wallops Island, VA, and demonstrated that local biogenic sources yield a mean of about 5 to 8 µg m^{-3} of SO_2 and 8 to 10 µg m^{-3} of SO_4^{2-} to the ground-level summertime atmosphere (4). These results were somewhat surprising, in that they suggested a more important role for natural sources than most workers using conventional techniques have inferred, but they are compatible with the results of others who have also measured stable sulfur isotope ratios.

Our results strongly suggested that there are important inland sources of atmospheric S, so the 1980 experimental season included a comparison of the ground-level H_2S concentrations at an inland site with those at the Virginia site previously studied. We chose a region of inland wetlands which appeared from previous studies to be a natural "hot spot" for sulfide emissions.

Experimental Results

Methods

Twenty-four hour H_2S samples were collected on cellulose filters treated with $AgNO_3$, and glycerol mounted in flow-controlled high volume samplers behind prefilters which removed particulates and SO_2. Preliminary tests showed that no H_2S was removed by the first two stages, and that SO_2 was selectively removed by the second stage at the flow rates employed. Silver sulfide formed on the filter was removed by boiling in an acid solution, and the H_2S thereby generated was precipitated as AgS and determined gravimetrically. The detection threshold for this system corresponds to a 24-hr mean concentration of 0.1 ug m^{-3}. Details of the system will be presented elsewhere (8).

Sites

One objective of this study was to determine the H_2S concentrations in the coastal marsh where previous studies had confirmed a significant biogenic contribution to SO_x levels, and the other was to see whether one candidate inland system also produced H_2S in abundance. A promising inland system in Monroe County, NY had been identified in a survey of air monitoring data (9). This study contains extensive wetlands (marshes, swamp forests, and bogs) where aquatic sulfate levels are naturally high (100 to over 300 ppm). The county lies on the southern shore of Lake Ontario. Wetlands occur in all parts, but they are most dense in the southwestern section, a generally rural area where the odor of H_2S is usually marked on calm summer nights and some cloudy afternoons. There are a total of nearly 13,000 km^2 of flat, well watered land in a belt along the southern shore of Lake Ontario extending to Syracuse in the east. Marsh systems are abundant in the whole area, and we judge that about 60% (8000 km^2) contains high densities (8 to 30%) of high-sulfate marshlands analogous to those in the southwestern section of Monroe County.

Twenty-four hour samples were simultaneously collected at four sites in the county (Figure 1). Three in the southwestern section included one (Churchville) immediately adjacent to a representative swamp-like pond whose shallow sediments (0.3 to 1.2 m) were strongly reducing, and two at points 5 to 12 km distant (Brockport and Griffen Road). The latter were between 1 and 3 km from wetlands similar to that at Churchville. A fourth site in a grassy meadow at Mendon Ponds in the southeastern part of the county in a region of gentle hills was used as a control site, although it was within 1 km of small systems. The main differences between the western sites and the control sites were higher levels of sulfate in surface waters (>300 ppm in the west and <180 ppm in the east) and the widespread presence of intense algal blooms and evidence of reducing sediments in the western systems (thick muck sediments which were apparently reducing at their surfaces were characteristic of most systems surveyed) apparently due to high nutrient loadings from farm fertilizer runoff. (The Mendon Ponds sampler was inside a large park where conditions remained more natural, the sediments in nearby systems appeared less reducing on visual inspection.)

The coastal samples were taken on the grounds of the NASA Flight Test Center at Wallops Island, VA. Samples were collected on the lee shore of a barrier island, and also on the edge of the mainland opposite (Figure 2). The island sampler was positioned about 3 m from a stand of *Spartina alterniflora*, and was about 5 meters from a small salt creek. The soil beneath the sampler was periodically wetted at high tide. The mainland sampler was at the edge of a small salt pond, and about 200 m from the nearest Spartina stand. The island and the mainland are separated by a 3-km-wide expanse of mixed salt marsh, meandering salt creeks, and salt pans.

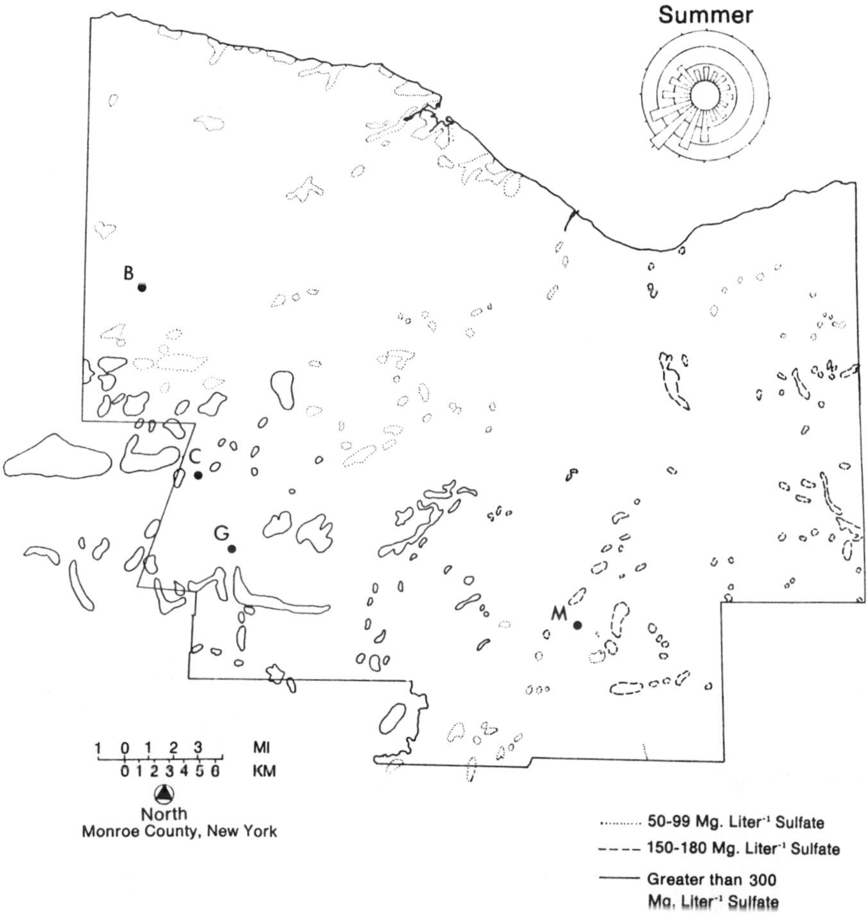

Fig. 1 Monroe County Sampling Sites. Sites are: B, Brockport; C, Churchville; G, Griffen Road; M, Mendon Ponds. Map shows major swamps and marshes coded for mean summertime inorganic sulfate concentration. Surface waters are not shown. Windrose shows mean summertime conditions. Wetlands and windrose taken from Hitchcock (1977).

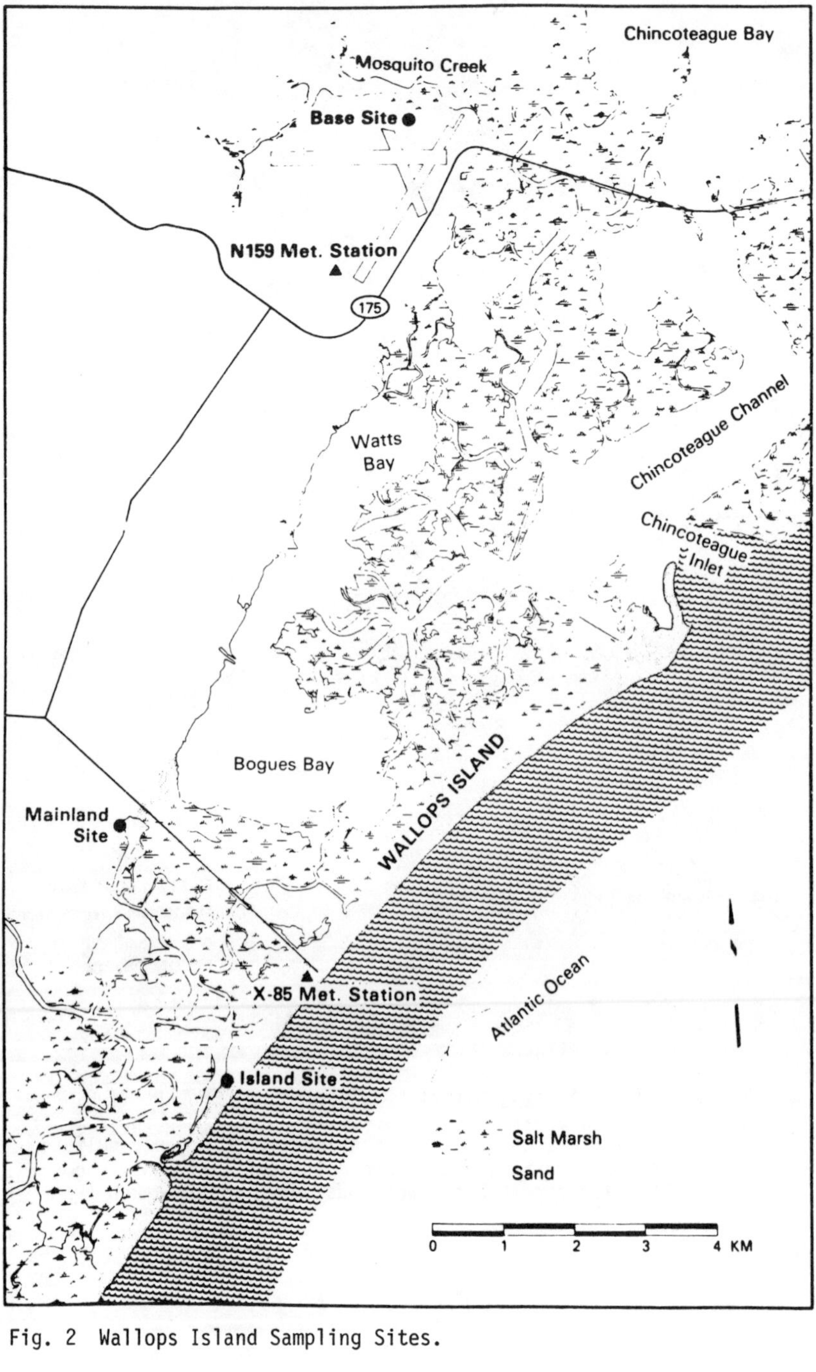

Fig. 2 Wallops Island Sampling Sites.

Results

Salt Marsh

Sampling began on September 6 and continued through September 21. Island H_2S concentrations ranged from below detection on one day to 5.5 μg m^{-3}, and averaged 1.4(±1.6) μg m^{-3}. Concentrations at the mainland were much lower, being below detection on 9 days (56%) and ranging up to only 0.7 μg m^{-3}. The average (calculated by assuming zero concentrations in samples below detection level) was 0.17(±0.24) μg m^{-3}. Levels at both sites varied inversely with wind speed (r = .51 at the island, $P < .05$, and .41 at the mainland, $P < .10$). On three days when the wind blew steadily from the ocean concentrations at the island ranged from 0.1 to 0.3 μg m^{-3}, but were below detection at the mainland, although air sampled there had passed over the entire marsh system. The data are summarized in Table 1.

Table 1. Concentrations and Meteorological Observations

A. Hydrogen Sulfide (μg m^{-3})

Wallops Island	Island	Mainland
Range	<0.1 to 5.5	<0.1 to 0.7
Mean*	1.4(±1.6)	0.2(±0.2)

Monroe County	Churchville	Brockport	Griffen Road	Mendon Ponds
Range	0.4 to 6.0	<0.1 to 0.5	<0.1 to 0.4	<0.1 to 0.2
Mean	2.6(±1.6)	0.2(±0.1)	0.2(±0.1)	0.05(±0.07)

B. Means of Meterological Observations (24-hour mean)

	Wallops Island	Monroe County
Temperature (°C)	22.9(±2.1)	23.0(±1.9)
Wind Speed (m s^{-1})	3.6(±1.1)	3.6(±1.2)
Relative Humidity (%)	76(±5)	61(±13)

* Mean (±1 SD).

Inland Wetlands

Samples were collected from August 7 to August 29. The highest levels were observed at Churchville, where they ranged from 0.4 to 6.0 μg

m^{-3} (mean 2.6 $\pm$ 1.5 μg m^{-3}). Concentrations at the two other sites in the southeast ranged from below detection (10% of days at Griffen Road and 29% at Brockport) to 0.4 and 0.5 μg m^{-3}. At both the means were 0.2($\pm$0.1). The lowest levels were observed at Mendon Ponds, where they were below detection in 63% of samples, never exceeded 0.2 μg m^{-3}, and averaged 0.05 ($\pm$0.07) μg m^{-3}. There was no significant correlation between wind speed and the levels at any site, or between levels at any pair of sites. These relationships are compatible with the irregular distribution of emitting sources around all sites, and they also suggest that the lowest detected levels did not reflect a widespread background, but rather local influences of specific emission sources.

Discussion

The island results at the coastal site agree well with those obtained by Goldberg et al. (10) who measured H_2S concentrations at the 7 m height at a point about 1 km north of our island site. They found extremely low concentrations during the day rising to between 3 and 20 μg m^{-3} under the nightime inversion, a diurnal pattern that has been observed by others in both atmospheric concentrations and emission rates (see, for example, 11, 12, and 13). Their 24-hour mean for one day reported in detail appears to be about 2 μg m^{-3}, which is about the same as the mean we obtained for the 16-day sampling period. The extremely low concentrations we observed at the mainland site confirm that emission strengths are extremely variable within this system, a finding compatible with the reported variability in sulfate reduction rates and rates of H_2S emissions for coastal zones (11-14).

The similarity in ambient sulfide concentrations at the coastal system and within the high-sulfate marsh zone in southwestern Monroe County was the most surprising finding of this study. It suggests that sulfide emission rates in very large areas of upstate New York equal those in the Virginia marsh system. Concentrations cannot readily be translated into emission rates, as they depend critically on local air movements. For example, Hansen et al. (13) sampled air at the 0.5 m height while measuring emissions from a shallow tide pool. Ambient levels ranged from 6 to 36 μg m^{-3} (varying inversely with wind speed) while emissions at the surface ranged up to 0.2 g $m^{-2}h^{-1}$. Nevertheless, it is reasonable as a first approximation to assume that mean emissions in upstate New York regions where high-sulfate marshes are dense equal those at Wallops Island, and a crude estimate of the summertime emissions can be based on the mean nighttime emission calculated by Goldberg et al. (10) for the region between the island and the mainland at Wallops. These authors report a range of 300 to 1800 μg $m^{-2}h^{-1}$ (estimated from the concentration in a column of air passing their sampling site and averaged over the total fetch of 3 km). These release rates are close to the median of those reported in the literature for coastal sediments (60 to 200,000 μg $m^{-2}h^{-1}$ (15)), and comparable to the nighttime rate of 300 μg $m^{-2}h^{-1}$ estimated for nightime flooded Ivory Coast tropical forest soils (16). Extrapolating these rates to the estimated 8000 km^2 of upstate New York of which the southeastern corner of Monroe County is representative yields a summertime total between 2.6 and 15.5 x 10^9 g, or between 20 and 125% of the local summertime anthropogenic S emissions. (These are estimated as 25% of the total summertime emissions in the state.) Biogenic sulfide emissions of this magnitude could influence local precipitation sulfate levels, and may contribute to the increased levels of sulfate in precipitation documented for summer months.

Although the crude estimates developed here are clearly in need of confirmation, there is little doubt that some inland systems release large quantities of sulfide into the atmosphere. This being the case, we may speculate that human activities can sometimes disrupt the sensitively balanced processes of H_2S production and consumption in aquatic systems, and promote release to the atmosphere. It seems likely that emissions from the Monroe County systems are already impacted by high nutrient loading and the emissions have been increased by the resulting availability of organic matter. It is also possible that these systems have always been strong sources of atmospheric S. Controlled studies are needed.

Research Needs in Florida

Very large areas of Florida comprise permanent or temporary wetlands, and one may surmise that at least some of them are potent sources of atmospheric S, so field work to identify existing sources is indicated. Some prime candidates can be identified in a fairly straightforward way. One simply looks for analogues of the extremely productive, shallow, high-sulfate marshes of Monroe County. The first logical step is a survey of the atmospheric sulfide concentrations in the vicinity of such systems.

A more ambitious--and scientifically inherently more interesting--program would systematically relate sulfide emission potential to variables such as sulfate concentration, trophic state, sulfide production in sediments (or possibly in anoxic micro-niches in seston), availability of iron or other metals that may form metal sulfides, and water depth and current and other factors which influence the delivery of sulfide to the air-water interface. Such a program must be an interdisciplinary one, and it requires that extremely diverse aquatic systems be available for study. Here, Florida would seem to be most richly endowed. A cursory survey of the available data (17-20) confirms that aquatic systems in Florida are extremely diverse. Surface water sulfate concentrations range from a few ppm up to 700 ppm and up more in brackish and marine systems on the coasts. All levels of trophic state appear to be represented, and pH ranges from about 4 to over 8. Cultural eutrophication occurs widely in varying degrees, and is under study. Other factors--such as the availability of iron which removes sulfide from solution--also vary widely.

There are some major instrumental problems which must be solved to clear the way for substantial progress in a broad program of the kind envisioned. One must be able to trace the formation and fate of sulfide in aquatic systems, and detect aquatic sulfide concentrations which are atmospherically relevent. Limnologists are accustomed to dealing with millimoles of sulfide in water, but one millimole is many orders of magnitude above equilibrium with 0.1 μg m^{-3} in air. For example, in freshwater at 20°C, this concentration ranges from about 10^{-8} mM l^{-1} at pH 5 up to only about 2×10^{-7} mM l^{-1} at pH 8.5.

Acknowledgments

The work reported here was performed under NSF grant no. PFR 79-08967, EPA grant no. CR 806628, and EPA contract no. 1D 539. Participation in the field studies by D.F.S. Natush and J. Dill, Chemistry Department, Colorado State University, is gratefully acknowledged.

Literature Cited

1. Howarth, R.W., and Teal, J.M., 1979. Sulfate Reduction in a New England Salt Marsh. Limnology and Oceanography 24:999-1013.

2. Jørgensen, B.B., 1982. Mineralization of Organic Matter in the Sea Bed--The Role of Sulphate Reduction. Nature 296:643-645.

3. Berner, R.A., 1971. Worldwide Sulfur Pollution of Rivers. Journal of Geophysical Research 76:6597-6600.

4. Hitchcock, D.R., and Black, M.S., 1983. $^{34}S/^{32}S$ Evidence of Biogenic Sulfur Oxides in a Salt Marsh Atmosphere. Atmospheric Environment _In Press_.

5. Ludwig, F.L., 1976. Sulfur Isotope Ratios and the Origins of the Aerosols and Cloud Droplets in California Stratus. Tellus 28: 427-433.

6. Grey, D.C., and Jensen, M.L., 1972. Bacteriogenic Sulfur in Air Pollution. Science 177:1099-1100.

7. Krouse, H.R., and Case, J.W., 1981. Sulfur Isotope Ratios in Water, Air, Soil, and Vegetation Near Teepee Creek Gas Plant, Alberta. Water, Air and Soil Pollution 15:11-28.

8. Dill, J., 1983. In prep.

9. Hitchcock, D.R., 1977. Biogenic Sulfur and Air Quality in the United States. Final Report NSF Grant No. AEN-7514571. Arthur D. Little, Inc., Cambridge, MA.

10. Goldberg, A.B., Maroulis, P.J., Wilner, L.A., and Bandy, A.R., 1981. Study of H_2S Emissions from a Salt Water Marsh. Atmospheric Environment 15:11-18.

11. Ingvorsen, K., and Jørgensen, B.B., 1981. Seasonal Variation in H_2S Emissions to the Atmosphere from Intertidal Sediments in Denmark. Atmospheric Environment 16:855-866.

12. Jaeschke, W., Claude, H., and Jerrman, J., 1980. Sources and Sinks of Atmospheric H_2S. Journal of Geophysical Research 85:5639-5644.

13. Jansen, M.H., Ingvorsen, K., and Jørgensen, B.B., 1978. Mechanisms of Hydrogen Sulfide Release from Coastal Marine Sediments to the Atmosphere. Limnology and Oceanography STET 28:70-82.

14. Howarth, R.W., and Giblin, A., 1983. Sulfate Reduction in the Salt Marshes of Sapelo Island, Georgia. Limnology and Oceanography 28:70-82.

15. Aneja, V.P., Aneja, A.P., and Adams, D.R., 1982. Biogenic Sulfur Compounds and the Global Sulfur Cycle. Journal of the Air Pollution Control Association 32:803-807.

16. Delmas, R., Baudet, J., Servant, J., and Baziard, Y., 1980. Emissions and Concentrations of Hydrogen Sulfide in the Air of the Tropical Forest and of the Ivory Coast and Temperate Regions of France. Journal of Geophysical Research 85:4468-4474.

17. Canfield, D.E., 1981. Chemical and Trophic State Characteristics of Florida Lakes in Relation to Regional Geology. Institute of Food and Agricultural Sciences, University of Florida. Gainesville.

18. Baker, L.A., Brezonik, P.L., and Kratzer, C.R., 1981. Nutrient Loading-Trophic State Relationships in Florida Lakes. Pub. 56, Florida Water Resources Research Unit, Gainesville.

19. Flannery, M.S., Snodgrass, R.D., and Whitmore, T.J., 1982. Deepwater Sediments and Trophic Conditions in Florida Lakes. Hydrobiologia 92:597-602.

20. Crisman, T.L., 1980. Chydorid Cladoceran Assemblages from Florida. PP. 657-668 in Kerfoot, W.C. (Ed) "Evolution and Ecology of Zooplankton Communities" ASLO Spec. Symp. 3. University Press of New England. Hanover.

INTERPOLATED ANALYTIC WIND ROSES FOR FLORIDA LOCATIONS

J.M. Schwartz, J.H. Miglietta, and A.E.S. Green
Interdisciplinary Center for Aeronomy and other Atmospheric Sciences
University of Florida, Gainesville, Florida 32611

Abstract

A three-dimensional joint frequency distribution (3DJFD) map of Florida was developed, based on wind rose data available for 10 calibration cities, by Green, Schwartz, Singhal, Bolch (GSSB) (1). An approximate 3DJFD can now be assigned to any site in Florida after determination of the location on a geographic polar coordinate system. An illustrative application of the GSSB interpolated analytic wind rose method is given for a particular site in Florida (Palatka). This includes an examination of environmental impact of a coal-fired plant when used with the Green, Singhal, Venkateswar (GSV) Analytic Gaussian Plume Model (2). Thus a concentration map of the expected incremental SO_2 levels near Palatka is given for an assumed annual SO_2 emission. A second way of using the analytic wind rose is to generate a discrete 3DJFD for use with standard Gaussian Plume Models. An illustration of this route is given. The possibility of generating a pollution transport function map of Florida and of connecting this to a mesoscale transport map is discussed. Finally, using the GSV and GSSB methodology in conjunction with climatological data we infer a summer 3DJFD for the Tampa region and calculate the distant concentrations associated with a representative SO_2 emission rate. The transport time, high ultraviolet levels, sea breezes, and summer rains are favorable to the return of SO_2 emissions from the center and southern part of the state as acid rain on the northern part of the state.

Introduction

Two previous works, [Green, Singhal and Venkateswar (2) (GSV),] and [Green, Schwartz, Singhal and Bolch (1) (GSSB)] provide a simple analytic technique for estimating the impact of a new emission source at any location in Florida. This work is an outgrowth of our prior work on SO_2 effects in Florida (3-5) in which standard air pollution models (6) proved very expensive and time consuming when coupled to a comprehensive benefit cost analysis.

The essential equations for GSV and GSSB are summarized in Table 1. To illustrate the methodology, we here consider an application of these two works to Palatka, Florida, where two 600-megawatt coal-burning plants are under construction. Since an accumulation of wind data is not available to assign a three-dimensional joint frequency distribution which is needed in standard pollution-dispersion calculations, the interpolation procedure of GSSB is as reasonable a procedure as is currently available.

The Interpolated Wind Distribution

The coordinates for Palatka as obtained from the geometric construction made on a Southern States AAA map, are $X_g = -.50$ and $Y_g = +.43$ (see

Fig. 1). When these parameters are inserted into the biquadratic interpolation formula given by Eq. 1 in Table 1 and used with the geographic coefficients given in Table II of GSSB we obtain the array of coefficients given in Table 2. These are now the basic smooth, interpolated, localized parameters for Palatka, Florida, which reflect, for the most part, the numerical data from the nearby stations but encompass information from all of the ten calibration sites used as inputs in the GSSB work. Here the index $\nu = 0$ denotes the omnidirectional or scalar component of the wind (U), whereas $\nu = 1, 2, 3, 4$ denote four vectored wind plumes in the N, E, S and W quadrants. We will refer to this model as a five vector model by analogy with the five vector model of the nuclear force (7). To complete the characterization of the three-dimensional joint frequency distribution (3DJFD) we use Eqs. 2 and 4 of GSSB together with the Palatka parameters in Table 2 to obtain the wind amplitudes $M_{\nu s}$ in Table 3. Here s = 1, 2,...6 denotes the stability classes A, B,...F respectively. To obtain the characteristic wind velocities $U_{\nu s}$ we use Eq. 5 together with the Palatka parameters in Table 2 to obtain the results given in Table 4.

The Polar Coordinate Gaussian Plume Model

The work of GSV shows that the GPM expression for the ground level concentration $\chi(r,\theta,0,H)$ at the distance r, angle θ from a source of effective height H is given by Eq. 6 where the vertical and transverse dispersion parameters are given by Eqs. 7, and a, L, q, K and p are parameters given in Table 1 of GSV. Figures 1 and 2 of GSV give the basic data for Z_s and Y_s and the fits with Eqs. (7) which are very good. The quantity Ω^s is a correction term given by Eq. 8. When the distribution of wind directions for various wind velocities and stability classes is represented analytically in the special polar coordinate form given by Eq. 9, the winds distribution can be convoluted with Eq. 6 to give a final result for the pollutant concentration given by Eq. 10 where $\delta_s(r)$ is given by Eq. 11. The factor $10^6 = (kM/M)^2$ corrects for the units conventionally used and the constant $C = 4(2)^{\frac{1}{2}}ek/\pi^{3/2} = 1.420$ (when speeds are in knots and angles in radians). Here $\Sigma_{s\nu}$ denotes the summation over the five vector and the six stability classes. Equation 10 expresses the final concentration at a point r,θ due to all angular, velocity and stability class components from a given source located at r = 0, and $\theta = 0$ which in this case is the new coal plant in Palatka.

The final result is a sum of products of factors which reflects the down-wind distance dependence of the GPM and modified angular wind rose components. Some insight into this feature is given by Taylor and Hernandez (8). Note that the characteristic velocity parameters $U_{\nu s}$ do not enter the results. Figure 2 gives the isopleths of SO_2 concentrations for a source with an annual average emission Q = 10,000 tons per year and for an effective stack height H = 200 meters. The results are quite smooth and reasonable with the dispersion plumes in the opposite direction of the wind plumes.

We suspect that the dispersion plumes might be expressed by a five vector of the type given by Eq. 10 with s replaced for each plume by an average s each with a total amplitude $T_\nu = \Sigma_s M_{s\nu}$. The adjustment of the five parameters could be carried out by non-linear least squares using the numerical output from Eq. 10 as input data. For this purpose it

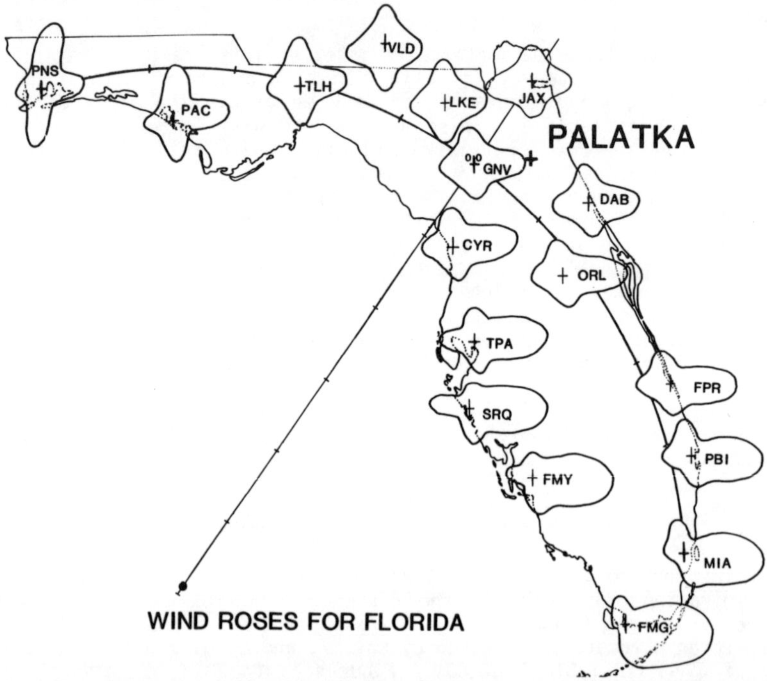

Fig. 1 Wind rose map of Florida showing location of Palatka. The cross corresponds to the geographic coordinates $X_g = -0.50$, distance southward along the arc shown through Gainesville, in units of 100 km and $Y_g = 0.43$ radial distance outward from this arc along a radius passing through Palatka.

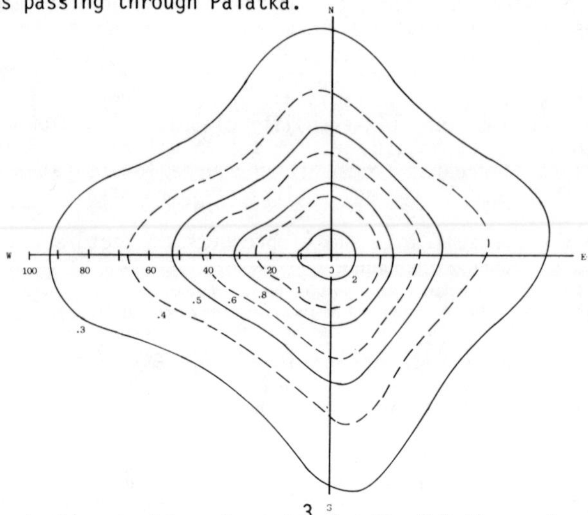

Fig. 2 Isoconcentration contours in $\mu g/m^3$ for the Palatka region due to an average source of 10,000 tons of SO_2 per annum. The distance scale is in kilometers.

Table 1. Equations for the Analytic Gaussian Plume Model

---

(1) $$P = \alpha + \beta X_g + \gamma X_g^2 + \delta Y_g + \phi X_g Y_g + (\phi^2 Y_g^2/4\gamma)$$

(2) $$M_{\nu s} = \frac{4\ D_\nu}{e\sqrt{\pi}\ U_\nu} \left(\frac{X_\nu}{\bar{V}_\nu}\right)^{\hat{s}-s} \left(\frac{\bar{Y}}{\bar{W}}\right)^{(\hat{s}-s)^2} \eta_\nu$$

(3) $$\eta_\nu = \frac{1}{1 + X_\nu \bar{Y} + X_\nu^2 \bar{Y}^4 + \ldots . X_\nu^{\hat{s}-1} \bar{Y}^{(\hat{s}-1)^2}}$$

(4) $$D_o = \left(\frac{2\pi}{16} - \sqrt{2\pi}\ \Sigma_j \sigma_j D_j\right) \frac{1}{2\pi}$$

(5) $$U_{\nu s} = U_\nu V_\nu^{\hat{s}-s}\ \bar{W}^{(\hat{s}-s)^2}$$

(6) $$\chi_s(r,\theta,0,H) = \frac{Q}{u\pi Y_s Z_s} \exp-\left(\frac{H^2}{2Z_s^2} + \frac{r^2\theta^2}{2Y_s^2} + \frac{\Omega_s\theta^2}{2}\right)$$

(7) $$Z_s = \frac{L_s r}{[1 + (r/a_s)]^{q_s}},\quad Y_s = \frac{K_s r}{[1 + (r/a_s)]^{p_s}}$$

(8) $$\Omega_s(r) = -2 + \frac{p_s r}{a_s + r} + \frac{q_s r}{a_s + r} + \left(1 - \frac{q_s r}{a_s + r}\right) \frac{H^2}{Z_s^2}$$

(9) $$F_s(\theta,u) = \sum_{y=0}^{4} eM_{s\nu} \left(\frac{u}{u_{\nu s}}\right)^2 \exp - \left\{\left(\frac{u}{u_{\nu s}}\right)^2 + \frac{2\sin^2[\frac{1}{2}(\theta - \theta_\nu)]}{\sigma_{\nu s}^2}\right\}$$

(10) $$\bar{X}(r,\theta,0,H) = C\ Q\ \Sigma_{s\ \nu} \frac{M_{s\nu}\ \delta_s}{Y_s\ Z_s} \cdot \frac{1}{[1 + (\delta_s/\sigma_{\nu s})^2]^{1/2}} \times$$
$$\exp - \left\{\frac{H^2}{2Z_s^2} + \frac{2\sin^2 \frac{1}{2}(\theta - \theta_\nu)}{\sigma_{\nu s}^2\ [1 + (\delta_s/\sigma_{\nu s})^2]}\right\}$$

(11) $$\delta_s = \delta_s(r) = [10^6 \frac{r^2}{Y_s^2} + \Omega_s(r)]^{-1/2}$$

Table 2. Analytic Windrose Parameter for Palatka

P	ν Dir	0 U	1 N	2 E	3 S	4 W
D_ν		.0478	.0270	.0492	.0202	.0254
X_ν		1.973	.163	1.511	.640	.0199
U_ν		3.205	2.973	3.667	3.089	2.927
V_ν		1.492	2.035	2.035	2.035	2.035
Θ_ν		---	.708	5.022	8.752	12.729
σ_ν		∞	.686	.811	.753	.810

For all ν, Y = 0.822 and W = 0.857.

Table 3. Wind Amplitudes for Palatka ($M_{\nu s}$ values)

CL.	s	ν Dir	0 U	1 N	2 E	3 S	4 W
A	1		.676 (-2)	.190 (-7)	.538 (-3)	.836 (-5)	.565 (-12)
B	2		.747 (-2)	.347 (-6)	.106 (-2)	.389 (-4)	.843 (-10)
C	3		.758 (-2)	.583 (-5)	.191 (-2)	.166 (-3)	.116 (-7)
D	4		.708 (-2)	.901 (-4)	.318 (-2)	.652 (-3)	.146 (-5)
E	5		.607 (-2)	.128 (-2)	.486 (-2)	.235 (-2)	.169 (-3)
F	6		.479 (-2)	.167 (-1)	.683 (-2)	.781 (-2)	.180 (-1)

Table 4. Characteristic Wind Velocities ($U_{\nu s}$) for Palatka

CL.	s	Dir	U	N	E	S	W
A	1		.500	2.193	2.704	2.279	2.159
B	2		1.343	4.320	5.328	4.489	4.252
C	3		2.653	6.251	7.709	6.500	6.154
D	4		3.847	6.643	8.193	6.903	6.541
E	5		4.097	5.185	6.396	5.399	5.106
F	6		3.205	2.973	3.666	3.089	2.927

would be helpful to have the basic parameters a, L, q, K and p expressed as continuous functions of the stability index s. Our preliminary attempts suggest that power series with two or three terms would give adequate accuracy and would complete the transformation of the 3DJFD and the corresponding five vector transfer function (5VTF) into continuous mathematical functions of all variables.

For some purposes such as regulatory applications it might be desirable to go the opposite way i.e. from the continuous to the discrete. We next describe how our work might be used in this regard.

The Numerical 3DJFD

The usual tabular wind rose data specify the probabilities P_{iks} for the wind to lie in various directions θ_k and various velocity classes u_i for each of s stability classes. The wind data are usually given in sixteen angular intervals ($\Delta\theta = 2\pi/16$) and six velocity classes. Thus, characteristically 96 numbers are given for each of six stability classes.

In the work of GSV these numbers are used to establish a set of two dimensional continuous analytic functions of direction and speed representing the probability densities $F_s(u,\theta)$. In GSSB, the continuous functions for 10 calibration sites were used to infer our wind rose map of Florida. Here having generated these $F_s(u,\theta)$ functions for Palatka we may also calculate the discrete probability values by reversing our initial procedure in GSV and GSSB. Thus we would use

$$P_s(u_i, \Theta_k) = F_s(u_i, \Theta_k)\,\Delta u_i \Delta\Theta$$

where the u_i and Θ_k denote the centers of the velocity and angle intervals and Δu_i and $\Delta\Theta$ denote velocity class and angular sizes. For standard regulatory use tables constructed in this way might reasonably be adopted as inputs in a standard dispersion model code (6) or the Industrial Source Dispersion Model (ISC)(9).

Research Needs on Pollution Dispersion

The research needs for establishing a definitive wind rose climatology for Florida, are manifold. In the first instance, the data from the original ten calibration sites used by GSSB were not of uniform quality and better data is now available. Since a procedure has been established for boiling down tabular 3DJFD into continuous functions $F_s(\theta, u)$. It would be desirable to redo our original work with a uniform treatment of stability classes and other details. Secondly, some effort to move from a 3DJFD at the wind measurement height (z) to a 3DJFD at the effective emission height (H) of the stack would be desirable. One might modify our characteristic velocities $u_{\nu s}$ by adapting the ISC power law device using Table 2-2 of (9). However in view of the insensitivity of our final 5VTF to $u_{\nu s}$ this might not be necessary. One also needs some rules for the turning of our four plume vectors in going from the boundary layer to the stack height altitude and to allow for surface roughness and unusual topographical features. It would also be highly desirable to place the stability index s on a more physical basis. Such an attempt has been made by Fahien and Theyunni (10).

The fact that our final 5VTF is characterized in terms of four concentration vectors and a uniform component suggest that it should not be too difficult to develop a transport function map of Florida and to connect it to physically reasonable fundamental flow field models. In this work we have illustrated a local SO_2 dispersion application of the analytic Gaussian plume model and our analytic wind rose map of Florida. However, the fact that our wind rose mapping system predicts the local 3DJFD for any point in Florida indicates that the system can be used to generate larger mesoscale patterns of average wind trajectories of the type that is needed for acid rain transport, both from Florida sources and from sources outside of Florida. Developing relationships between our work and the approaches of Pielke (11), Kreitzberg (12) and of Chen (13) is an important research need. It should be noted that with the high humidity, high ultraviolet sunlight levels, and land-sea breeze circulation patterns, distances for acid rain conversion should be smaller in Florida than elsewhere (14).

As discussed in GSSB it would be useful to develop a time dependent 3DJFD which can be used to provide daily, monthly, or seasonal wind rose maps of Florida, which could then be used with Equation 10 to predict maximum pollution concentrations. To illustrate the utility of a seasonal 3DJFD in acid rain applications we have made a qualitative attempt to disaggregate our interpolated annual average wind rose for Tampa into seasonal components giving special attention to the summer winds. Climatological data (15, 16) indicate that the prevailing winds in the summer are southerly since Florida sits on the western edge of the Bermuda high. This suggests that the southerly component of our annual 3DJFD primarily arises from the summer winds. To construct a reasonable summer wind rose, we set $D_3(S) = 0.12$, $D_2(E) = 0.03$, $D_1(N) = D_4(W) = 0$ to give a southeasterly wind and adjusted $D_0(U) = 0.038$ to restore the total wind probability to unity. The formulas for concentration, Eqs. 10 and 11 of Table 1 can then be applied directly.

Let us now choose as a nominal SO_2 emission rate for the Big Bend plant near Tampa in the summer, Q = 100,000 tons/yr. Using an effective stack height, H = 200 m. we calculate isopleths of pollutant concentration at various distances from Tampa as shown in Figure 3. It is seen that substantial concentrations, which are mostly contributed by the high stability components of the summer 3DJFD, reach several hundred kilometers to the north and somewhat into the Gulf of Mexico. The characteristic velocities for F stability class winds (see Table 4) are approximately 3 knots so these component winds would take 3 or so days to travel to the northern border of the state. With high ultraviolet sunlight levels, high humidities, the regular afternoon cumulus build up and rainfall, lightning production of OH and H radicals and NO_x, the SO_2 conversion to SO_3 and then H_2SO_4 followed by rainout should proceed quite rapidly. The scenario of SO_2 conversion and transport suggested above might account for the observations of Brezonik (17) that (a) the summer rains are more acidic than the winter rains and (b) the northern lakes in Florida are more acidic than the southern.

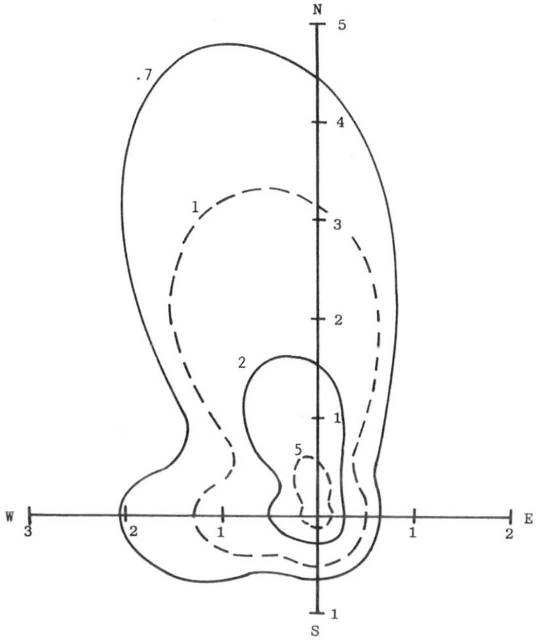

Fig. 3 Isoconcentration contours (in μg/m^3) centered around Tampa for an illustrative summer 3DJFD and an SO_2 source of strength of 100,000 tons/yr. The distances are in 100 kilometers.

The prevailing sea breezes in the summer in Florida might be expected to modify the northward flow into a well defined east coast right handed helical vortex up the peninsula and a weaker west coast left handed vortex also up the peninsula. The combined motion would tend to channel the SO_2 contributions from all power plants located in Florida into the northward flow so that 1,000,000 Ton/yr effective emission rate might be more appropriate. Thus Florida may be a substantial exporter of SO_2 to the north and midwest (in the summer anyway) rather than an importer as has been conjectured in the past. The possibility of lightning generated NO_x contributing to this flow also warrants study (18). While more detailed calculations are clearly needed before this hypothesis can be regarded as quantitative, nevertheless our qualitative conclusions provide a heuristic explanation of several observations and illustrate the applicability of the GSV and GSSB methodologies to acid rain questions. The need for further studies should be self evident.

ACKNOWLEDGEMENTS

We would like to thank Larry A. George of the Department of Environmental Regulation for helpful information on the available meteorological data and for suggestions on the applicability of this work, and to Professor Kreitzberg for a helpful conversation on sea breezes.

Literature Cited

1. A.E.S. Green, J.M. Schwartz, R.P. Singhal and W.E. Bolch, "Wind Roses of Florida", JAPCA, 32, 8, 822, 1982.

2. A.E.S. Green, R.P. Singhal and R. Venkateswar, "Analytic Extensions of the Gaussian Plume Model", JAPCA, 30, 7, 773, 1980.

3. ICAAS, An Interdisciplinary Study of the Health, Social and Environmental Economics of Sulfur Oxide Pollution in Florida, ICAAS Report, A. E. S. Green, P. I., February, 1978.

4. Loehman, E. T., S. V. Berg, A. A. Arroyo, R. A. Hedinger, J. M. Schwartz, M. E. Shaw, R. W. Fahien, V. H. De, R. P. Fishe, D. E. Rio, W. F. Rossley, and A. E. S. Green, "Distributional Analysis of Regional Benefits and Cost of Air Quality Control", J. Environ. Econ. and Managem., 6, pp. 222-243, 1979.

5. Green, A. E. S., E. T. Loehman, S. V. Berg, R. W. Fahien, M. E. Shaw, M. J. Jaeger, H. Wittig, A. A. Arroyo, J. M. Schwartz, R. A. Hedinger, V. H. De, D. E. Rio, T. J. Buckley, R. P. Fishe, W. F. Rossley, and D. Trimble, "Pollution Dispersal Modeling and Regional Public Policy Implications", Proceedings of Symposium held in Williamsburg, April 17-19, 1979, Environmental and Climatic Impact on Coal Utilization, V. V. Singh and A. Deepak Eds., Academic Press, 1980.

6. Englund, H. M. and W. T. Bury, Eds., "Atmospheric Dispersion Modeling", APCA Reprint Series, Air Pollution Control Association, Pittsburg, PA, March 1980.

7. Green, A. E. S., Phys. Rev. 75, 1926; 76, L870; A460 (1949), Rev. Mod. Phys. 39, 495 (1967); Science 169, 933 (1970).

8. Taylor, K. E. and E. Hernandez, "Estimating Average Deposition Pattern in Data Sparse Regions of Florida", Proceedings of Acid Deposition Science Workshop, March 23-24, 1983, University of Florida, Gainesville, Florida.

9. Bower, J. F., J. R. Bjorklund and C. S. Cheney, "Industrial Source Complex (ISC) Dispersion Model, User's Guide", Vol. 1, EPA-450/4-79-030, PB 80-133044, Dec. 1979.

10. Fahien, R. W. and M. Theyunni, "Air Pollution Dispersion Problems of Florida", Chapter 9, The Impact of Increased Coal Use in Florida (IICUF), ICAAS University of Florida, Gainesville, Florida, November 1980.

11. Pielke, R. A., "A Three Dimensional Numerical Model of the Sea Breeze over South Florida", Mar. Wea. Rev. 102, 115-139, 1974.

12. Kreitzberg, C. W., "Transport of Pollutants on the Scale of Hundreds of Kilometers", Proceedings of Acid Deposition Science Workshop, March 23-24, 1983, University of Florida, Gainesville, Florida.

13. Chen, E. Y., "Florida Mesoscale Climatology Bearing upon Acid Deposition", Proceedings of Acid Deposition Science Workshop, March 23-24, 1983, University of Florida, Gainesville, Florida.

14. Taylor, K. E., W. L. Chameides and A. E. S. Green, "Atmospheric Modifications", Chapter 11, Coal Burning Issues, ICAAS, University Presses of Florida, Gainesville, Florida, 202-230, 1980.

15. Environmental Science Services Administration, Climatic Atlas of the United States, U.S. Department of Commerce, Environmental Data Services, 1968.

16. Record, F. A., D. V. Bubenick and R. J. Kindya, Acid Rain Information Book, Noyes Data Corporation, Park Ridge, N.J., 1982.

17. Brezonik, P. L., "Acid Precipitation and Its Effects on Florida Lakes; Where Do We Stand After Seven Years of Research", Proceedings of Acid Deposition Science Workshop, March 23-24, 1983, University of Florida, Gainesville, Florida.

18. Green, A. E. S. and M. A. Uman "Lightning in Florida as a Source of Acid Rain", Proceedings of Acid Deposition Science Workshop, March 23-24, 1983, University of Florida, Gainesville, Florida.

ESTIMATING AVERAGE DEPOSITION PATTERNS IN DATA-SPARSE REGIONS OF FLORIDA

Karl E. Taylor & Elisa Hernandez
Department of Physics
University of Florida, Gainesville, FL 32611

Abstract

The meteorological data needed to predict annual-average deposition around a point source with Gaussian plume models are contained in a joint frequency distribution that indicates how stable the atmosphere is and how often the wind blows with a given speed from a particular direction. Data in this form have been collected at relatively few sites in Florida, but data with less information content (e.g., with no stability class characterization) are available for many more locations. The aim of this study is to use the detailed data collected at the few locations to improve estimates of climatological pollution levels in the data-sparse regions.

Certain parameters that characterize the joint frequency distribution have been identified and are found to be nearly invariant for all data-rich sites. If we assume that these parameters are indeed constants, we can use them for any site in Florida even if the detailed data needed to compute them are locally unavailable. A series of calculations is carried out to determine how the accuracy of the estimates depends on the quality and quantity of data available.

We find that the wind speed data are similar for all Florida sites and that the fraction of days in the year that a particular stability will exist is also nearly the same for all locations. Only the wind directions (wind roses) vary from place to place and are different for each stability class. It is this data then that must be collected for individual sites if pollution concentrations are to be estimated accurately.

In the next section we derive the formula for calculating average concentrations near a point source. We then show how pollution estimates depend on the data available. We discuss how to make reasonable estimates even when the full three dimensional frequency distribution is not known for a particular site. In the final section we discuss our conclusions and indicate what further research is needed in Florida to improve our understanding of acid deposition patterns.

Theory

According to the Gaussian plume model [1], the ground concentration in direction θ at a distance r from a point source of effective stack height H is given by

$$\chi(r,\theta-\phi,H) = \frac{Q}{\pi\ \sigma_z\ r\sigma\ u} \exp\left[-\frac{H^2}{2\sigma_z^{\ 2}}\right] \exp\left[-\frac{(\theta-\phi)^2}{2\sigma^2}\right] \qquad (1)$$

where $r\sigma$ and σ_z are the dispersion parameters which vary with distance and stability class, u is the adjusted wind speed, θ the wind direction, and Q the source strength. As shown by Green, Singhal, and Venkateswar [2], henceforth referred to as GSV, this polar coordinate form is quite similar to the familiar cartesian coordinate form, except $r\sigma$ defined below, replaces the traditional cross-wind dispersion parameter σ_y and differs from it by the generally small correction term Ω, defined in GSV.

$$\frac{1}{\sigma^2} = \frac{r^2}{\sigma_y^{\ 2}} + \Omega(r,s)$$

GSV also provide useful analytical fits to the functions σ_y and σ_z that simplify the problem.

The adjusted wind speed u which appears in (1) is nearly equal to the true wind speed u* in the parameter range of interest ($\sigma^2 << 1$) and is defined by the following expression:

$$u = u^* \exp(-\tfrac{1}{2}\,\sigma^2)$$

This small adjustment did not appear in GSV but has been added here primarily for aesthetic reasons since without it minor amounts of mass disappear and are regenerated as the plume moves downwind. A similar argument is invoked to justify replacing σ_y with $r\sigma$ where it appears in the denominator of the polar coordinate Gaussian plume model derived by GSV. These minor changes are small and make little practical difference.

We should note further that because χ should be cyclic in $\theta-\phi$, the $(\theta-\phi)^2$ appearing in the exponential function is actually an approximation to the function $4\sin^2[\frac{1}{2}(\theta-\phi)]$. In practice this function is small enough so that the approximate form appearing in (1) can be used. (See GSV for further details.)

Our aim here is to derive a simple formula for calculating average concentrations over long time periods. To compute the average ground concentration for a year, we need daily measurements of wind speed and direction and also a daily determination of the Pasquill stability classification. This data can then be described by a three-dimensional joint frequency distribution in which the probability density $F_s(u,\phi)$ is defined such that F_s du dϕ is the probability that the atmosphere will be found in stability class s (where s = A, B, C, D, E, or F) <u>and</u> the wind will be found within $\pm\,\frac{1}{2}$du of speed u <u>and</u> $\pm\,\frac{1}{2}$dϕ of direction ϕ. The total probability must be unity:

$$\sum_s \int_0^{2\pi} \int_0^{\infty} F_s(u,\phi)\; du\; d\phi = 1$$

It is now possible to determine the long-term average concentration ($\bar{\chi}$) by performing a weighted average of the individual daily plumes:

$$\bar{\chi}(r,\theta,H) = \sum_s \int_{\theta-\pi/2}^{\theta+\pi/2} \int_0^{\infty} \chi(r,\theta-\phi,H)\ F_s(u,\phi)\ du\ d\phi \qquad (2)$$

Substitution of (1) into (2) yields

$$\bar{\chi} = \sum_s \int_{\theta-\pi/2}^{\theta+\pi/2} \frac{Q}{\pi\ \sigma_z\ r\sigma} \exp\left[-\frac{H^2}{2\sigma_z{}^2}\right] \exp\left[-\frac{(\theta-\phi)^2}{2\sigma^2}\right] \int_0^{\infty} \frac{F_s(u,\phi)}{u} du d\phi \qquad (3)$$

It is clear that $\bar{\chi}$ does not depend on the details of the probability distribution in u, only the average value of 1/u. This point can be emphasized by rewriting (3) in an alternate form in which the average value of the reciprocal of u (represented by U^{-1}) and the two-dimensional joint frequency distribution $\Phi_s(\phi)$ appear explicitly:

$$\bar{\chi} = \sum_s \int_{\theta-\pi/2}^{\theta+\pi/2} \frac{Q}{\pi\ \sigma_z\ r\sigma} \exp\left[-\frac{H^2}{2\sigma_z{}^2}\right] \exp\left[-\frac{(\theta-\phi)^2}{2\sigma^2}\right] U^{-1}(s,\phi)\ \Phi_s(\phi)\ d\phi \qquad (4)$$

where
$$U^{-1} = \frac{1}{\Phi_s(\phi)} \int_0^{\infty} \frac{F_s(u,\phi)}{u} du$$

and
$$\Phi_s(\phi)\ d\phi = \int_0^{\infty} F_s(u,\phi) du\ d\phi$$

The probability (integrated over speed) that stability class s applies <u>and</u> that the wind is within ± ½dϕ of direction ϕ is equal to $\Phi_s(\phi)d\phi$.

Following GSV, we wish to approximate the directional dependence of the probability density $F_s(u,\phi)$ and $\Phi_s(\phi)$ as analytical functions that will allow the ϕ integration in (3) and (4) to be carried out analytically. To be specific, $F_s(u,\phi)$ will be approximated by a sum of equally spaced Gaussian functions of equal width σ_I:

$$F_s(u,\phi) = \sum_{i=1}^{I} A_{si}(u) \frac{1}{\Delta\phi} \exp\left[-\frac{(\phi-\phi_i)^2}{2\sigma_I{}^2}\right] \qquad (5)$$

where $\Delta\phi = 2\pi/I$ is the distance between the Gaussian peaks and $\phi_i = \Delta\phi(i-1)/I$. In what follows, the number of directions (I) will be either 16 or 8 depending on what degree of smoothing is desired.

The total probability of being in stability class s and within $\pm\frac{1}{2}du$ of wind speed u should be conserved under the above approximation, so we require

$$\int_0^{2\pi} F_s(u,\phi)d\phi = \int_{-\infty}^{\infty} \sum_{i=1}^{I} A_{si}(u) \frac{1}{\Delta\phi} \exp\left[-\frac{(\phi-\phi_i)^2}{2\sigma_I^2}\right] d\phi$$

$$= \sum_{i=1}^{I} A_{si} \frac{\sqrt{2\pi}\ \sigma_I}{\Delta\phi} \tag{6}$$

In evaluating the integral in (6) the integration limits were extended to $\pm\infty$ following GSV. To facilitate interpretation of (6) and also its application to real data, we further require that the Gaussian amplitudes (A_{si}) be the probability that the wind direction is within $\pm\ \Delta\phi/2$ of ϕ_i. This means that the factor multiplying A_{si} in (6) must equal 1 and the width of the Gaussians must be

$$\sigma_I = \frac{\Delta\phi}{\sqrt{2\pi}} = \frac{\sqrt{2\pi}}{I} \tag{7}$$

With (7) it is now possible to simplify (3) by substituting (5) and analytically evaluating the integral (after extending the integration limits to $\pm\infty$):

$$\bar{\chi} = \sum_s \frac{Q \exp\left[-\frac{H^2}{2\sigma_z^2}\right]}{\pi\ \sigma_z\ r\sqrt{\sigma^2+\sigma_I^2}} \sum_{i=1}^{I} \exp\left[-\frac{(\theta-\theta_i)^2}{2(\sigma^2+\sigma_I^2)}\right] \int_0^{\infty} \frac{A_{si}(u)}{u} du \tag{8}$$

In practice the integral over u is replaced by a sum over the observed speed classes.

If the alternate formula (4) is to be used, analytical approximations for the angular dependence of $\Phi_s(\phi)$ and $U^{-1}(s,\phi)$ must be found. As we demonstrate in the next section, the average value of the reciprocal of u is very nearly independent of direction (for major cities in Florida), so we shall replace it with its angular average value for each stability class:

$$U^{-1}(s,\phi) \cong \overline{(1/u)}_s \equiv \frac{\int_0^{2\pi} \Phi_s(\phi)\ U^{-1}(s,\phi)d\phi}{\int_0^{2\pi} \Phi_s(\phi)\ d\phi} \tag{9}$$

An approximation for $\Phi_s(\phi)$ similar to (5) can be found but with the amplitudes now independent of u and defined by

$$\Lambda_{si} \equiv \int_{-1/2\,\Delta\phi}^{1/2\,\Delta\phi} \Phi_s(\phi)\, d\phi = \int_0^\infty A_{si}(u)\, du$$

Presuming the approximation for U^{-1} given in (9) is valid, (4) becomes

$$\bar{\chi} = \sum_s \frac{Q \exp\left[-\frac{H^2}{2\sigma_z^2}\right]}{\pi\, \sigma_z\, r\sqrt{\sigma^2+\sigma_z^2}}\; \overline{(1/u)}_s \sum_{i=1}^{I} \Lambda_{si} \exp\left[-\frac{(\theta-\theta_i)^2}{2(\sigma^2+\sigma_z^2)}\right] \qquad (10)$$

Besides the familiar dispersion parameters that appear in (10), the average concentration depends on the wind and stability climatologies that determine $\overline{(1/u)}_s$ and Λ_{si}. It is clear from the above formula that if the average value of the wind speed increases, then the dillution of pollutants increases and χ decreases. The directional variation of concentration is determined solely by the coefficients Λ_{si} which indicate the probability that the wind is in direction θ_i and that the stability class is s.

Discussion and Applications

In this section we shall show that in Florida the average ground concentration of pollutants emitted from point sources can be estimated with reasonable accuracy even if some of the wind rose climatological data is not available. We shall consider first some of the annual wind rose characteristics for four cities in Florida (Jacksonville, Pensacola, Tallahassee, Tampa) and one in southern Georgia (Valdosta), all of which will henceforth be referred to as "Florida" sites. Data from a few other cities in Florida are available but have not yet been processed.

In Fig. 1 the probability (Ψ_s) that the stability class on a randomly chosen day will be s is plotted for each of the five cities. The figure indicates that to a good approximation, Ψ_s is independent of location. This suggests that for locations where this stability class information is unavailable we might use a typical Florida value as given by the solid curve in Fig. 1. It should be noted that at some locations seven stability classes were reported instead of the normal six. It was found that at these sites class D had been subdivided into nighttime D and daytime D. To recover the traditional classification we added these together before processing the data.

In Fig. 2 $\overline{(1/u)}_s$ is plotted in a similar way. Again the close grouping of points at each stability class suggests that for locations in Florida where wind speed information is missing, we might use typical Florida values as given by the solid curve in Fig. 2.

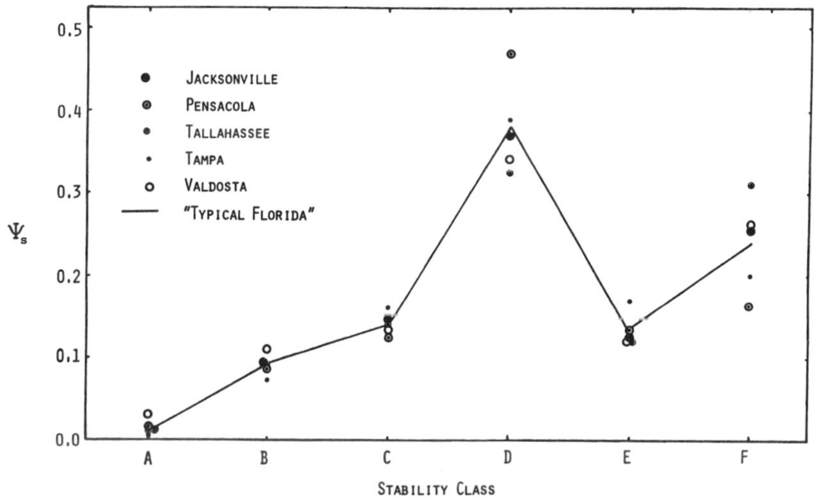

Fig. 1 Fraction of time that each stability class is observed at five cities. The line segments connect the average ("typical Florida") values for each stability class.

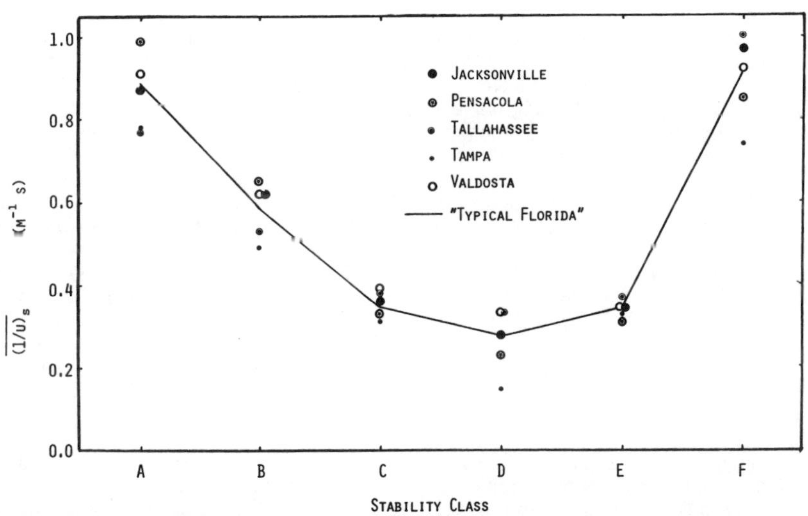

Fig. 2 Wind speed parameter $\overline{(1/u)}_s$ shown as a function of stability class for five cities. The line segments connect the average ("typical Florida") values for each stability class.

The two "typical-Florida" curves in Figs. 1 and 2 provide enough information to calculate immediately the long term axially averaged concentration at various distances from a point source in Florida. The formula for estimating these concentrations is obtained by averaging (10) over all angles:

$$\langle\bar{\chi}\rangle = \sum_{s} \frac{Q}{\sqrt{2\pi^3}\ \sigma_z\ r} \exp\left[-\frac{H^2}{2\sigma_z^2}\right]\ \overline{(1/u)}_s\ \Psi_s \tag{11}$$

where the angular brackets mean axially-averaged, (averaged over all angles). When no local meteorological data is available, values from Figs. 1 and 2 may be used in (11).

Fig. 3 shows how concentration depends on radial distance for a typical Florida site when the stack height is 100 meters and the source strength Q is 1 kg/s. The relative importance of the different stability classes can also be determined since the contribution by each one to the total is plotted separately. The classes representing relatively unstable conditions dominate near the source whereas during stable conditions (E and F) ground concentrations are low everywhere. Fig. 3 shows that the maximum average ground concentration appears at about 1 km distance from the source and most of the pollution there has been contributed on days when the stability was classified as A, B or C.

To this point we have not shown the directional dependence of pollution levels. To illustrate the importance of the wind direction data, we shall consider a set of sample calculations using meteorological data from Tampa. The annual-average pollution concentrations at a distance of 1 km from a hypothetical source of height 100 meters and strength 1 kg/s were calculated. Tampa was chosen arbitrarily and the distance of 1 km was chosen because the maximum ground concentration is found in that vicinity. Other stack heights and distances were also studied and these further support the conclusions based on the cases discussed here.

A series of five cases incorporating various amounts of meteorological data were analyzed to determine which data are most important for obtaining good estimates of pollution levels. The control case represents the best estimate of pollution concentration near the Tampa point source because the complete three-dimensional joint frequency distribution is used in (8) and the model derived concentrations are therefore the same as those that would be obtained with the traditional ("brute force") Gaussian plume climatological models [3]. The solid curve in Fig. 4 shows the directional dependence of the ground concentration for the control case. The two maxima due to the prevalence of westerlies and easterlies are clearly displayed.

Tampa is a "data-rich" location since all the meteorological data needed to perform a traditional calculation is available. We can illustrate how pollution dispersion calculations could be made for data-sparse regions by eliminating some of the data used in calculating the control case. For example, if all the wind rose information is eliminated, a reasonable estimate can still be obtained with (11) by inserting the typical Florida values for $\overline{(1/u)}_s$ and Ψ_s and averaging over all angles.

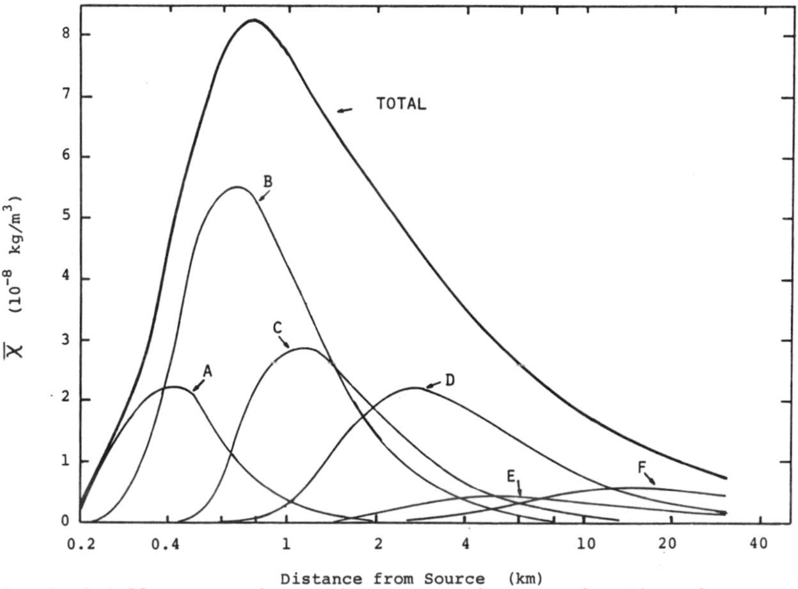

Fig. 3 Axially averaged ground concentration as a function of source distance calculated using "typical Florida" meteorological data. The contribution of each stability class to the total concentration is also shown. The source strength (Q) is 1 kg/s and the stack height (H) is 100 m.

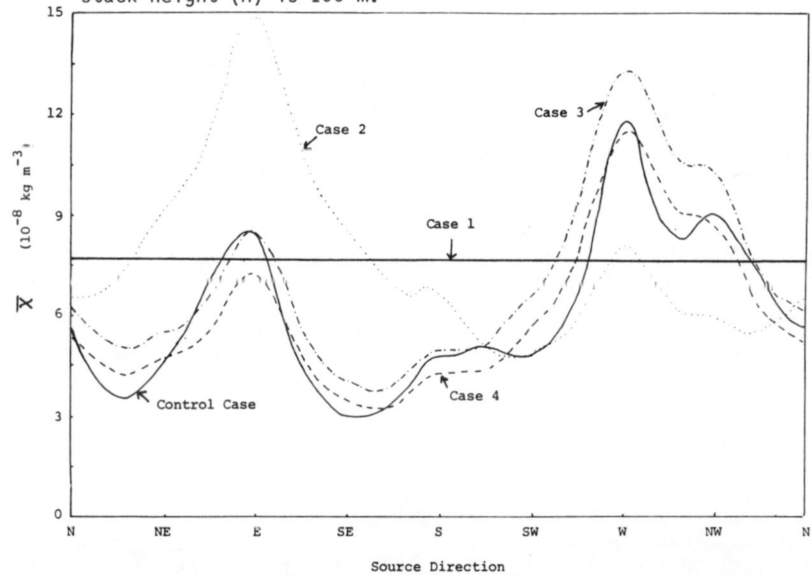

Fig. 4 Annual average pollution levels at a distance of 1 km from a point source in Tampa Bay of strength Q = 1 kg/s and stack height 100 meters. The different cases discussed in the text show how estimates depend on the meteorological data available.

The concentration at 1 km can then be read from Fig. 3 and is plotted as case 1 in Fig. 4 (unbroken, horizontal straight line). For the most part the errors are less than a factor of 2.

Case 2 incorporates a minimal amount of directional information from Tampa, but still no speed or stability class information. The wind directional data was smoothed to 8 instead of the original 16 directions. This type of calculation could therefore be done for sites where a crude, annual average wind rose is available, but without a breakdown according to stability class or speed. This case is plotted as a dotted curve in Fig. 4. The two peaks appear as they did in the control case but their relative importance is reversed. This is primarily because the wind roses for individual stability classes differ to a certain extent, and this difference cannot be accounted for without separate wind roses for each stability class. In particular, when the atmosphere is unstable at Tampa, westerlies dominate whereas stable conditions commonly occur during easterlies. Since stable conditions are more common, the total wind rose is also dominated by easterlies, but it is the westerlies during unstable conditions that give the highest pollution levels at 1 km downwind. If the simple (stability class independent) wind rose is applied, the importance of the westerlies is not apparent whereas the effect of the easterlies is exaggerated.

If six wind roses are available, one for each stability class, the estimates improve dramatically. For this third case represented by a dash-dot curve in Fig. 4, no wind speed information was used from Tampa and the wind roses were smoothed as above. As is shown in Fig. 4, the errors discussed in the previous paragraph were greatly reduced and the pollution estimates were therefore significantly improved. Indeed there is probably no practical difference between the accuracy of pollution concentrations predicted by this case and the control case in light of the many well known deficiencies of the Gaussian plume model [4, 5].

A final case (case 4) was examined to ascertain the importance of wind speed information. This case differs from case 3 in that Tampa values for $\overline{(1/u)}_s$ were used instead of typical Florida values. It differs from the control case in that smoothed wind roses were used and $U^{-1}(s,\phi)$ was assumed to be independent of direction (see Eq. 9). The dashed curve in Fig. 4 shows that case 4 is nearly the same as the control case so that the assumption that $U^{-1}(s,\phi)$ is independent of direction is a good one and that smoothing the wind roses into eight sectors is also justified. The differences in the ground concentrations between this case and case 3 are minor for the most part and simply indicate that $\overline{(1/u)}_s$ for Tampa is about the same as the typical Florida values for this parameter as shown in Fig. 2.

The annual-average concentration at 1 km is mainly due to contributions during less stable conditions (A, B and C). Further downwind, on the other hand, concentrations are highest when the atmosphere is stable as is indicated by Fig. 3. To further illustrate the accuracy of the present approach, we consider pollution concentrations around the hypothetical Tampa source already discussed above, but now at distances of 10 km and 100 km. Fig. 5 shows a comparison between the traditionally calculated (control case) concentrations and estimates made with limited wind rose information for Tampa at these two distances. The dotted curves in Fig. 5 are based on the wind rose data that was used in case 2 of Fig. 4: a single, smoothed, annual wind rose (without a breakdown according to stability class or speed), and typical Florida values for $\overline{(1/u)}_s$ and Ψ_s. It is clear that at these greater distances, typical Florida meteorological data and a single smoothed wind rose (eight directions) give a reasonably accurate estimate of average pollution concentrations. The agreement between the case 2 curves and the control case at these distances is much better than it was at 1 km. The marked similarity indicates that Tampa is very nearly a typical Florida site and that the wind roses for stability classes D, E and F are approximately the same as the total, annual-average wind rose at Tampa.

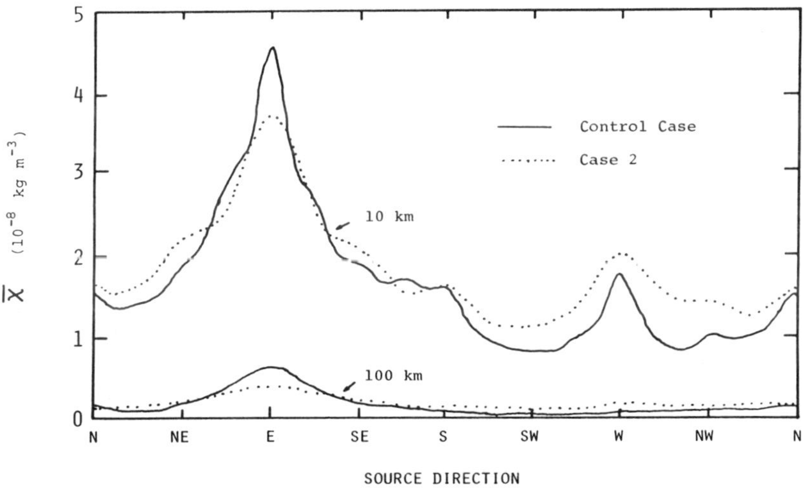

Fig. 5 As in Fig. 4 but for distances of 10 km and 100 km. Note that the ordinate has a different scale.

We conclude that for the example discussed here, the only local information needed to predict pollution concentrations beyond a few kilometers from a 100 meter stack is a single smoothed wind rose; no speed or stability class information is needed locally since typical Florida values may be used.

Research Needs in Florida

We have found that in Florida the potential impact of proposed point pollution sources can be estimated with reasonable accuracy even if some of the meteorological data normally required by traditional Gaussian plume climatological models is unavailable. For example, our study indicates that typical Florida values for the wind speed parameter $\overline{(1/u)}_s$ can be substituted for the actual data without significantly affecting the estimates. More serious errors (factor of two), however, can result at some distances if only a single wind rose with no stability class information is available. Even when no local meteorological data is available, crude estimates of average ground concentration can be made by constructing a set of curves like those in Fig. 3, but they could be in error by a factor of five or more.

Our results indicate that the most valuable meteorological data for air pollution dispersion modeling in Florida is wind direction information, preferably broken down into separate wind roses for each stability class. For most places in Florida this wind direction and stability class joint frequency distribution is not presently available. For reasonable predictions to be made, this wind data must either be collected in the future or, alternatively, be inferred by some other means.

One approach would be to simply interpolate between the nearest data-rich sites or by methods discussed by Green et. al in [6]. Another would be to analyze GOES data to deduce stability class and sea breeze circulation climatologies which could then perhaps be used to generate the needed wind roses (cf., Chen [7]). In addition, seasonal average calculations should be made with existing meteorological data to determine whether a more natural basis could be found for making pollution estimates in data-sparse regions of Florida.

Literature Cited

1. Hanna, S. P., G. A. Briggs, and R. P. Hosker, "Handbook on Atmospheric Diffusion", Technical Information Center, U. S. Department of Energy, 1982.

2. Green, A. E. S., R. P. Singhal, and R. Venkateswar, "Analytical Extensions of the Gaussian Plume Model", JAPCA, 30, 773-776, 1980.

3. Busse, A. D. and J. R. Zimmerman, "User's Guide for the Climatological Dispersion Model", U.S. Environmental Protection Agency, EPA-R473-024, 1973.

4. Turner, D. B., "Atmospheric Dispersion Modeling: A Critical Review", JAPCA, 29, 502-519, 1979.

5. Egan, B. A., J. D. Spengler, and V. A. Mirabella, "Atmospheric Dispersion Modeling: A Critical Review", JAPCA, 29, 927-941, 1979.

6. Green, A. E. S., J. M. Schwartz, R. P Singhal, and W. E. Bolch, "Wind Roses for Florida", JAPCA, 32, 822-825, 1982.

7. Chen, E. Y., "Florida Mesoscale Climatology Bearing Upon Acid Deposition", Proceedings of Acid Deposition Science Workshop, University of Florida, Gainesville, Florida, March 23-24, 1983.

FLORIDA MESOSCALE CLIMATOLOGY BEARING UPON ACID DEPOSITION

Ellen Y. Chen
Fruit Crops Department
University of Florida, Gainesville, FL 32611

Abstract

Geostationary (GOES) satellite-detected surface infrared temperatures indicate the presence of mesoscale thermal regimes in the flat terrains of peninsular Florida. The thermal regions were found to be related to soil type and drainage class, specifically to soil thermal inertia, $(\lambda C)^{\frac{1}{2}}$, where λ is the thermal conductivity, and C is the volumetric heat capacity of the soil. While synoptic weather systems produce the overall weather in the peninsula, the difference in soil property is one of the principal factors governing the observed temperature difference at the local level. Since soil property is a comparatively stable quantity, thermal regions are persistent and more predictable. The GOES temperatures have confirmed that regions with the largest diurnal and annual temperature range were associated with soils of lowest thermal inertia. Areas in the peninsula with the lowest winter nocturnal temperatures have the highest summer daytime temperatures. The high daytime temperatures provide low level convergence for sea breeze circulation for both central and southern Florida. Sea breeze circulation is a dominant mechanism for mesoscale transport and deposition.

Introduction

Summertime mesoscale weather and climatology in the Florida peninsula are dominated by sea breezes which form along both the east and west coasts of the peninsula. The mesoscale sea breeze circulation is generally superimposed on moist prevailing synoptic southeasterly to southwesterly flows. Convective rainfall in the peninsula is frequently associated with sea breezes (1, 2). The sea breeze therefore is a vehicle for transport of atmospheric particles (3) and deposition. Low level convergence generated by a land-water temperature difference is a principal factor for the initiation of the sea breeze (4). The initial temperature field was modeled and found to be an important parameter to subsequent forecast (5). A three-dimensional model (6) of the sea breeze of South Florida indicated that sea breeze circulations are the dominant control on the location of thunderstorms and convective rainfall. Watson and Holle (7) found correlation coefficients between convergence and rainfall were about 0.5 to 0.8. They characterized Florida convective systems as slow moving, relatively simple, and triggered primarily by land-water temperature difference. The influence of mesoscale surface parameters on convection, and hence, the sea breeze systems, were also studied. Using aircraft photographs, initial cumulus formation was related to surface features such as non-swampy areas, and to perimeters of lakes and rivers (8). Influence of albedo, soil moisture and thermal inertia to sea breeze was investigated by Gannon (9), using a two-dimensional model over South Florida. A numerical model was used to study convection over a forested surface (10). Research indicates a positive influence of the surface on convective motion. It is apparent that a set of data whose resolution is capable of defining mesoscale surface thermal information is needed to study the sea breeze, and in turn, transport and deposition. Beside surface parameters, upper level (800-500 mb) instability and moisture content frequently determines whether rainfall will be associated with sea breeze convection (11, 7). The greater the instability and the larger the moisture content, the greater the precipitation.

Infrared temperatures obtained from meteorological satellites were found to contain useful mesoscale thermal information of Florida. The data were applied to study the winter nocturnal climate of Florida (12). The following describes the data, and with supplementary meteorological information (13), can be used to study factors affecting mesoscale transport and deposition. Research on this set of data is still at the early stage, but some of the results have direct bearing upon acid transport and deposition, for the present and especially in the future.

Mesoscale Climate Information Derived from Meteorological Satellites

The data. - Infrared temperatures (10.5-12.6μ) detected by the Geostationary Operational Environmental Satellite/Synchronous Meteorological Satellite (GOES/SMS) were used to study surface condition of Florida during clear winter nights (1976-1982). The purpose was to develop a system to use GOES-detected temperature for cold protection of citrus in central and southern Florida (14, 15). A spin-off of the projects was the use of the data to derive surface (16, 17) and climate information (12). Because of the relatively short time-span of the satellites in operation (beginning 1975) the conventional meaning of climate (30-50 year period) is not implied here.

The digital infrared temperatures originated from GOES-east, stationed at the equator and 75°W longitude (18). Data were either purchased from Environmental Data and Information Service (EDIS) or were transmitted by computer to computer linkage (1200 Baud telephone lines) between the National Meteorological Center (NMC) and Fruit Crops Department (Hewlett-Packard computer, 1000E). From 1981-1983 the data were available through a direct downlink from satellite to antenna (Fruit Crops Department). The infrared data were sectored over Florida (24° to 31°N latitude, 80° to 85°W longitude) and have an average resolution of 8km (northsouth) by 6km (eastwest). Digital data obtained from direct satellite to antenna downlink have 8 km by 3 km resolution (Fig. 1). Data are available half-hourly; for our purpose, hourly data are adequate.

Mesoscale climate information. - Satellite infrared images of Florida for clear winter nights were manually overlaid to obtain average, cold season, low temperature maps of Florida (Fig. 2). For detail information on the construction of the maps, see (12). It was found that under winter nocturnal conditions images can be used without emissivity or atmospheric moisture correction (19). Errors introduced were small, especially if the analysis was carried out over large mesoscale regions. However, to further minimize error, references were made only to relative temperatures; references to absolute temperature values were avoided. The average low temperature map for 1979-80 (Fig. 2A) and 1980-81 (Fig.

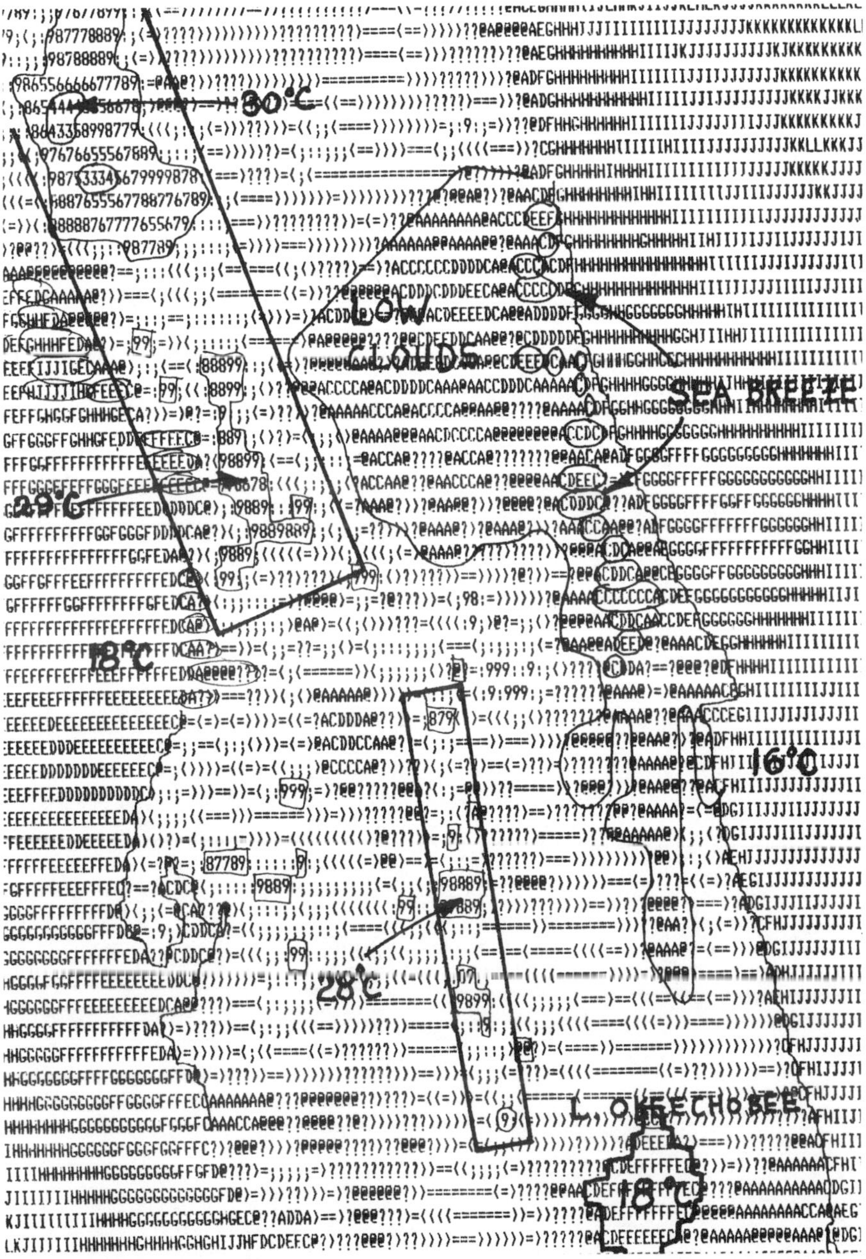

Fig. 1 GOES IR digital image of Florida for 0200 EDT, 8 June 1982, warmest land temperature, coastal water temperature, and sea breeze are identified. Boxed areas are part of The Central Florida Ridge.

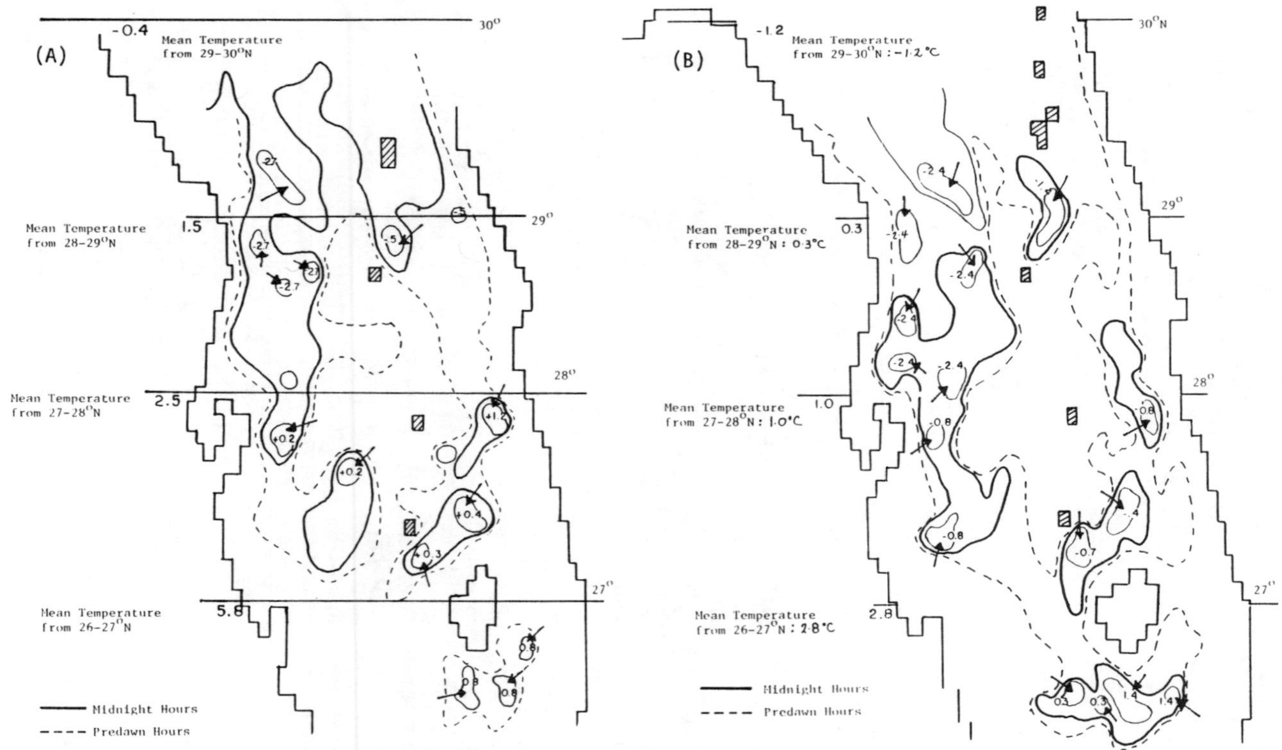

Fig. 2 Average low temperature map constructed from GOES IR images of Florida for 1979-80 (A) and 1980-81 (B) winter. Temperatures are in °C. Lines are time-expansion of cold areas initiated from cold pockets (arrows) appearing early in the night (9-10 pm). Thick lines indicate average coldest conditions during midnight hours, dashed lines indicate average coldest conditions during predawn hours. (A) was compiled from 8 nights, two images per night. (B) was compiled from 10 nights, three images per night.

2B) showed cold areas were consistently located over soils of the sandy ridge of Florida (Fig. 3). The similarity between the two maps is apparent. The persistence of the cold areas were found to be present (qualitatively) in the five years GOES images were examined (12). During the initial phase of the research, it was not possible to machine process the data to obtain quantitative results. The persistence of the cold areas were found to be related to soil radiative flux. One quantity which best describes the condition observed is thermal inertia, $(\lambda C)^{\frac{1}{2}}$, where λ is the thermal conductivity, and C is the volumetric heat capacity of the soil. The quantity appears as a coefficient in the heat flux equation. Considering only vertical heat flux, soil heat flux (G) at the surface is,

$$G(t,0) = \theta\ (\lambda C)^{\frac{1}{2}}\ \cos\ (Wt + \pi/4),$$

where t is the time, θ is the amplitude, and W is the frequency.

The GOES images show strong indications that the persistence of the thermal pattern changes with antecedent rainfall (17). GOES-detected infrared temperatures can differentiate moisture contents of the top few centimeters of the surface. This was tested by plotting GOES diurnal temperatures for a dry (Newberry Sand Hills) and a wet (Waccasassa Flats) area in North Central Florida (Fig. 3). The dry area showed a larger diurnal temperature range for both a cold day (January) and a warm day (April) (12).

GOES data covering the warm season (April to June) were examined to locate warm thermal patterns in the dry sandy ridge areas in an attempt to obtain a satellite-derived annual temperature range to confirm the claim that satellite temperatures can be used to delineate mesoscale climate information. Ten days (April to June) in 1981 and 1982 were examined. The hottest areas were also consistently located in the Florida sandy ridge area (20) (Fig. 1). Figure 1 also shows the landwater temperature difference along the coast of Florida. The sea breeze was also located with the help of higher resolution GOES visible image of the same hour. Hourly GOES visible data indicate that they can be used to track the hourly progression of the sea breeze front inland. Sea breezes were found along both east and west coast of central Florida.

Conclusion

GOES-detected surface temperatures can be used to identify mesoscale thermal regions in the topographically flat terrains of Florida. The thermal regions were detected in both the cold and the warm season of the climatic temperature spectrum. The thermal patterns were related to soil type through soil thermal property and to the moisture contents of the top layer. Temperature information obtained from GOES images can be used as model inputs. They can be used to study effects of surface temperatures on convective systems and sea breeze formation, and bearing upon acid transport and deposition. GOES images can be used to locate and track sea breezes, and to obtain land-water temperature difference. Finally, the Ad Hoc Panel on Mesoscale Processes (21) pointed out the lack of observation for the study of mesoscale meteorological processes; and the importance of mesoscale meteorological phenomena such as sea

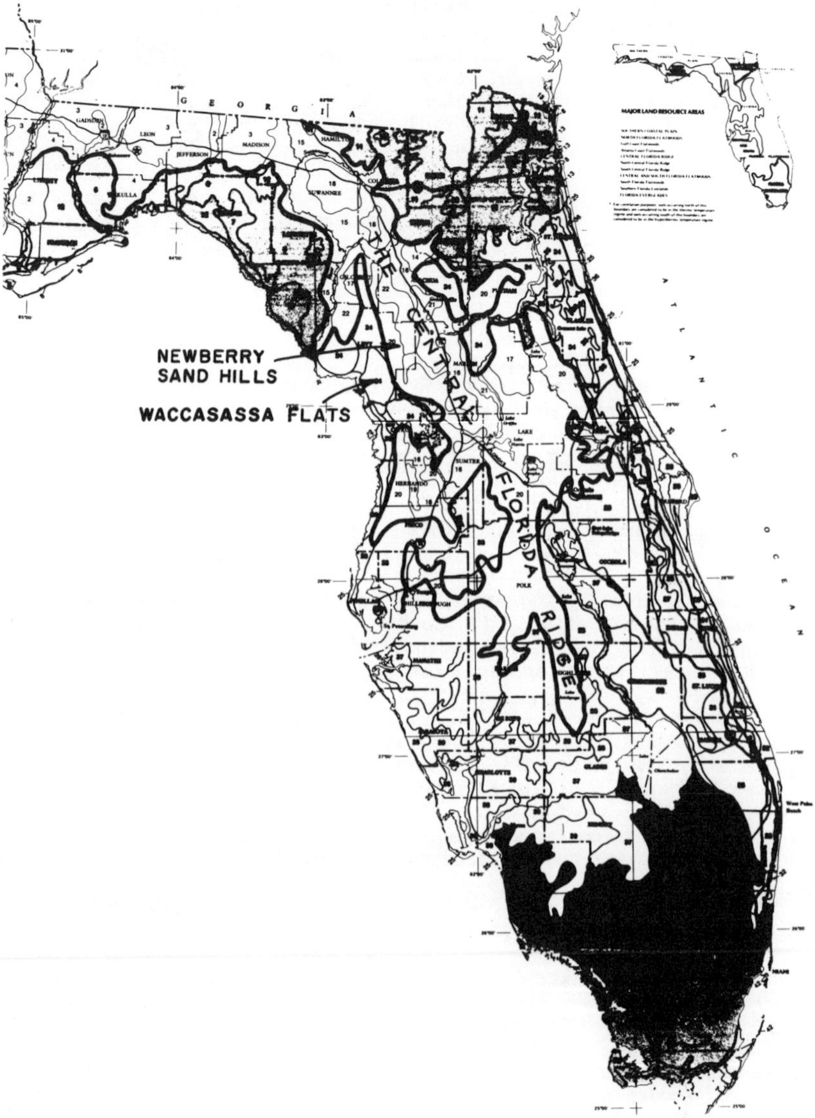

Fig. 3 The Central Florida Ridge (thick line) is a region of thick to thin sandy soils, excessively drained to well drained, level to strongly sloping. General Soil Map of Florida, R.E. Caldwell and R.W. Johnson, USDA - Soil Conservation Service, IFAS - Agricultural Experimental Station and Soil Science Department, 1982.

breezes, mesoscale convective complexes, rainbands, to air-quality, model input, etc. it appears that infrared and visible images from GOES can satisfy some of this need.

Research Needs in Florida

Initial research has shown positive results and show potentials of the infrared and visible images of the meteorological satellites when used to study mesoscale processes. Results are valuable not only to acid transport and deposition, but should contribute significantly to other mesoscale processes, to climatology, and to hydrology. The tools (satellite to antenna downlink; IFAS Computer Network; FAST (Florida Agriculture Service and Technology) and the Control Data Corporation Inc., and the Image-100 system for image analysis) for computer and machine processing of satellite-derived information are available at the University of Florida. It remains that research goals for the state and priority be established, and efforts directed towards these goals be coordinated.

Literature Cited

1. Byers, H. R., and Rodebush, H. R., 1948. Causes of Thunderstorms of the Florida Peninsula, J. Meteor. (5):275-280.

2. Trewartha, G. T., 1981. The Earth's Problem Climates, The University of Wisconsin Press, 371 pp.

3. Tingle, G. A., and Dieterle, D. A., 1976. A Numerical Study of Pollutant Distributions in Sea Breeze Circulations (114-117), Preprints Volume, Conference on Coastal Meteorology, Virginia Beach, Va., AMS.

4. Defant, F., 1951. Local Winds (655-662), in Compendium of Meteorology, T. F. Malone (Ed.), AMS, Boston, Mass.

5. Carpenter, K. M., and Lowther, L. R., 1982. An Experiment on the Initial Conditions for a Mesoscale Forecast, Quart. J. Royal Meteor. Soc. (108):643-660.

6. Pielke, R. A., 1974. A Three-Dimensional Numerical Model of the Sea Breezes over South Florida, Mon. Wea. Rev. (102):115-139.

7. Watson, A. I., and Holle, R. L., 1982. The Relationship Between Low Level Convergence and Convective Precipitation in Illinois and South Florida, Office of Weather Research and Modification, Environmental Research Laboratories, NOAA, Boulder, Colorado, Technical Report 7, NSF Grant ATM 78-08865.

8. Plank, V. G., 1965. Cumulus Convection Over Florida - A Preliminary Report (109-118), Projects 5038 and 8620, Air Force Cambridge Research Laboratory, Bedford, Mass.

9. Gannon, P. T., 1977. On the Influence of Surface Thermal Properties and Clouds on the South Florida Sea Breeze, Ph.D. Dissertation, The University of Miami, Coral Gables, Florida.

10. Garrett, A. J., 1982. A Parameter Study of Interactions Between Convective Clouds, the Convective Boundary Layer, and a Forested Surface, Mon. Wea. Rev. (110):1041-1059.

11. Burpee, R. W., 1978. Peninsula Scale Surface Convergence and Area Averaged Rainfall in South Florida (363-368), Preprints, Conference on Weather Forecasting and Analysis and Aviation Meteorology, Silver Springs, Md., AMS.

12. Chen, E., and Martsolf, J. D., 1982. The Development of Nocturnal GOES Infrared Data as a Source of Climate Information, National Climate Program, Office, NOAA, USDC, 106 pp. J. D. Martsolf, Co-Principal Investigator.

13. Green, A. E. S., Schwartz, J. M., Singhal, R. P., and Bolch, W. E., 1982. Wind Roses for Florida, APCA Journal, (32):822-825.

14. Sutherland, R. A., Langford, J. L., Bartholic, J. F., and Bill, Jr., R. G., 1979. A Real-Time Satellite Data Acquisition, Analysis and Display System - A Practical Application of the GOES Network, J. App. Meteor. (18):355-360.

15. Martsolf, J. D., 1981. A Satellite Frost Forecast System (143-165), in Computer Techniques and Meteorological Data Applied to Problems of Agriculture and Forestry: A Workshop, Anaheim, Ca. AMS/ASAE/ASA.

16. Chen, E., Allen, Jr., L H., Bartholic, J. F., Bill, Jr., R. G., and Sutherland, R. A., 1979. Satellite-Sensed Winter Nocturnal Temperature Patterns of the Everglades Agricultural Area, J. App. Meteor., (18):992-1002.

17. Chen, E., Allen, Jr., L. H., Bartholic, J. F., and Gerber, J. F., 1982. Delineation of Cold-Prone Areas Using Nighttime SMS/GOES Thermal Data: Effects of Soils and Water, J. App. Meteor., (21): 1528-1537.

18. Office of System Engineering, 1979. Geostationary Operational Environmental Satellite/Data Collection System, NOAA Technical Report NESS 78.

19. Chen, E., Allen, Jr., L. H., Bartholic, J. F., and Gerber, J. F., 1983. Comparison of Winter Nocturnal Geostationary Satellite Infrared Surface Temperature with Shelter-Height Temperature in Florida, Remote Sensing of Environment, in press.

20. Chen, E., 1982. Applications of Geostationary Satellite Surface Infrared Radiance Temperature to Provide Mesoscale Climate Temperatures, Final Report. NOAA Post Doctoral Grant, Contract No. NA81AA-D-00098, 91 pp.

21. Ad Hoc Panel on Mesoscale Processes, 1981. Current Mesoscale Meteorological Research in the United States, Committee on Atmospheric Sciences, National Academy Press.

PRECIPITATION AND CLOUD CLIMATOLOGIES OF FLORIDA

James A. Henry
Department of Geography
University of Florida
Gainesville, Florida 32611

Abstract

The rainfall climatology of Florida as established by a recent updating of earlier studies based upon climatic data from 62 weather stations is described. Sources of recent data are given. Temporal variations of cumulus and stratus cloud cover for May to September periods for 1973, 1974, and 1977 are also given for the seven climatic divisions of Florida.

Introduction

The geographic location and physical shape of Florida produce a very unique climate in the state. Two research projects were recently completed that analyzed two aspects of this unique climate. The first one produced an updated set of statewide rainfall maps, including mean annual, monthly, and seasonal maps using 1949-1978 data. The second study was a cloud climatology of Florida. High-resolution satellite data were used to overcome many of the problems inherent in ground observations of cloud cover.

Rainfall Climatology

Climatic data from 62 weather stations for the period 1949-1978 were used in conjunction with the SYMAP computer graphics program to generate various rainfall maps for Florida. The stations were chosen for geographic location and continuous records. Averages were calculated for months, years and seasons. Seasonal values were derived by using the appropriate monthly means (e.g., December, January and February values were used to create the winter season data). Corrections were applied for station changes in location and for missing observations.

A sample rainfall map (mean annual rainfall) is shown in Figure 1. Stations in the extreme northwest portion of the panhandle recorded the highest 30-year averages. This results because of the two precipitation maxima that exist in the summer and late winter. Figure 2 shows the mean rainfall patterns for the four seasons.

The climate in the region of Lake Okeechobee is strongly affected by the lake. Isolines were not drawn through the lake because it has been shown that rainfall over the lake is about 15% less than the surrounding land (1). According to this calculation, rainfall over the lake would be less than 45.0 inches annually.

The lowest annual precipitation values in Florida occur in the Keys. The Keys receive most of their rainfall in summer and early fall, but average less than mainland sites partly because of the very small amount

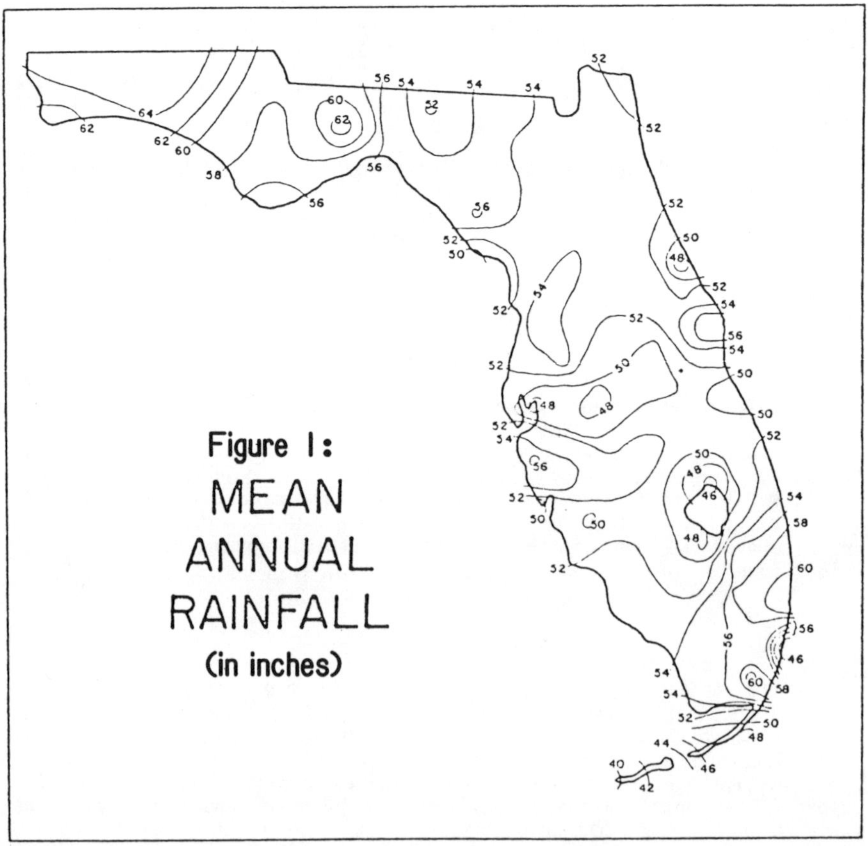

Fig. 1 1949-1978 Mean Annual Rainfall Map of Florida (in inches)

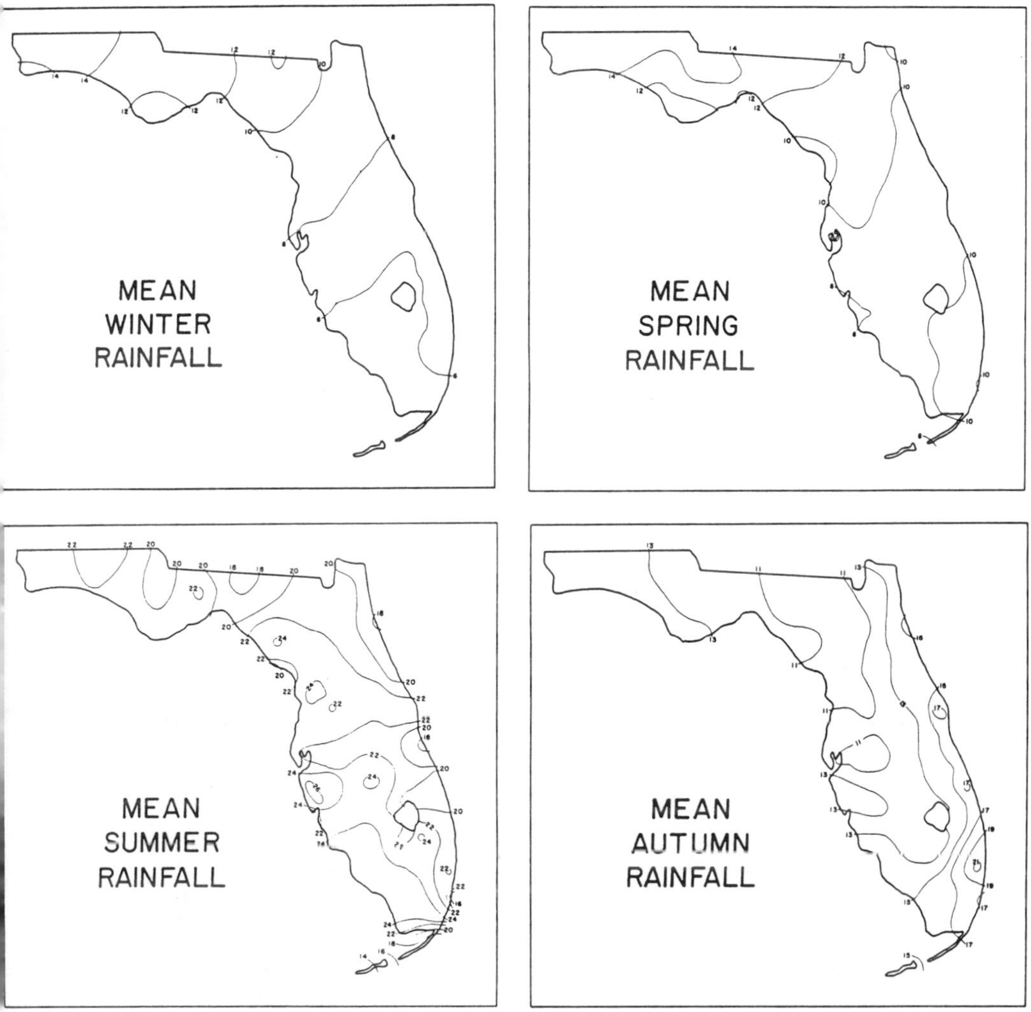

Fig. 2 1949-1978 Mean Seasonal Rainfall Patterns in Florida (in inches)

of land surface and resultant minimal convection. This area, along with the extreme southeast coast, has the highest hurricane potential for the state, and therefore precipitation values can vary dramatically from year to year.

The second phase of the rainfall study involved a search for periodicities in the data of 8 Florida stations (at least one in each of the 7 climatic divisions of the state) utilizing power spectrum analysis in an effort to ascertain whether cycles appear in the rainfall regimes of Florida. Fort Myers had the longest major cycle (7.5 years) as well as a notable 2.0-year cycle. Key West shows a 6.0-year cycle. A 5.5 and 6.0-year periodicity appears at Tampa, as does a 2.0-year cycle. Periodicities of 5.5 years occur at Fort Lauderdale and Apalachicola, while a 5.0 to 5.5-year cycle is dominant at Orlando. The latter station also displays a 2.0-year cycle. Jacksonville exhibits a 3.3-year cycle and Pensacola has the shortest major periodicity (2.7 years). No simple pattern emerged for the state as a whole.

Cloud Climatology

Cloud cover is one of the basic, routinely observed climatic elements. The amount of clouds plays a significant role in many atmospheric processes. For example, a modeling scheme indicated that temperatures are very sensitive to cloud cover: a change of 0.1 in cloud cover produced a change of several degrees in surface temperatures (2). Clouds also directly influence sunshine, visibility, turbulence and winds. The importance of the effects of clouds on the atmospheric and surface radiation budgets and the implication for climatic change have been shown by several authors (3, 4).

Knowledge of cloud cover in Florida is particularly important as spatial and temporal variations of clouds affect the location, structure and intensity of the sea breeze in southern Florida by influencing the surface heating pattern (5). The sea breeze is an important mechanism in initiating precipitation over peninsular Florida.

This section presents results of a summertime cloud climatology of Florida based on data obtained from a satellite system that reveals the amount of cloud cover over the state in a more representative manner than has appeared in climatological records and the general literature. The satellite data that were found to be most suitable for this study were produced by the Defense Meteorological Satellite Program (DMSP) satellites.

Daily noontime coverage of the three May-September periods of 1973, 1974 and 1977 were chosen for analysis. The May-September period was selected for two reasons: first, this is generally the most critical time for water supplies and cloud seeding operations in Florida; second, cloud cover estimation by ground observers is generally least accurate during summer, when vertically developed clouds are most common. The years 1973, 1974 and 1977 were chosen because these were relatively normal, wet and dry summers, respectively, over most of Florida.

Florida was divided into cells of 0.5 degrees latitude by 0.5 degrees longitude (approximately 55.4 x 48.2 km) for data recording purposes. For each cell on every image the percentage of total cloud cover, cumuliform cover and stratiform amount were visually determined. Figure 3 presents the May-September total cloud cover. Generally, total cloud cover is greatest in the northeastern and southern portions of the peninsula. This is true for the individual years as well as for the total period. The highest value is 61 percent, which occurred in the south in 1973. Values are higher in coastal regions than in the interior, as revealed by the trough of low values in the central portion of the peninsula--particularly in 1973 and 1974. Lowest cloud cover values are most common over the Keys.

In terms of cumulus clouds most of the state has values of 21-25 percent. This category covers all of the panhandle and most of the interior of the peninsula. Maximum values occur in the northeast and scattered areas in the central portion of the state. Minimum values are mostly associated with coastal areas. The land/water contrast is especially significant in the distribution of cumulus clouds. It has been noted that during May-September there is an almost daily occurrence of convective activity in southern Florida caused by the sea breezes on the east and west coasts (6). This system produces more cumulus clouds inland, away from the coasts, and accounts for many occurrences on the satellite images when the water portions and immediate coastal sections of the data cells were relatively cloudfree, while the inland portions were very cloudy or completely covered. The islands of the Keys are mostly unaffected by the sea breeze effect, and the two data cells in this region are consistently the least cloudy.

The only noncoastal cell with a very low cumulus cloud cover value is associated with Lake Okeechobee. On many of the satellite images the lake region appeared as a hole in the cloud cover, often being completely cloudfree when surrounding areas contained numerous clouds. It has been suggested that the lake-induced cooling of the overlying atmosphere suppresses convergence and, therefore, cumulus cloud development (7). The effect is apparently strong enough to inhibit convergence even when the strong sea breeze convergence zone passes over the lake.

Figure 4 shows the changes in total, cumulus and stratus cloud cover for the seven climatic divisions of Florida. A qualitative assessment of the monthly changes revealed by the general shapes of the curves shows several marked similarities among the northern four divisions and among the southern three. For example, the total cloud cover curves for all three years for the three southernmost divisions are very similar in shape and general trend, and are significantly different than the equivalent curves for the northern four divisions--further highlighting the contrasting synoptic controls dominant in northern and southern Florida in summer.

A more detailed account of Florida's cloud climatology has been given (8). No specific consideration has been given to the implications of this work to the acid rain question. However, several of the other papers at this workshop have already described these relationships.

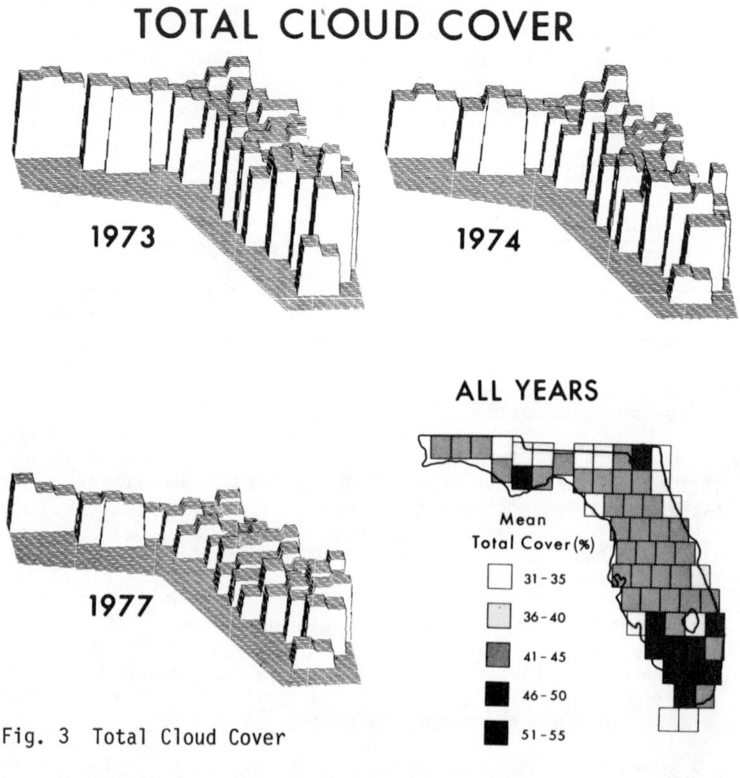

Fig. 3 Total Cloud Cover

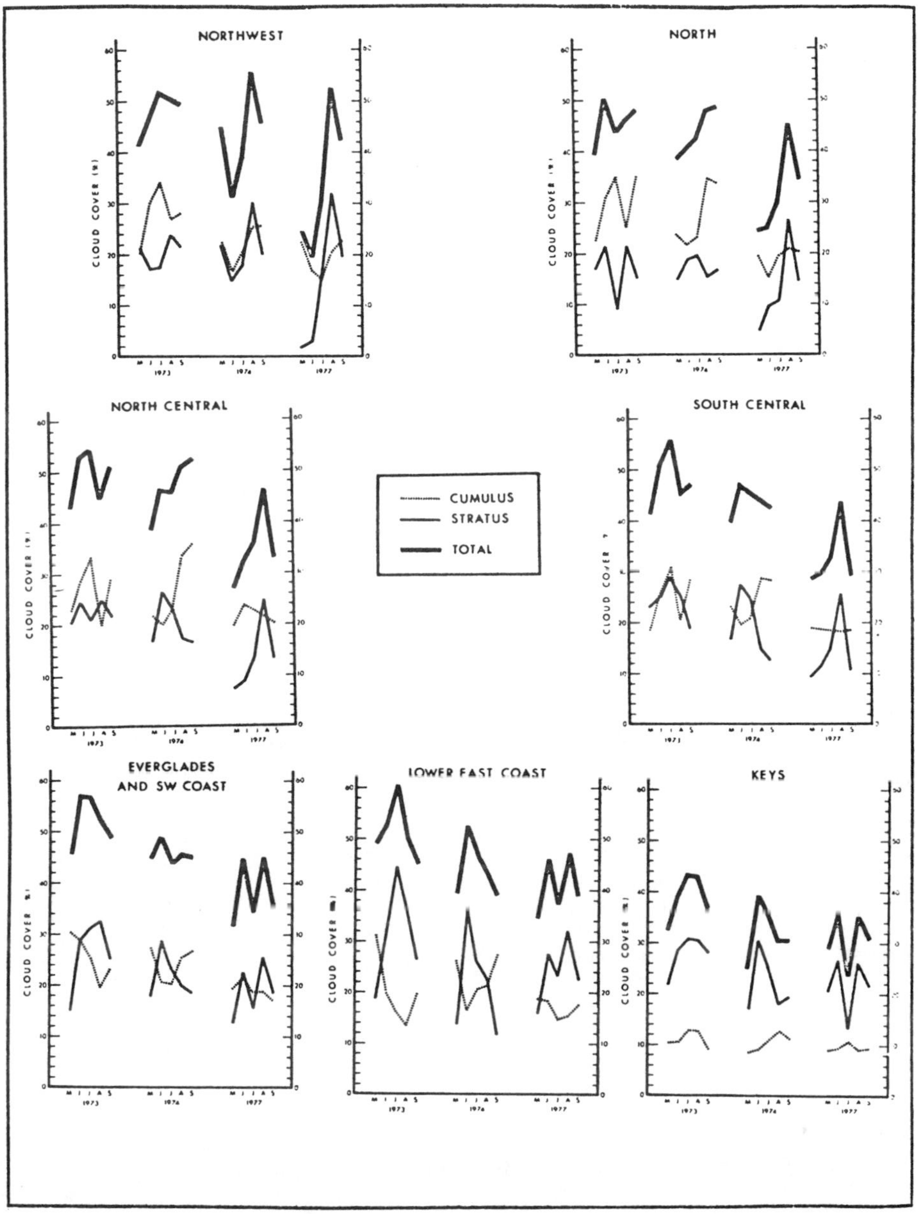

Fig. 4 Cloud Cover Change in the Seven Climatic Divisions of Florida

Research Needs in Florida

As the population and resultant urbanization in Florida increase there is a real possibility that rainfall patterns may change. An increase in particulate matter may increase precipitation in certain areas. Therefore, monthly and seasonal maps of Florida should be continually updated. This is also applicable to the time-series analyses; additional sites could be included in this portion of the analysis.

Statistical correlations between peak-rainfall years and hurricane occurrences would be a fruitful avenue of future research. Some qualitative associations have been attempted, but more rigorous approaches are needed.

Acknowledgment

Appreciation is extended to Jeffrey A. Isaacs who performed the rainfall climatology analysis for this report.

Literature Cited

1. Riebsame, W. E., Woodley, W. L., and Davis, F. E., 1974. Radar Inference of Lake Okeechobee Rainfall for Use in Environmental Studies, Weatherwise 27 (5):206-11.

2. Adem, J., 1964. On the Normal Thermal State of the Troposphere-Ocean-Continent System in the Northern Hemisphere, Geofisica Internacional 4:3-32.

3. Schneider, S., 1972. Cloudiness as a Global Climatic Feedback Mechanism: The Effects on the Radiation Balance and Surface Temperature of Variations in Cloudiness, Journal of the Atmospheric Sciences 29:1413-22.

4. Ohring, G., and Clapp, P., 1980. The Effect of Changes in Cloud Amount on the Net Radiation at the Top of the Atmosphere, Journal of the Atmospheric Sciences 37:447-54.

5. Gannon, P. T. Sr., 1978. Influence of Earth Surface and Cloud Properties on the South Florida Sea Breeze, National Hurricane and Experimental Meteorology Laboratory, NOAA Tech. Rept. ERL 402-NHEML 2.

6. Sax, R. I., and Keller, V. W., 1980. Water-Ice and Water-Updraft Relationships near -10C within Populations of Florida Cumuli, Journal of Applied Meteorology 19 (5):505-14.

7. Pielke, R. A., 1974. A Three-dimensional Numerical Model of the Sea Breezes over South Florida, Monthly Weather Review 102:115-39.

8. Henry, J. A., and Isaacs, J. A., 1981. Florida Summertime Cloud Cover from Satellite Data, National Weather Digest, 6, 3, 39-46.

LIGHTNING IN FLORIDA AS A SOURCE OF ACID RAIN

Alex E. S. Green
Department of Mechanical Engineering
and
Martin A. Uman
Department of Electrical Engineering
University of Florida, Gainesville, FL 32611

Abstract

Research grade lightning location systems of the type described in (1) are available in the Tampa Bay, the Keys and the Kennedy Space Center areas of Florida to locate and measure the location and other salient properties of cloud-to-ground lightning. From these systems the duration, area, total number of ground flashes, flash density, average flashing rate, maximum flashing rate, average number of strokes per flash, and initial peak magnetic field intensity can be obtained for each storm analyzed. Such data provide an excellent quantitative basis for attempts to establish detailed local correlations between lightning activity, NO_x, and acid rain production. On a global bases Chameides has estimated the nitrogen fixed in the production of NO_x to be 50 MT/yr of which lightning contributes 10 and fossil fuel combustion contributes 20. The unusual concentration of lightning in Florida suggests that this natural source be investigated in detail in acid rain and NO_x measurements coordinated with lightning measurements. In addition, the effects of intracloud lightning and corona in clouds must be considered. The results would not only be of interest in resolving the acid rain issue in Florida but would also be valuable in achieving a better scientific understanding of these relationships.

Lightning and Other Charged Particle Deposition Mechanisms

Various charged particle deposition mechanisms serve as natural sources of NO_x in the earth's atmosphere including 1) cloud to ground lightning; 2) intracloud lightning; 3) coronal discharges within clouds; 4) cloud to air lightning; 5) discharges under clouds associated with pointed vegetation, e.g., pine tree needles, grass etc.; 6) natural radioactivity; 7) photoelectrons produced by the extreme ultraviolet radiation of the sun; 8) auroral electrons; 9) auroral protons; and 10) cosmic rays. Deposition mechanisms (7-10) primarily occur at high altitudes and contribute to the global production of NO_x. The first six may have special Florida involvements because of our intense cumulus activity in the summer and because phosphate fields are sources of radioactivity.

A cloud to ground lightning flash typically consists of several large current surges or strokes [see Uman (2)]. It begins with a faint stepped leader, which proceeds rather slowly from cloud to ground in a series of short, intermittent steps. When the stepped leader contacts the ground, a very energetic and bright return stroke propagates rapidly back up the ionized path established by the leader. After a pause of 30-50 ms, a dart leader often propagates down the previous channel and is followed by another bright return stroke propagating upward. A typical flash to ground contains three or four leader-return stroke combinations, which almost always transfer negative charge to ground. The currents in return strokes typically rise to peak values of 10-40 kA in 1-10 μs or less. A typical flash produces a cylinder of air heated to about 30,000° K roughly 5 cm in diameter and 8 km long. It is estimated by optical and electrical techniques that 50 to 1000 MJ (megajoules) are dissipated in an average cloud to ground flash (3-7). For order of magnitude of estimation we will here use the nominal value of 500 MJ for the thermal yield.

Few estimates have been made of the energy dissipated in intracloud lightning, intracloud coronal discharges, cloud to air lightning, and discharges under clouds associated with pointed vegetation. Estimates of high altitude charge particle energy deposition are generally low compared to lightning energy dissipation. Accordingly in this work we will confine our quantitative discussion mostly to lightning phenomenology as determined by recent satellite and ground based measurements.

Satellite Detection of Lightning

Satellite observations provide the opportunity to study global lightning distributions without the geographical bias associated with sparcely spaced ground observations. Optical sensors designed for airglow measurements on the Orbiting Solar Observatory satellites (8,9) were used to establish lightning distributions between 35°N and 35°S. Unfortunately these sensors were saturated by reflected sunlight or moonlight and thus gave lightning data only near midnight local time during new moon periods. Orville and Spencer (10) have more recently counted lightning streaks seen in the cloud imagery data from the Defense Meteorgological Satellite Program (DMSP) and thus estimated global lightning flash frequency at dusk and midnight. Good qualitative comparisons are possible with this technique, which recorded latitude variations and seasonal dependence of lightning activity.

The Piggyback Experiment Sensor (PBE) designed by Sandia Laboratories, Albuquerque, New Mexico, to detect moderate intensity lightning flashes from an altitude of 830 km during day or night has been flown on DMSP flights 2 and 3. From their sun-synchronous orbits these two lightning detectors have recorded terrestrial lightning activity at dawn and dusk, accumulating a data base of approximately 30,000 lightning triggers.

Turman and Edgar (11) using two PBE lightning sensors on DMSP satellites have provided a consolidated set of global lightning activity measurements. Seasonal lightning rates have been calculated within geographical cells of size 10° x 10° which occasions a location uncertainty of 680 km radius. An example of trigger patterns are mapped in Figure 1 for dusk in August 1977. The totality of maps indicate that ocean storms contribute only a samll fraction of the total global lightning activity. At dawn, about 37% of global lightning originates over the oceans, and at dusk the percentage is 15%. A comparison of PBE trigger rate with previously estimated lightning flash densities for 16 geographical regions

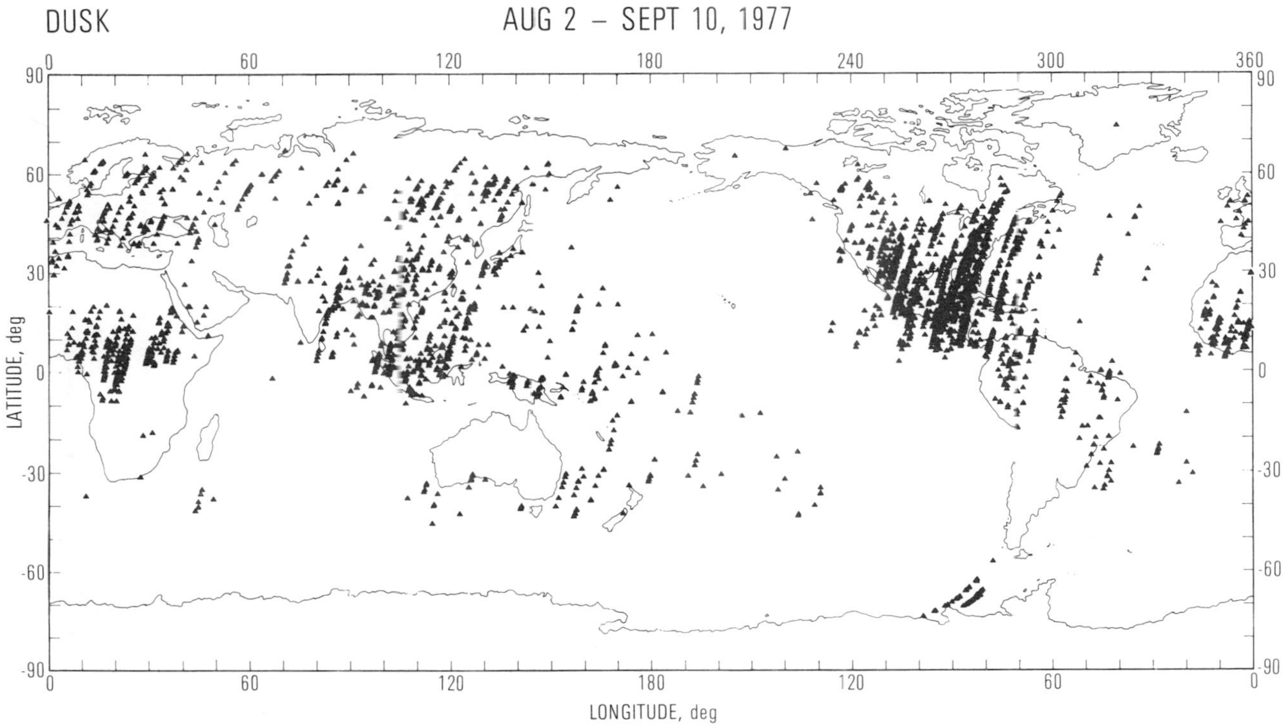

Fig. 1 Subpoint Locations of all Triggers Recorded at Dusk

indicates that PBE detects about 2% of the lightning flashes within its field of view. Accepting this detection fraction one can estimate the total number of flashes for various regions. Turman and Edgar estimate that the mean global lightning flash rate is 40-120 flashes per second, with seasonal variation of about 10%.

Ground Based Detection of Lightning

Extensive networks of magnetic direction-finding (DF) stations have been installed throughout the United States, particularly western United States, Alaska and Florida (1, 12,13). Their primary purposes are to facilitate early detection of lightning-caused fires and causes of damages to electrical distribution systems. Each station contains a new wideband direction-finder that responds primarily to cloud to ground lightning and discriminates against cloud discharges and background noise. Good angle accuracy is obtained by measuring the lightning direction at just the time the return-stroke magnetic field reaches its initial peak. Lightning locations are calculated from the intersections of direction vectors and/or from the ratio of signal strengths recorded simultaneously at two, three, or four DF sites. The development of these systems has proved to be a significant aid in 1) the detection of lightning-caused fires; 2) in fire weather forecasting; 3) determining the effects of lightning on electrical distribution systems, and 4) developing scientific data on the physics of lightning strokes to distribution lines.

Figure 2 is a map giving the mean annual ground flash density (number of flashes per unit area per unit time) for the United States which was assembled by Meier (14) from lightning direction finding and thunderstorm data. Note that on this map annual flashing densities in Florida range from 10 to 18. Granting an average of 14 and a land area of 1.4 x 10^5 km^2 we estimate about 2 million cloud to ground lightning flashes per year in Florida from data based upon ground based lightning locations. One might multiply the ground based Florida annual lightning flash rate by 4 to allow for intracloud lightning which occurs in approximately 3:1 proportions to cloud to ground lightning (14). This would give approximately 8 million lightning flashes per year in Florida.

On the other hand satellite measurements (11) suggest a global flashing rate of 100/s or about 3000 million per year. Thus we estimate a ratio of Florida to global annual flashes of 1/400 which seems rather low if Figure 1 is representative of summer season lightning flash densities in various geographic locations. The question is important as will be illustrated in the NO_x - HNO_3 lightning source strength calculation which follows.

Fixation of Nitrogen by Lightning in Florida

A direct calculation of the mass of nitrogen in NO_x produced by the air heated by lightning bolts is very sensitive to 1) the assumed dimensions and temperature of the average bolt; 2) the number of flashes over the area, and 3) the conversion efficiency from energy to NO_x production.

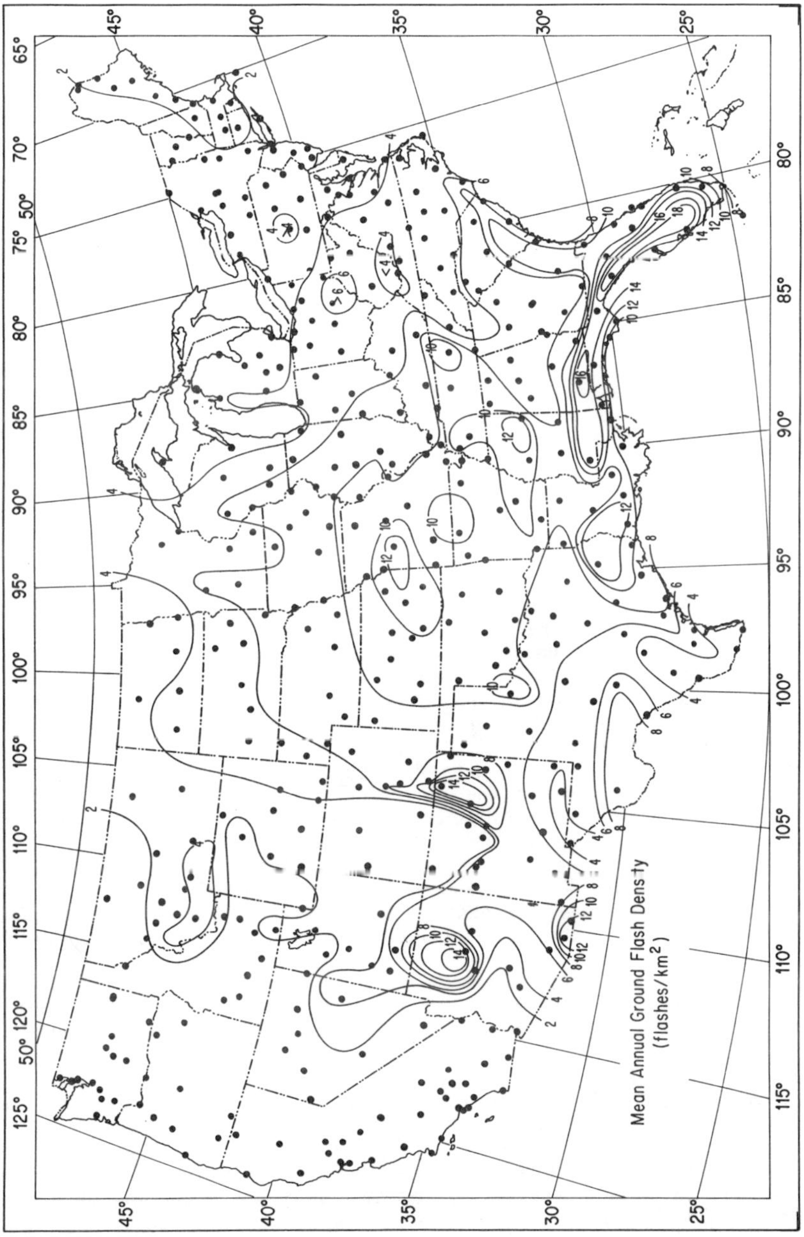

Fig. 2 Mean Annual Ground Flash Density (flashes/km^2)

Thus the production of fixed nitrogen over an area may be given by the obvious formula

$$P = A \times F \times E \times Y \times M/N_a \qquad (1)$$

where P = production rate in metric tons/yr, A is the area, F = flashing density (number per unit area), E = average energy per flash in Joules, Y = the thermal yield in nitrogen atoms/Joule of an average flash, M = atomic weight of the product [in gr/mole], N_a = Avogadros number.

The energy E is approximately 5×10^8 Joules/fl, Y is approximately 5×10^{16} atoms/Joule (15-17). Allowing for atomic weight of nitrogen (14 gr/ mole) we get for the mass of fixed nitrogen

$$\frac{8 \times 10^6 \times 5 \times 10^8 \times 5 \times 10^{16} \times 14 \text{ gr/mole}}{6 \times 10^{23}} = 4.7 \text{ kT/yr}$$

This is small compared to the 500 kT/yr estimated for anthropogenic production of NO_x in Florida.

The University of Florida Tampa Bay Area Lightning Project

A stroke by stroke analysis of lightning electric and magnetic field records and lightning optical records is being carried out by University of Florida researchers (18) to characterize the basic physical parameters of lightning in the Tampa Bay area. A Lightning Locating System (LLS) (1) was used to measure lightning phenomenological parameters such as flash rates and flash density as a function of such variables as types or locations of storms (e.g., off-shore vs over land), time of occurrence within a storm, and time of occurrence during the day. A second project objective was to provide, via direct measurement, data on lightning-induced current surges on the distribution systems owned by the participating power utilities. These data will be used to characterize the induce transients and to improve the analytical models for simulating the effects of lightning on distribution systems. Table 1 lists the instrumentation employed during the summer of 1979.

Livingston and Krider (19) and Piepgrass et al. (2) have earlier measured the ground flash density at the Kennedy Space Center (KSC), Florida. They used the KSC electric field mill network to detect lightning over KSC water, an effective area of approximately 550 km^2. The mean ground flash density estimated by Livingston and Krider (19) for data obtained during the months of June and July of 1974 and 1975 was 6.0 km^{-2} $month^{-1}$. Piepgrass et al. (2) found a mean ground flash density of 3.9 ± 2.9 km^{-2} $month^{-1}$ for the months of June and July of 1976 through 1980. Both of these monthly flash densities correspond to annual flashing densities which are higher than the annual flashing densities assigned by Meier (14) (see Fig. 2).

Livingston and Krider (19) have given total (ground flash and cloud flash) flashing rate statistics for storms at KSC occurring on 22 days during 1974 and 1975. They found, for most large storms which produce 100 discharges or more, that 70 percent of the lightning occurred during

Table 1. Recording equipment and data sources used for lightning research in the Tampa Bay Area during Summer 1979

Equipment and Data Sources	Recording Medium
Lightning Instrumentation	
lightning locating system	hard copy and magnetic digital tape
electric and magnetic field measurements*	35mm film records of oscillograms, strip charts, analog instrumentation magnetic tape
TV lightning monitors (4 + 1*)	video-tape
thunder ranging and visual observations*	strip chart and audio tape
still and streak photography (2)*	film
Power System Instrumentation	
transient surge recorders (27)	magnetic tape cassette
magnetic strip recorders (66)	analog magnetic tape
lightning flash counters (8)	counters and loggers
TECO data acquisition system	hard copy and magnetic tape
circuit breaker operations records	hard copy
FPC circuit-breaker operations records	hard copy
TECO and FPC line outages and equipment damage records	hard copy

* equipment installed in the University of Florida's Mobile Lightning Laboratory.

a period, which they called the active period of the storm, which lasted about 30 percent of the storm duration. The total number of flashes per storm ranged from 8 to 1,987. A log-linear relationship was found relating the total number of lightning flashes during the active period of the storm, N, to the duration of the active period, D i.e.

$$\text{Log}_{10} N = a D + b \tag{2}$$

where a = 0.1 and b = 1.9. The average total flashing rate for the storms studies was between 0.3 and 9.3 flashes per minute. The maximum flashing rate averaged over five minutes was found to exceed 20 flashes per minute. Statistics for 79 storms which occurred at KSC during the summers of 1976 through 1980 (20) indicated total number of flashes which ranged from 1 to 3,687 flashes. The average total flashing rates ranged from 0.1 to 12.2 discharges per minute, while the maximum flashing rates averaged over five minutes ranged from 0.2 to 30.6 flashes per minute. Lhermitte and Krehbiel (21) found for a "normal" storm over KSC a maximum flashing rate of about 60 cloud discharges per minute over a several minute interval.

Peckham and Uman studies in the Tampa Bay area suggest that Livingston and Krider may have observed a combination of individual storm entities, multiple-peak storms, and storm systems. Table 2, a summary of their results, shows that the average number of flashes produced by a multiple-peak storm is more than three times greater than the number of flashes produced by an individual storm entity. The average area of a multiple-peak storm is almost two and a half times greater than that of an individual storm entity, while the average duration of a multiple-peak storm is about twice as long as the duration of an individual storm entity. This supports the postulate that multiple-peak storms consist of more than one, typically three, individual storm entities which are overlapping in time and space. Peckham and Uman's correlation between numbers of flashes and storm duration have the parameters a = 0.02 and b = 0.57. The marked differences with the parameters of Livingston and Krider (a = 0.1, b = 1.9) may be due to differences in definition of storm duration used by the former and duration of active period of the storm used by the latter.

Table 2. Summary of means observed in Tampa Bay area during Summer 1979

	Individual Storm Entities	Multiple Peak Storms	Storm Systems
Average Number of Flashes	52	169	595
Average Duration, minutes	41.2	76.6	127
Average Area, km^2	106	256	900

Discussion

These recent detailed studies of lightning in the Tampa Bay area provide a natural opportunity to resolve many issues related to lightning production of NO_x and HNO_3. The duration, area, total number of ground flashes, flash density, average over five minute intervals, average number of strokes per flash, and average initial peak electric field intensity can be obtained for each storm analyzed. Such data provides an excellent quantitative base for attempts to establish detailed local correlations between lightning activity and NO_x and acid rain production. The result would not only be of interest in resolving the acid rain issue in Florida but would also be valuable in achieving a global understanding of the relationships between lightning and coronal type discharge and nitrogen fixation.

On a global basis Chameides (22) has estimated the fixation of nitrogen to be 50 MT/yr (million metric tons/yr) of which lightning contributes 10 MT/yr and fossil fuel combustion contributes 20 MT/yr. If

1/400 of the estimated global annual number of flashes occurred in Florida, the fixed nitrogen would be 25 kT/yr. This is an order of magnitude smaller than estimated anthropogenic emissions in Florida. However a few factors of 2 or 3 could bridge this gap.

Thus far our calculations have not considered the fate of all the electrical energy stored in a cloud or the energy lost in the complex charging and recharging processes. A typical cloud to ground lightning flash reduces the electrical field strength of a cloud to 1/2 its initial value which implies a reduction of the electrical energy of the cloud by 3/4 (23). The remainder of the electrical energy might still be dissipated by other channels such as coronal discharges within the cloud, (24) cloud to air discharges or coronal type ground to cloud discharges. If the dissipation of the residual cloud energy and charging losses proceeds via non-thermal or electrical discharge processes the nitrogen fixed per joule would rise considerably with respect to the yield of 5 x 10^{16} N atoms per joule estimated for shock heated air. Charged particle deposition theory (25) gives a yield of about 25 x 10^{16} N atoms per joule which is about 5 times more efficient than the shock heating mechanism (15).

We have noted earlier (26) that the prevailing summer southerly winds channel and transport most of the anthropogenic emissions in the heavily populated areas of Florida to the northern part of the peninsula. The fact that lightning activity peaks during this period suggests that we should investigate the possibility that lightning flashes might activate these anthropogenic emissions of NO_x and SO_x and influence their time history. Indeed lightning acting on the large water content of clouds would be expected to be an important source of OH and H radicals (27) which play central roles in atmospheric chemistry. The origin of proteins and primitive life forms have been explained in terms of lightning flashes in the primordial atmosphere of the earth. The origin of acid rain in north Florida might well find an explanation in lightning flash activation of the atmosphere of the south and central Florida megapolis. Florida has the lightning detectors, air pollution monitors and acid rain monitors all in place and hence is certainly in a good interdisciplinary position to undertake the next scientific steps needed to resolve the many open questions.

Acknowledgements

One of us (A.E.S.G.) would like to acknowledge helpful conversations with B. C. Edgar, M.W. Meier, B.R. Vonnegut, A.A. Few, W.M. Chameides and J.M. Schwartz.

Literature Cited

1. Fisher, B., and E.P. Krider, "On-line lightning maps lead crews to trouble", Electrical World, May 1982.

2. Uman, M.A., "Lightning", McGraw-Hill Book Co., New York, 1969.

3. Krider, E.P., G.A. Dawson, and M.A. Uman, "Peak power and energy dissipation in a single-stroke lightning flash", J. Geophys. Res., 73, 3335-3339, 1968.

4. Guo, C., and E.P. Krider, "The optical and radiation field signatures produced by lightning return strokes", J. Geophys. Res., 87, 8913-8922, 1982.

5. Berger, K., "The earth flash", in Lightning, edited by R.H. Golde, pp. 119-190, Academic Press, New York, 1977.

6. Hill, R.D., R.G. Rinker, and H. Dale Wilson, "Atmospheric nitrogen fixation by lightning", J. Atmos. Sci., 34, 179-192, 1980.

7. Turman, B.N., "Analysis of lightning data from the DMSP satellite", J. Geophys. Res., 83, 5019-5024, 1978.

8. Vropahl, J.A., J.G. Sparrow, and E.P. Ney, "Satellite observations of lightning", Science, 169, 860, 1970.

9. Sparrow, J.G., and E.P. Ney, "Lightning observations by satellites", Nature, 232, 540, 1971.

10. Orville, R.E., and D.W. Spencer, "Global lightning flash frequency", Mon. Weather Rev., 107, 934, 1979.

11. Turman, B.N., and B.C. Edgar, "Global lightning distributions at dawn and dusk", J. Geophys. Rev., 87, 1191-1206, 1982.

12. Krider, E.P., R.C. Noggle, A.E. Pifer, and D.L. Vance, "Lightning direction finding systems for forest fire detection", Bulletin American Meteorology Society, 61, 980-986, 1980.

13. Beaty, H.W., "Researchers gather lightning data", Electrical World, August 1, 1978.

14. Meier, M.W., private communication, Lightning Location and Protection, Inc., Tuscan Arizona, 85719.

15. Levine, J.S., R.E. Hughes, W.L. Chameides, and W.E. Howell, "N_2O and CO production by electric discharge: Atmospheric implications", Geophys. Rev. Lett., 6, 557-559, 1979.

16. Levine, J.S., R.S. Rogowski, G.L. Gregory, W.E. Howell, and J. Fishman, "Simultaneous measurements of NO_x, NO, and O_3 production in a laboratory discharge" Atmospheric implications", Geophys. Res. Lett., 8, 357-360, 1981.

17. Chameides, W.L., D.H. Stedman, R.R. Dickerson, D.W. Rusch, and R.J. Cicerone, "NO_x production in lightning", J. Atmos. Sci., 34, 134-149, 1977.

18. Peckman, D.W., M.S. Thesis Dept. of Electrical Engineering, University of Florida, May 1982 (M.A. Uman, Chairman).

19. Livingston, J.M., and E.P. Krider, "Electric fields produced by Florida thunderstorms", J. Geophys. Res., 83, 385-401, 1978.

20. Piepgrass, M.W., E.P. Krider, and C.B. Moore, "Lightning and surface rainfall during Florida thunderstorms", J. Geophys. Res., 87, 11,193-11,201, 1982.

21. Lhermitte, R., and P.R. Krehbiel, "Doppler radar and radio observations of thunderstorms", IEEE Trans. Geoscience Electronics, GE-17, 162-171, 1979.

22. Chameides, W.M., Acid Deposition Workshop Paper, 1983.

23. Vonnegut, B., J. Geophys. Res., 88, 3911-3912, 1983.

24. Green, A.E.S., T. Sawada, B.C. Edgar, and M.A. Uman, J. Geophys. Res., 78, 5288, 1973.

25. Porter, H.S., C.H. Jackman, and A.E.S. Green, J. Chem. Phys., 65, 154-167, 1976.

26. Schwartz, J.M., J.H. Miglietta, and A.E.S. Green, Acid Deposition Workshop Paper, March 1983.

27. Green, A.E.S., J.J. Olivero, and R.W. Stogat, "Microdosimetry of low energy electrons". IN: Biophysical Aspects of Radiation Quality, IAEC, Vienna, 79-97, 1971.

TERRESTRIAL AND AQUATIC EFFECTS OF ACID DEPOSITION: A FLORIDA OVERVIEW

David R. Barker
Florida Department of Environmental Regulation
2600 Blairstone Road, Tallahassee, FL 32301

We cannot dispute the fact that acid deposition is a basic air pollution control problem. We must also concede that the most immediate and obvious impacts are in the aquatic environment.

Florida is a state, almost entirely surrounded by the marine environment, whose topography protrudes intermittently and sometimes impertinently from a surface water table which itself permeates through a generally porous strata of various soils, clay, limerock and sand. The coastal regions entertain a diversity of brackish water situations where nutrient laden fresh waters flow from the land and enrich estuarine areas.

The state is a spectrum of diverse and unique aquatic systems. The sensitivity of these systems to acid deposition is in question.

Acidified waters may prevent fish eggs from hatching, damage fish larvae, and lower immunity factors in adult fish. Through a secondary effect, that of the release of metals (notably aluminum) from lake sediments, acid rain causes a rapid biochemical deterioration of fish gills and brings about death of the animals via suffocation. Equally severe effects on the fresh water organisms which compose the aquatic food web have also been recorded.

The estimated 8,500 Florida lakes have a very wide natural range of annual average pH values which are critically dependent on highly specific regional geology. The range extends from about 4.0 pH units where naturally occurring organic rich soil leachates predominate, to about 8.8 pH units where suspended limerock is the main source of alkalinity. In south Florida the pH of rainfall is commonly in the pH range of 4.8 to 5.1. The lakes and surface waters of south Florida which have high suspended limestone alkalinities could be considered well buffered compared to the clay and silt lined lakes of north Florida where higher rainfall acidities in the pH range of 3.8 to 4.3 are commonly encountered. Florida lakes with naturally acidic watersheds and substrates are inhabited by somewhat more acid tolerant aquatic organisms. Some University of Florida researchers have reported that there has been a measurable lowering of pH, a loss of alkalinity, and an increase in sulfate in a study group of 20 softwater north Florida lakes over the last 25 years. These changes were reported to be accompanied by decreases in both average chlorophyll *a* levels and phytoplankton species diversification. It is worth noting that the nitrates and sulfates associated with acid rain are also available, once deposited in surface waters as nutrients, to increase primary productivity and hasten lake eutrophication processes. Some estimates place the nitrogen and phosphorous input to lakes in Florida from atmospheric deposition at close to 30% of the total overall nutrient input to these aquatic systems. Some researchers point to low nutrient concentrations as being primarily responsible for the reduced animal and plant populations reported in acidic Florida lakes.

No evidence is currently available to clearly show that degradation of aquatic systems or adverse human health effects have occurred as a direct result of acid precipitation in Florida. Our concerns with the common and traditional acid rain effects issues in Florida may not now contain the same elements of urgency found in northeastern U.S. situations. We may, however, be in a position to look into some of the potentially more serious and long term side effects of acid rain such as: increased mobilization of heavy and trace metals in leachate effluents from landfills and hazardous waste dump sites into surface and drinking water systems; potential effects of increased acidity on the half life (reactivity and stability or persistence) of trace organics and commercial pesticides in the environment; and the direct effects of acid rain on the epiphyton communities, which play such a large role in the purification of wastes in sheet and surface waters as they flow towards the marine organism nurseries along the coast. The process of increased environmental mobilization of toxic materials by acid rain and their subsequent availability for food chain biomagnification may be of concern.

Today we will hear discussions regarding laboratory and field study comparisons, pH versus total alkalinity concepts, calcium hardness, calcite saturation index, cation exchange capacity, base saturation, hydraulic conductivity, soil nutrient loss, and heterotrophic community comparisons in the various studies conducted in acid deposition research. For some the challenge will be found in correlations of the many temperature regime lake studies with those now already completed for subtropical Florida situations. It is important that this latter correlation be made, if at all possible, in order to fully utilize the enormous amount of data and information already available at the national and international levels on this subject. For others, original Florida research will be discussed and yet others will attempt to put together the pieces of a puzzle here in Florida which stems from a number of independent university, governmental and private industry studies. Variances in recent acid precipitation data from the Florida Power Coordinating Group Phase II Study and past research data from the University of Florida may be discussed in the light of different methodologies, quality assurance controls, or real and striking differences which need to be fully explained.

The Department of Environmental Regulation is concerned that national studies have been slow and uncoordinated and that the Southern Appalachian Study which we were eagerly anticipating is stalled and not proceeding aggressively. Here in Florida and at this conference, we have the makings of some very cooperative and productive research projects which will certainly help us in forging our policies on acid rain, regardless of the uncertainties and indecisions of studies beyond our control. If there are real problems in the aquatic environment which are attributable to acid precipitation, we need your help in identifying and defining these and in delineating a course of action which will lead to the appropriate solutions.

THE FLORIDA ACID DEPOSITION STUDY, PHASE II: DISTRIBUTION AND ANALYSIS OF LAKE AND SOIL BUFFERING CHARACTERISTICS IN FLORIDA

Curtis D. Pollman and Charles D. Hendry
Environmental Science and Engineering, Inc.
Gainesville, FL 32602

Abstract

The ecological effects program of the Florida Acid Deposition Study is currently in Phase II of a 4-phase study initiated in January 1981. Phase I consisted of a review of the literature regarding the effects of acidic deposition on abiotic and biotic processes in terrestrial and aquatic ecosystems. Indices quantifying soil and aquatic sensitivity were evaluated for applicability to Florida, and the results were subsequently extended to Phase II to delineate the spatial distribution of lake and soil buffering characteristics within Florida.

Results show that the majority of lakes surveyed within Florida may be classified as acidic (i.e., pH <7), softwater, and, unlike temperate lakes of similar chemistry, productive systems. Of the lakes surveyed, 12 percent had average pH levels in excess of 8; 25 and 31 percent ranged between pH 7 to 8 and 6 to 7, respectively; and 32 percent had average pH levels less than 6. Several physical and chemical factors contribute to enhanced levels of primary production in Florida lakes, including: a subtropical environment; naturally occurring phosphatic deposits; and the shallow, well-mixed nature of most of the lakes.

Objectives

In the Florida Acid Deposition Study (FADS), a comprehensive literature review and analysis were made during Phase I to determine those receptors (i.e., soils, materials, vegetation, and surface waters) that are potentially affected by acidic deposition and to identify areas for further study. Interest with respect to evaluating the potential effects of acidic deposition on soils and aquatic ecosystems has focused on sensitivity indices as predictive tools; however, an evaluation conducted during Phase I of several indices to determine the sensitivity of Florida soils indicates that the tools are not very well defined and that there is some variance between the results (1). For example, McFee (2) has presented the most finely delineated regional indicator of sensitivity to date; his model, however, is based on cation exchange capacity (CEC) and does not implement all the available data for Florida soils for pH and base saturation. Conversely, Klopatek et al. (3) have presented a conceptually more elaborate and complete model to predict sensitivity; however, the degree of resolution is rather poor and tends to obscure potentially sensitive areas that will manifest themselves only at higher levels of detail.

Similar problems exist with delineating lakes and lake districts sensitive to acidic deposition. Although considerable information exists on the chemical characteristics of a variety of Florida lakes, the information has not been consolidated and evaluated to determine the spatial distribution of Florida lakes of similar buffering capacity. Furthermore, the relationship between watershed soils characteristics and lake chemistry has not been adequately investigated in Florida. This causal linkage between soils and receptor aquatic systems has recently been the subject of considerable interest (4, 5, 6, 7). However, studies specific to Florida have been restricted to date to the Lake McCloud watershed (8, 9).

Several objectives thus describe the intent of Phase II of the FADS. The initial objective was to establish the geographical distribution of lakes and lake districts with respect to buffering capacity. A parallel objective was to construct groupings of soils with equivalent buffering characteristics using a more refined index, incorporating base content, to supersede the spatially well-defined but conceptually incomplete model of McFee (2). A final objective was to evaluate the influence of soil characteristics on the buffering capacity of receptor aquatic systems.

Methodology

A data base comprising 465 samples obtained by Calhoun et al. (10) and Carlisle et al. (11, 12) was used to evaluate the spatial distribution and variability of soil buffering capacity to acidic deposition. This data base includes samples from 26 of Florida's 67 counties. Of these counties, 17 correspond to counties which have recently published soil surveys mapping the distribution of major soil associations. The data base encompasses all of the major soil orders found in Florida: entisols, inceptisols, mollisols, spodosols, alfisols, ultisols, and histisols. The most widely distributed orders are entisols, spodosols, ultisols, and histosols (Figure 1). General characteristics of these soils have been summarized previously (1, 13) and are discussed in detail by SCS (14).

Soil sensitivity generally has been assessed by calculating soil buffering capacity and assigning somewhat arbitrary threshold levels to define varying degrees of susceptibility (2). In this study, soil study, soil buffering capacity was calculated by integrating CEC, base saturation, and bulk density over the surficial 25 centimeters (cm) of the soil column. Perhaps even more important than the nutrient status of the soil column in defining sensitivity is the rate at which nutrients are recycled between the soil and the vegetation it supports. In many terrestrial systems, nutrient recycling rates control primary production; consequently, if acidic deposition results in a concomitant reduction in nutrient recycling, the system may be considered sensitive. Other variables, including weathering rates and sulfate adsorption, may be critical in defining the ability of a particular soil to assimilate acidic deposition; however, the lack of data currently precludes incorporation of these variables into an index and further illustrates the need for additional study in quantifying soil dose/response relationships to define sensitivity.

In order to map areas of the state containing lakes potentially sensitive to acid deposition, a computerized data set was compiled that includes physical and chemical parameters for Florida lakes. Lake data were obtained from recently published reports which included the 165 lake data base collected by Canfield (15). Additional data sources included the data base of softwater lakes developed by Brezonik et al. (16) during their investigation of the chemical aspects of Florida rainfall as well as data from the limnological survey conducted by Milleson (17) of the seven lakes in the Lake Istokpoga drainage basin.

Because a number of lakes were included in more than one of the above surveys, the resulting data base totaled 179 lakes in 46 of the 67 counties. Statistical and data reduction calculations were performed using programs included in the Statistical Package for the Social Sciences (SPSS) (18).

The water quality chemical parameters generally used in assessing potential lake sensitivity to acidic precipitation are pH, alkalinity, and calcium (Ca^{+2}). The EPA Acidification Chemistry Information Database (ACID) (19) has used an alkalinity level of $\leq$ 5 mg/L as $CaCO_3$ (100 microequivalents per liter (100 ueq/L)) as the criterion for identifying lakes potentially sensitive to acidic deposition. It should be noted, however, that this is simply a tool for identifying softwater lakes and does not imply an acidification rate or even if acidification will occur.

The calcite saturation index (CSI) also has been used as a sensitivity index (20, 21, 22). The CSI is a multivariate parameter incorporating pH, alkalinity, and Ca^{+2} into a single index that relates these parameters to the thermodynamic equilibrium solubility condition for calcite ($CaCO_3$). CSI is calculated by:

$$p\ (Ca^{+2}) + p\ (ALK) - pH + 2 = CSI$$

where p (x) = -LOG (MOLES X/L).

If a water is saturated relative to $CACO_3$, the CSI = 0; if supersaturated, the CSI < 0; and if undersaturated, the CSI > 0. Thus, CSI can be used as a tool to characterize lakes or lake districts which contain carbonaceous (e.g., limestone) deposits in their watershed based on three water quality chemical parameters. The implication is that lakes containing such deposits have virtually an infinite capacity to buffer acidic inputs, while lakes that do not contain carbonate minerals potentially may be sensitive to acid precipitation or at least more so relative to carbonaceous lakes. As with the alkalinity index, the accuracy with which lake sensitivity (or nonsensitivity) can be predicted by CSI is, of course, unknown because many other physical, chemical, and hydrological factors can affect the response of a lake to acid loading. These factors include, but are not limited to, sediment buffering capacity, trophic conditions, thermal stratification regime, and lake water residence times. Alkalinity and CSI are useful, however, to the extent that they can identify lakes containing low levels of the principal acid-mediating substances, and, therefore, these indices can be used as screening tools to select lakes for more detailed analysis of the other physical, chemical, and hydrological factors described above.

Results and Discussion

Soil Sensitivity

Average buffering capacity

Averaging buffering capacity of Florida soils (Figure 1) as a function of soil order, suborder, and great group is presented in Table 1. Of the four major orders that constitute Florida soils (entisols, ultisols, histosols, and spodosols), the greatest average buffering capacity was observed in the histosols. These soils, which are organic in nature, have extremely high buffering capacities by virtue of the large CEC afforded by weak acid functional groups (e.g., carboxylic acid groups). Buffering capacities of the other principal orders are substantially lower, with ultisols the most weakly buffered order. Buffering capacity for ultisols (Figure 1) averaged 54 kiloequivalents per hectare (keq/ha). Comparison with an annual deposition rate of 279 to 702 eq/ha acidity currently observed in Florida (based on 140 cm rainfall annually and rainfall volume-weighted pH ranging from 4.3 to 4.7 at all 14 sites of the Florida Acid Deposition Study monitoring network during Phase II; see ref. 23) infers sufficient soil buffer capacity to neutralize 77 to 193 years' deposition. This analysis is presented to offer some perspective towards the significance of soil buffering capacity relative to annual deposition rates of rainfall acidity. It is assumed that each milliequivalent of acidity percolating through the surficial 25 cm of the soil column readily exchanges (on a one-to-one basis) with base cations; in reality, the replacement efficiency of protons for base cations decreases as the base content is depleted (24). Furthermore, this analysis assumes that cation input is H^+ only and that no countering forces (including weathering and nutrient recycling) are present to influence soil acidification. The likelihood that all these assumptions hold is not realistic, and the reader is cautioned that this analysis is a qualitative oversimplification of a process not well understood.

The ultisol suborders represented by the most abundant number of samples are grossarenic paleudults and typic paleudults, with average buffering capacities of 46 and 57 keq/ha, respectively. Base content in these soils is typically low.

The available data indicate that entisols are somewhat more highly buffered than either spodosols or ultisols. However, some of the most poorly buffered soils in Florida are classified as entisols; for example, typic quartzipsamment soils averaged only 43 keq/ha buffering capacity. These soils constitute the most frequently sampled entisolic subgroups and are characterized as quartz sands with virtually no weatherable minerals. Base saturation and CEC are both low, accounting for the observed buffering capacity. Conversely, aquentic entisols have substantially higher buffering capacities, averaging 686 keq/ha, with average values as high as 2,930 keq/ha observed for the hydraquentic great group. Hydraquentic soils comprise, for the most part, clayey soils of tidal marshes that are permanently saturated with water. Base saturation and CEC are correspondingly high.

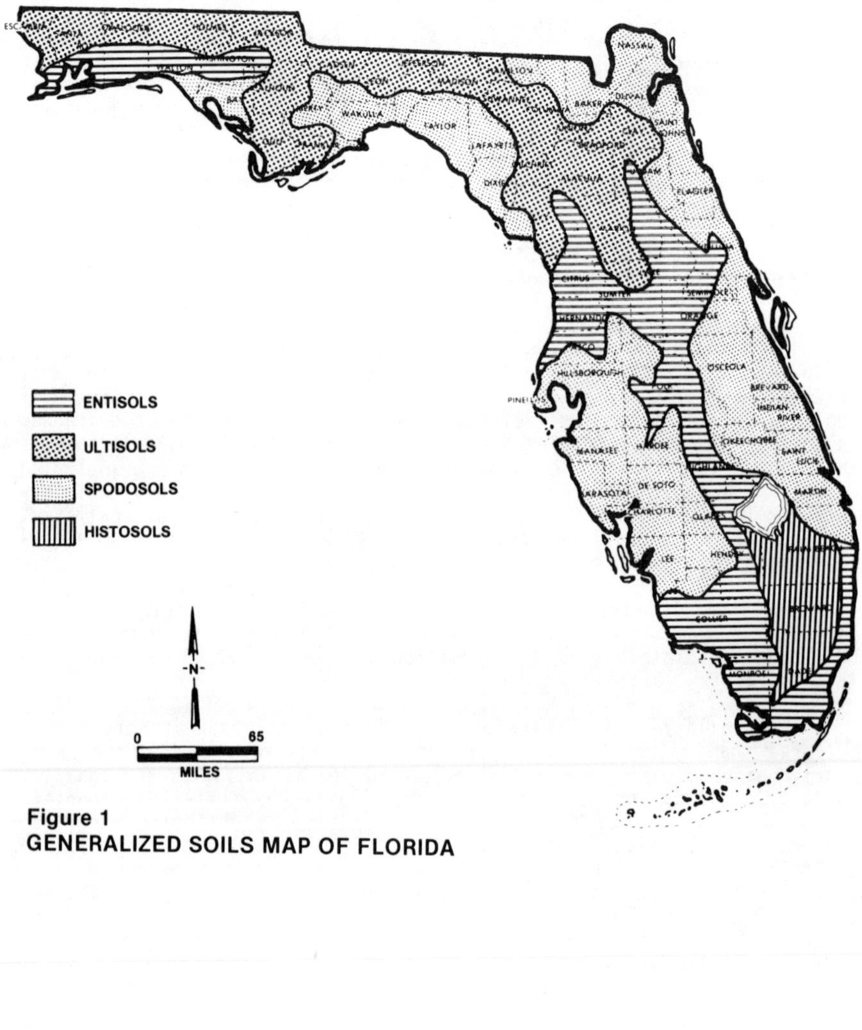

Figure 1
GENERALIZED SOILS MAP OF FLORIDA

Table 1. Summary of Available Buffering Capacity of Florida Soils (All Values in keg/ha)

Order	Average Buffering Capacity (keq/ha)	Suborder	Average Buffering Capacity (keq/ha)	Great Group	Average Buffering Capacity (keq/ha)
Entisol	203	Psamment	63	Quartzipsamment	52
				Udipsamment	240
		Aquent	686	Psammaquent	307
				Hydraquent	2,930
				Haplaquent	2,790
				Sulfaquent	2,510
Ultisol	54	Udult	54	Paleudult	55
				Hapludult	34
		Aquult	54	Ochraquult	60
				Paleaquult	52
Spodosol	68	Aquod	78	Halaquod	78
		Humod	25	Haplohumod	25
Alfisol	206	Aqualf	228	Albaqualf	222
				Glossaqualf	323
				Ochraqualf	193
				Umbraqualf	218
		Udalf	107	Hapludalf	130
				Paleudalf	65
Histosol	765	Saprist	765	Medisaprist	540
				Medifibrist	439
				Medihemist	167
				Sulfhemist	2,780
Mollisol	565	Aquoll	565	Argiaquoll	393
				Haplaquoll	757
Inceptisol	134	Aquept	78	Humaquept	78
		Ochrept	294	Dystrochrept	186
				Eutrochrept	510
		Umbrept	151	Haplumbrept	151

Spodosols, perhaps the most widely distributed soil order in Florida, are represented by two suborders: aquods and humods, with average buffering capacities of 78 and 25 keq/ha, respectively. These soils have received considerable attention regarding the effects of acidic deposition (25, 26, 27).

Inclusion of base saturation to define available soil buffering capacity results in calculated values considerably lower than the results reported by McFee (2). McFee defined several categories of soil susceptibility based on total CEC: nonsensitive soils encompass soils with greater than 500 keq/ha; slightly sensitive soils range between 200 and 500 keq/ha; and sensitive soils constitute all soils less than 200 keq/ha. Only 15 percent of the soils comprised by the data base included in this study had available buffering capacities in excess of 200 keq/ha, which illustrates the importance of base saturation in developing soil sensitivity indices. McFee's (2) results indicate 60.7 percent of the surface area of Florida is dominated by nonsensitive soils, whereas sensitive soils constitute between 14.7 and 29.6 percent. It is important to note that McFee's results are not directly comparable to this study because his results were reported by soil association on an areal basis. For this study, however, soil buffering capacity was calculated as a function of soil subgroup and a statewide map delineating soil types by this taxonomic structure was unavailable. However, as illustrated by the close relationship between lake buffering characteristics and soil buffering capacity, the general spatial relationships in terms of relative buffering capacity developed by McFee still remain valid.

Results from this study indicate that soil sensitivity is extremely difficult to delineate accurately within Florida. The complex, heterogeneous nature of soil distribution within Florida complicates any analysis of spatial distribution and raises the likelihood that relatively small but significant areas of unique buffering characteristics may be obscured during the mapping process. The heterogeneity of buffering characteristics within a particular soil great group or even subgroup contributes to the complexity of delineating the spatial distribution of soil sensitivity. Finally, the broad disparity in numerical results from application of McFee's index for determining soil sensitivity and calculation of available buffering capacity using both base saturation and CEC also illustrates that soil sensitivity is poorly understood and conceptually premature at this stage. Sensitivity presupposes that a dose/response relationship has been established, which would enable a prediction of system response given a certain acidic deposition rate, and that the chemical changes induced by this perturbation yield a known biological response. Additional research is required to elucidate the exchange characteristics of soils as a function of both acidic loadings and inherent exchange capacities.

Lake Sensitivity

Spatial distribution of lake water chemical characteristics

The lakes in the data base display a wide range of chemical characteristics. For example, specific conductance is a gross indicator of total ionic content, and mean conductance values for the lakes ranged over 2.5 orders of magnitude, from 11 to 5,600 microsiemens per centimeter (uS/cm). The lakes also exhibit widely differing limnological conditions. For example, based on nitrogen (N) and phosphorus (P) concentrations (28), the lakes represent vastly differing trophic states, ranging from ultraoligotrophic [100 micrograms of nitrogen per liter (ugN/L); < 5 micrograms of phosphorus per liter (ugP/L)] to hypereutrophic (> 4,000 ugN/L; > 800 ugP/L).

The lakes also display a large variation in values for the water quality chemical parameters typically associated with assessing sensitivity relative to acidic deposition (i.e., pH, alkalinity, and Ca^{+2}). For example, mean pH levels for the lakes ranged from 4.1 to 9.2, total alkalinity ranged from 0 to 204 mg/L as $CaCO_3$ (0 to 4,080 ueq/L), and Ca^{+2} ranged from < 0.5 to 86 mg/L.

Frequency distribution histograms were constructed for the lakes as functions of pH, alkalinity, and CSI (Figure 2). Of the lakes surveyed, 12 percent had average pH levels in excess of 8; 25 and 31 percent ranged between pH 7 to 8 and 6 to 7, respectively; and 32 percent had average pH levels less than 6. On an areal basis, 63 percent of the total lake surface areas represented by the lakes surveyed averaged pH 8 or greater; 17 and 15 percent averaged between pH 7 to 8 and 6 to 7, respectively; and only 5 percent had an average pH less than 6. It is important to note, however, that Lake Okeechobee, with a surface area of 448,000 acres and an average pH of 8.38, constitutes 51 percent of the total surface area of lakes surveyed during Phase II.

The majority of the lakes contain low levels of alkalinity (Figure 2b). An alkalinity level of < 5 mg/L as $CaCO_3$ (100 ueq/L) has been used by EPA's ACID for identifying lakes potentially sensitive to acidification (19). As shown by Figure 2b, roughly half (44 percent) of the lakes would be identified as potentially sensitive by the criterion. Likewise, the distribution of CSI values indicates that many Florida lakes and their watersheds do not contain carbonate minerals. For comparison, it has been hypothesized that lakes with CSI values less than 3 are generally stable relative to acid deposition, while lakes with CSI greater than 4 may be susceptible to acidic inputs (20, 21).

The above analysis of existing lake water data indicates that a number of Florida lakes are acidic and contain relatively low levels of buffering substances. Similarly, previous limnological studies (29, 30, 15) have concluded that Florida lakes can be classified as acidic, softwater, productive systems. Thus, although most Florida lakes are acidic and dilute in their ionic composition, productivity levels are generally greater than temperate lakes of similar chemistry. The productivity of Florida lakes is enhanced for several reasons: a subtropical environment (e.g., light, temperature); naturally occuring phosphatic deposits; and the shallow, well-mixed nature of most of the lakes.

In order to map the distribution of softwater lakes in the state and compare them with soil characteristics, the lakes in the data set were located on a 1:500,000 scale topographic map. Mean alkalinity, pH, and CSI values for each lake were then plotted on the maps and corresponding isopleths (i.e., lines of equivalent values) were drawn. These maps were subsequently combined with McFee's soil sensitivity map of Florda (2). (Contact the authors for details on these renderings.)

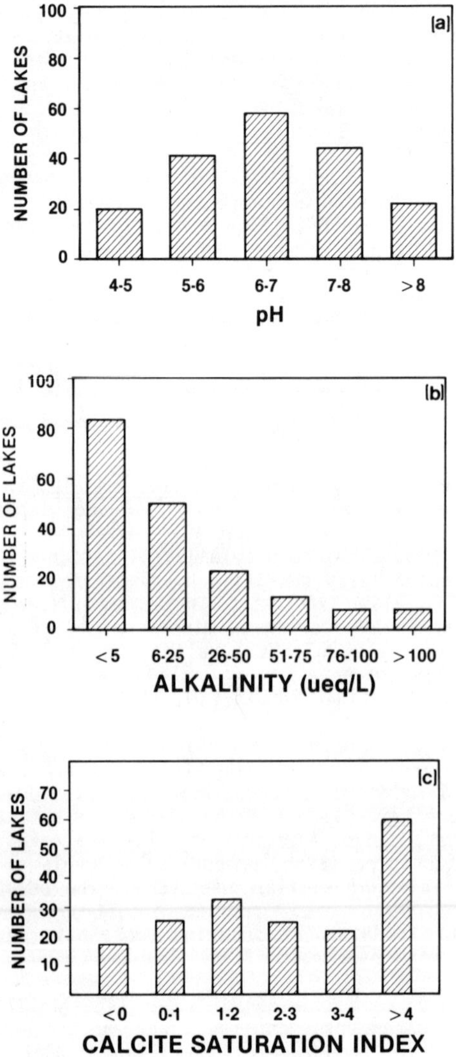

Figure 2
FREQUENCY DISTRIBUTIONS OF FLORIDA LAKES INCLUDED IN THE STUDY (N = 179) AS A FUNCTION OF (a) pH, (b) ALKALINITY, AND (c) CALCITE STATURATION INDEX

The distribution of acidic, soft-water lakes, and alkaline lakes in Florida is not random, but strongly correlates with soil exchange characteristics. A general increase in lake water pH and alkalinity and decrease in CSI are observed in a general north-northwest to south transect of the state. Furthermore, these gradients correspond to a general change in soil buffering capacity as mapped by McFee. Lakes exhibiting low pH and alkalinity are associated with the highlands and ridge physiographic regions of the state. Most prominent are the Western Highlands of the panhandle and the Central Ridge, which includes the Trail Ridge and Lake Wales Ridge lake districts in north and south-central Florida, respectively. Exceptions to this general model do occur, primarily because of cultural influences within the watershed, hydrogeological characteristics (e.g., artesian supply), or a combination of both. Central Florida has the greatest variability in lake pH, alkalinity, and CSI, due in part, to a diverse surficial geology and physiography (15). For example, this region contains several discontinuous sandy ridge formations (e.g., Lake Wales Ridge, Istokpoga Ridge, Mount Dora Ridge) separated by relatively low-lying, flat valleys. Lakes located along the ridges exhibit softwater acidic conditions, while lakes in the valley areas tend to be more alkaline. Additionally, lakes in the valley regions are generally more culturally developed (e.g., the Oklawaha chain), which increases nutrient (N and P) loadings to these lakes resulting in increased productivity and alkalinity.

Lake water chemical relationships

Alkalinity (the sum of the titratable bases with strong acid) is commonly referred to as the acid neutralizing capacity (ANC) of a water. In natural waters, alkalinity is largely a function of the carbonate species, but other dissolved substances such as borate, silicates, ammonia, organic bases, phosphate, and aluminum hydroxide are among the bases that may be titrated in an acidimetric titration of alkalinity. Mathematically, alkalinity is defined (31) by a charge balance involving species that are reactive with protons (H^+).

Carbonate (CO_3^{-2}) and bicarbonate (HCO_3^-) alkalinity are believed to arise in lakes due to weathering of carbonaceous minerals (e.g., $CaCO_3$, $MgCO_3$) in the watersheds. To test this hypothesis for Florida lakes, alkalinity data for the lakes were plotted, as a function of non-marine (i.e., "excess") Ca^{+2} plus Mg^{+2}. As shown by Figure 3, alkalinity is strongly correlated ($r^2 = 0.80$) with the sum of these two base cations. Thus, 80 percent of the variance in lake water alkalinity is attributable to variations in lake water "excess" Ca^{+2} plus Mg^{+2}. This indicates that alkalinity in Florida lakes is derived from Ca^{+2} and Mg^{+2} containing minerals, presumably carbonates. Therefore, lakes containing low levels of these base cations likely are located in drainage basins containing soils exhibiting relatively low levels of base cations. Conversely, lake watersheds consisting of limestones or limestone-derived soils are expected to have relatively alkaline lakes. An analysis of these soil-lake linkages is developed in a subsequent section.

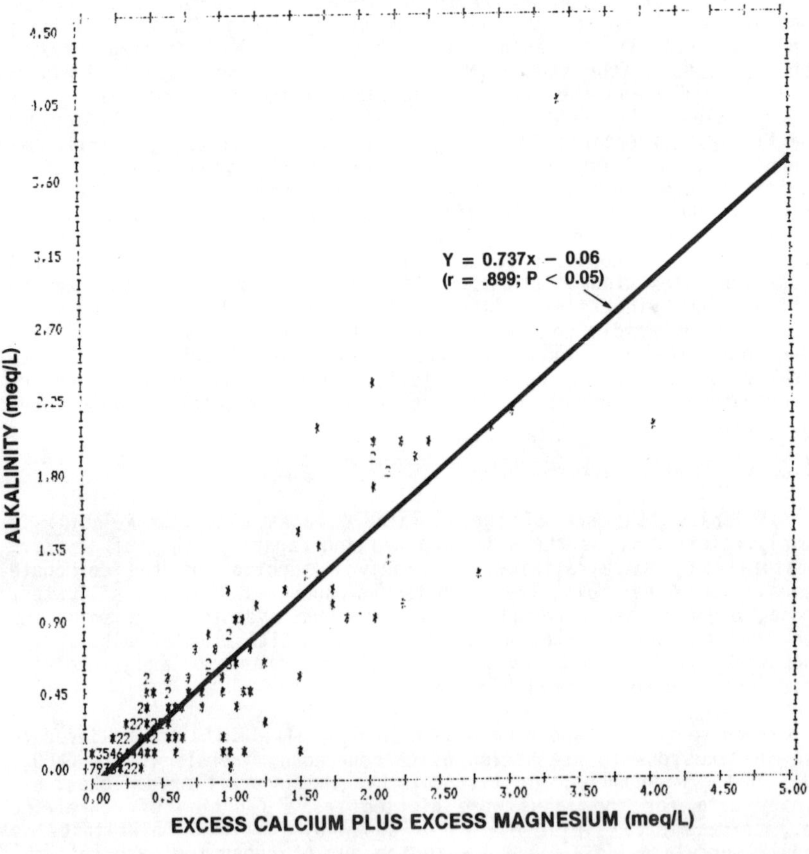

Figure 3
ALKALINITY OF FLORIDA LAKES INCLUDED IN THE STUDY (N = 179) "EXCESS" [CALCIUM PLUS MAGNESIUM] SOLID LINE SHOWS THE 1:1 RELATIONSHIP

Assuming a natural aquatic system is at equilibrium with atmospheric CO_2 ($PCO2 = 10^{-3.5}$ atm) with no additions of strong base or acid, the theoretical relationship between lake water pH and the CO_2- bicarbonate system can be calculated. Factors other than the CO_2- bicarbonate system may, however, control pH in lakes. These can include, for example, organic acids and/or metal (Al, Fe) hydroxides. To gain some insight into the degree with which only the CO_2-bicarbonate system controls the pH of Florida lakes, alkalinity data from the lake data set were plotted as a function of pH. As shown in Figure 4, pH is a strong function of alkalinity. The solid line in Figure 4 shows the theoretical relationship between pH and alkalinity, where the CO_2- bicarbonate system controls pH at atmospheric pressure and 25°C. The data indicate that for roughly 75 percent of the lakes, the pH is near the equilibrium value defined by the CO_2- bicarbonate system. Twenty-five percent of the lakes, however, exhibit pH values less than that expected for pure water in equilibrium with atmospheric CO_2 (pH 5.6). This indicates that the pH of these lakes is controlled by acid-base systems other than the CO_2-bicarbonate system. Based on acidity constants, it is likely that organic and/or metal (e.g., Al, Fe) weak acid buffer systems dominate the acid-base chemistry of these lakes. Unfortunately, a lack of chemical data specific for these compounds precludes an assessment of their importance in controlling the pH of these lakes . Additional work would be needed to assess this hypothesis accurately.

Interaction Between Soil Buffering Capacity and Lake Water Alkalinity

Aquatic ecosystems serve as receptors for the outputs of terrestrial ecosystems. Only a small fraction (5 to 10 percent) of precipitation falls directly, and hence unchanged, onto the surfaces of lakes and rivers (32). Surface runoff and interflow, the other principal components of the hydraulic budget for aquatic systems, are chemically modified as the aqueous phase comes into contact with vegetative surfaces and soils. Aquatic systems are strongly influenced by these diverse fluxes originating outside the lake but within the watershed; consequently, the linkage between these two systems is of considerable interest and importance.

The linkage between terrestrial and aquatic system characteristics was evaluated by quantifying lake alkalinity as a function of soil buffering capacity. Several counties were selected for analysis, including Osceola County, located in central Florida, and Leon and Jackson Counties in the panhandle region.

Osceola County

A linear model was used to test the hypothesis that soil buffering capacity dictates lake alkalinity. Estimates of watershed areal buffering capacities were estimated from plotted distributions developed for each county. Notably, individual drainage basins were not delineated, resulting in a somewhat qualitative estimate of buffering capacity afforded by each watershed; however this type of analysis is useful in

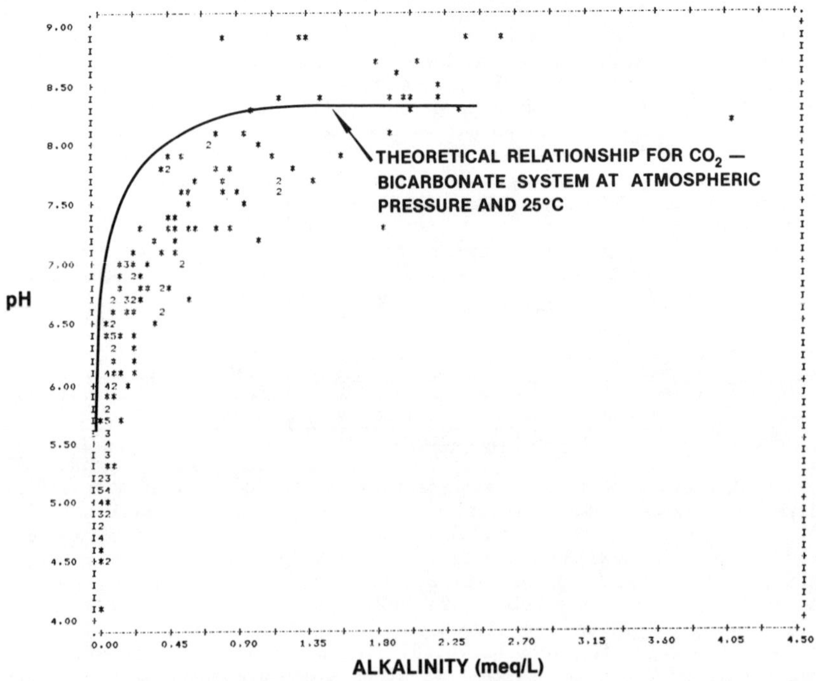

Figure 4
ALKALINITY OF FLORIDA LAKES INCLUDED IN THE STUDY (N = 179) AS A FUNCTION OF pH. SOLID CURVE DEPICTS THE THEORETICAL RELATIONSHIP WHERE pH IS CONTROLLED BY THE CO_2 - BICARBONATE SYSTEM AT ATMOSPHERIC CO_2 PARTIAL PRESSURE AND 25°C

establishing the general relationship between the two systems. Using data from 14 lakes within Osceola County for which a sufficient soil data base exists to estimate watershed buffering capacity, the following model was derived:

$$\alpha = -162 + 5.1\beta \qquad (r = 0.78,\ p < 0.10)$$

where α equals lake alkalinity (ueq/L) and β equals available soil buffering capacity (keq/ha). This model infers that 61 percent of the variability in lake water alkalinity within Osceola County is directly related to soil association distribution. These results conceptually agree with Kaplan et al. (7), who demonstrated the dominant role of soils in mediating water quality in Middle Atlantic and New England lakes. The remaining variability may result from the extent of cultural influence within each lake watershed (e.g., land use and fertilization practice, use of septic tanks, etc.) and the geochemical role of underlying bedrock in mediating the flux of alkalinity through the watershed.

The concept of soil buffering capacity as a major forcing function controlling lake water alakalinity in Florida is particularly striking in the case of East Lake Tohopekaliga and Lake Tohopekaliga. Although the two lakes are interconnected by a small channel only 5.8 km in length draining East Lake Tohopekaliga, alkalinites in the two lakes are quite different; an average alkalinity of 540 ueq/L (27 mg/L as $CaCO_3$) is observed for Lake Tohopekaliga contrasted with an average alkalinity of 140 ueq/L (7 mg/L as $CaCO_3$) in East Lake Tohopekaliga. The soils of the East Lake Tohopekaliga watershed are very poorly buffered and approximate only 32 keq/ha. These soils, which comprise primarily quartzipsamments and haplohumods are characterized in the region by low CEC and low to moderate base saturation. As a result, virtually all of the lakes within this region for which reported concentrations are available have alkalinities of the order of 100 ueq/L or less. Conversely, haplaquods generally dominate the watershed of Lake Tohopekaliga. CEC is somewhat higher, resulting in concomitant increases in lake alkalinity as evidenced, for example, by Lakes Hatchinea, Kissimmee, and Marian.

Leon and Jackson Counties

The limited number of lakes and the tightly clustered nature of the data preclude the development of a model (through regression analyses) to describe the causality of soils in controlling lake alkalinity. In general, lakes within these two counties have had very low alkalinities which, with the notable exceptions of Merritt's Mill Pond in Jackson County and Lake Munson in Leon County, do not exceed 80 ueq/L. Correspondingly, soils within these two counties also tend to be depleted in exchangeable base cations; for example, available buffering capacity in the sand ridge region within Leon and Jackson Counties averaged only 10 and 33 keq/ha, respectively.

Merritt's Mill Pond and Lake Munson are both characterized by alkalinities considerably in excess of _a priori_ expectations based on the predominant soils. These apparent deviations are readily explained by evaluating the hydrologic (and hence geochemical) sources to each lake. Lake Munson is an alkaline lake (alkalinity equals 1,560 ueq/L) which receives treated domestic wastewater from the City of Tallahassee. The wastewater is orginally potable water derived from the limestone of the Floridan Aquifer; consequently, these waters are highly alkaline. As a result, Canfield (15) concluded that the chemical composition of Lake Munson was not representative of other lakes within the same region. Merritt's Mill Pond, which may also be classified as alkaline with an alkalinity of 1,980 ueq/L (99 mg/L as $CaCO_3$), is spring-fed and thus receives inputs of water in equilibrium with the limestone of the underlying aquifer (15). Hydraulic connection between the underlying aquifer and other lakes within the same region is generally prevented by a clay layer and/or extensive sand deposits; as a result, these lakes tend to reflect to a stronger degree the buffering characteristics of the soils and are more acidic and softer in nature.

Summary and Research Needs in Florida

Insofar as the data allowed, the geographical distributions of both soil buffering capacity and lake buffering characteristics were delineated; in addition, the inherent influence of soil characteristics on the buffering capacity of the receptor aquatic system was investigated. Results of this study show that even soils within the same taxonomic subgroup are quite heterogeneous with respect to buffering capacity. Consequently, delineation of the areal distribution of available buffering capacity (and implicitly sensitivity) in Florida soils currently is difficult to assess.

In addition to CEC, other factors, including nutrient recycling and weathering, influence the susceptibility of Florida soils to acidic deposition. Additional research is required to elucidate the exchange characteristics of soils as a function of both acidic loadings and inherent exchange capacities.

Results of the study show that the majority of lakes surveyed within Florida may be classified as acidic (i.e., pH < 7), softwater, and, unlike temperate lakes of similar chemistry, productive systems. Several physical and chemical factors contribute to enhanced levels of primary production in Florida lakes, including: a subtropical environment; naturally occurring phosphatic deposits; and the shallow, well-mixed nature of most of the lakes. Lake water pH correlated strongly with alkalinity, indicating that the pH of the majority of Florida lakes is controlled by the CO_2- bicarbonate system. For some relatively acidic lakes this relationship does not apply, and other acid-base systems are hypothesized as controlling lake pH. Further studies are required to elucidate potential mechanisms which include Al-Fe and/or organic acid systems.

Lake water alkalinity and the sum of excess (non-marine) calcium and magnesium were strongly correlated for Florida lakes, implying that surficial geological characteristics (or soil buffering capacity) within a particular watershed controls lake water alkalinity and, hence, pH. However, several notable exceptions to this relationship occur. Research needs to further refine the chemical relationships between aquatic and terrestrial systems.

Acknowledgements

Assistance of K. Ingram, H. Prentice, and T. Gillespie in data reduction, V. Swasey for typing, and K. Kinsley-Momberger for illustrations is appreciated. Work reported in this paper was supported by the Florida Electric Coordinating Group (FCG) under ESE Contract No. 81631600.

Literature Cited

1. Environmental Science and Engineering, Inc., 1982. Florida Acid Deposition Study: A Literature Review of the Ecological and Material Effects of Acid Deposition, Gainesville, Florida, January 1982.

2. McFee, W., 1980. Sensitivity of Soil Regions to Acid Precipitation, U.S. Environmental Protection Agency, EPA-600/3-80-013, Corvallis, Oregon.

3. Klopatek, J.M., Harris, W.F., and Olson, R.J., 1980. A Regional Ecological Assessment Approach to Atmospheric Deposition: Effects on Soil Systems, In: Atmospheric Sulfur Deposition: Environmental Impact and Health Effects, D.S. Shriner et al., (Eds.), Ann Arbor Science Publishers, Inc., Ann Arbor, Michigan.

4. Gorham, E., and McFee, W., 1980. Effects of Acid Deposition Upon Outputs from Terrestrial to Aquatic Ecosystems, In: Effects of Acid Precipitation on Terrestrial Ecosystems, T.C. Hutchinson and M. Havas (Eds.), pp. 465-480, Proceedings of the NATO Conference on Effects of Acid Precipitation on Vegetation and Soils, Toronto, Ontario, Canada, May 21-27, 1978, Plenum Press, New York.

5. Galloway, J.N., and Bache, B.W., 1980. I. Effect of Acid Precipitation on Water Quality and on Effects via Soil into Water Bodies, In: Effects of Acid Precipitation on Terrestrial Ecosystems, T.C. Hutchinson and M. Havas (Eds.), pp. 609-610, Proceedings of the NATO Conference on Effects of Acid Precipitation on Vegetation and Soils, Toronto, Ontario, Canada, May 21-27, 1978, Plenum Press, New York.

6. Electric Power Research Institute, 1981. The Integrated Lake-Watershed Acidification Study (ILWAS)--Contributions to the International Conference on the Ecological Impact of Acid Precipitation, EA-1825, Research Project 1109-5, Palo Alto, California.

7. Kaplan, E., Thode, H.C., and Portas, A., 1981. Rocks, Soils and Water Quality Relationships and Implications for Effects of Acid Precipitation on Surface Water in the Northeastern United States, Environ. Sci. and Technology 15(5):5390.

8. Brezonik, P.L., Hendry, C.D., Jr., Edgerton, E.S., Schulze, R.L., and Crisman, T. L., 1982. Acidity, Nutrients, and Minerals in Atmospheric Precipitation Over Florida: Deposition Patterns, Mechanisms, and Ecological Effects, U.S. Environmental Protection Agency, Office of Research and Development, Corvallis Environmental Research Laboratory, Corvallis, Oregon.

9. Volk, B.G., Bitton, G., Byers, G.E., and Graetz, D.A., 1982. Effects of Acid Precipitation on Selected Soils of the Southeastern United States, Presented at the EPA Acidic Deposition Effects Program, Raleigh, North Carolina, February 9-12, 1982.

10. Calhoun, F.G., Carlisle, V.W., Caldwell, R.E., Zalazny, L.W., Hammond, L.C., and Breland, H.L., 1974. Characterization Data for Selected Florida Soils, University of Florida, Institute of Food and Agricultural Sciences, Gainesville, Florida.

11. Carlisle, V.W., Caldwell, R.E., Sodek, F., Hammond, L.C., Calhoun, F. G., Granger, M. A., and Breland, H. L., 1978. Characterization Data for Selected Florida Soils, University of Florida, Gainesville, Florida, Soil Science Research Report Number 78-1.

12. Carlisle, V.W., Hallmark, C.T., Sodek, F., Caldwell, R.E., Hammond, L.C., and Berkheiser, V. E., 1981. Characterization Data for Selected Florida Soils, University of Florida, Gainesville, Florida, Soil Science Research Report Number 81-1.

13. Brady, N.C., 1974. The Nature and Properties of Soils, 8th Edition, Macmillan, New York.

14. U. S. Soil Conservation Service, 1975. Soil Taxonomy, Washington, D.C.

15. Canfield, D.E., 1981. Chemical and Trophic State Characteristics of Florida Lakes: Relation to Regional Geology ,Institute of Food and Agricultural Sciences, University of Florida, Gainesville, Florida.

16. Brezonik, P.L., Baker, L.A., Ogburn, R.W., Edgerton, E.S., and Crisman, T.L., 1981. Ecological Effects of Acid Precipitation on Sensitive Softwater Lakes in Florida, EPA-NCSU Project No. APP-0007-02-1980, University of Florida, Gainesville, Florida.

17. Milleson, J.E., 1978. Limnological Investigations of Seven Lakes in the Istokpoga Drainage Basin, South Florida Water Management District Technical Publication Number 78-1, West Palm Beach, Florida.

18. Nie, N.H., Hull, C.H., Jenkins, J.G., Steinbrenner, K., and Bent, D.H., 1975. Statistical Package for the Social Sciences, Second Edition, McGraw-Hill Book Company, New York.

19. Gmur, R., 1981. General Information on the Acidification Chemistry Information Database, Brookhaven National Laboratory, New York.

20. Conroy, N., Jefferies, D.S., and Kramer, J.R., 1974. Acid Shield Lakes in the Sudbury, Ontario Region, Proc. Symp. Wat. Poll. Res. in Canada 9:45-61.

21. Kramer, J.R., 1976. Geochemical and Lithological Factors in Acid Precipitation, In: Proceedings of the First International Symposium on Acid Precipitation and the Forest Ecosystem, L.S. Dochinger and T. A. Seliga (Eds.), USDA Forest Service, Columbus, Ohio.

22. Galloway, J., Cowling, E., Gorham, E., and McFee, W., 1978. A National Program for Assessing the Problem of Atmospheric Deposition (Acid Rain), National Atmospheric Deposition Program Report to the Council on Environmental Quality, Fort Collins, Colorado.

23. Environmental Science and Engineering, Inc., 1983. Florida Acid Deposition Study: Phase II Report, Gainesville, Florida.

24. Wiklander, L.,1980. The Sensitivity of Soils to Acid Precipitation, In: Effects of Acid Precipitation on Terrestrial Ecosystems, T.C. Hutchinson and M. Havas (Eds.), pp. 553-567, Proceedings of the NATO Conference on Effects of Acid Precipitation on Vegetation and Soils, Toronto, Ontario, Canada, May 21-27, 1978, Plenum Press, New York.

25. Abrahamson, G., Bjor, K., Hornvedt, R., and Tveite, B., 1976. Effects of Acid Precipitation on Coniferous Forests, In: Impact of Acid Precipitation on Forest and Freshwater Ecosystems in Norway, F. H. Brakekke (Ed.), SNSF Project, RF 6/76.

26. Peterson, L., 1978. Podzolization: Mechanism and Possible Effects of Acid Precipitation, In: Effects of Acid Precipitation on Terrestrial Ecosystems, T.C. Hutchinson and M. Havas (Eds.), pp. 223-237, Proceedings of the NATO Conference on Effects of Acid Precipitation on Vegetation and Soils, Toronto, Ontario, Canada, May 21-27, 1978, Plenum Press, New York.

27. Roberts, T.M., Clarke, T.A., Ineson, P., and Gray, T.R., 1978. Effects of Sulphur Deposition on Litter Decomposition and Nutrient Leaching in Coniferous Forest Soils, In: Effects of Acid Precipitation on Terrestrial Ecosystems, T.C. Hutchinson and M. Havas (Eds.), pp. 381-393, Proceedings of the NATO Conference on Effects of Acid Precipitation on Vegetation and Soils, Toronto, Ontario, Canada, May 21-27, 1978, Plenum Press, New York.

28. Vollenweider, R.A., 1968. Scientific Fundamentals of the Eutrophication of Lakes and Flowing Waters, with Particular Reference to Nitrogen and Phosphorus as Factors in Eutrophication, Technical Report DAS/CSI/68.27, Organization for Economic Cooperation and Development (OECD), Paris.

29. Shannon, E.E., and Brezonik, P.L., 1972. Limnological Characteristics of North and Central Florida Lakes, Limnology and Oceanography 17:97-110.

30. Bakers, L.A., Brezonik, P.L., and Kratzer, C.R., 1981. Nutrient Loading-Trophic State Relationships in Florida Lakes, Water Resources Research Center Publication Number 56, University of Florida, Gainesville, Florida.

31. Stumm, W., and Morgan, J.J., 1981. Aquatic Chemistry, 2nd Edition, Wiley-Interscience, New York.

32. Seip, H.M., 1980. Acidification of Freshwater Sources and Mechanisms, In: Ecological Impact of Acid Precipitation, D. Drablos and A. Tollan (Eds.), pp. 242-243, SNSF Project, Sandefjord, Norway.

ATMOSPHERIC ACID DEPOSITION AND FRUIT PRODUCTION

W. J. Kender and C. A. Anderson
University of Florida, IFAS
Agricultural Research and Education Center
Lake Alfred, FL 33850

R. H. Biggs and P. G. Webb
University of Florida, IFAS
Fruit Crops Department
Gainesville, FL 32611

Gaseous Air Pollutants

From 1972 to 1982, a major research effort was placed on determining the incidence and control of air pollutants on temperate fruit crops in New York State (17). Initial studies demonstrated that ozone was the cause of an unknown disorder of grapevines later designated as "oxidant stipple" (28). Grape varieties were shown to have a wide range of susceptibility to ozone at ambient concentrations (16). Chronic symptoms of ozone injury were described under field conditions (16, 18, 28) with soluble solids of berries reduced as a result of leaf area destruction by oxidant stipple (18, 25). Under ambient conditions, damage to grapevines was shown to result in as high as a 20% reduction in yield (18, 19, 25). Vines in field plots protected from ambient air pollutants with carbon filters showed marked reduction in symptom development (25). The amount of summer rainfall and ambient ozone levels influenced the severity of oxidant stipple (25).

Methods to control air pollution injury have, in general, not been effective, although certain soil and physiological factors which influence plant responses to air pollutants afford flexibility in controlling such damage (15). With Concord grapes, ambient levels of ozone caused significantly more oxidant stipple at low N leaf levels (18). Total N concentration of tissue was negatively correlated to oxidant stipple severity. Optimum N fertilization, use of own-rooted vines and flower cluster thinning were found to have an ameliorating effect on oxidant stipple (18). Also, foliar applications of benomyl fungicide was shown to protect grape leaves against oxidant stipple (19).

Sulfur dioxide was also shown to influence grapevines (26, 29) and apple trees (30). Continuous SO_2 fumigations at 0.10 and 0.05 ppm reduced growth and yield and delayed maturity of Concord grapes (26). Sulfur dioxide in combination with O_3-containing air increased oxidant stipple (26) and resulted in reduced yields and soluble solids. Concern has been expressed over the effects of SO_2 from coal-fired industrial plants on crops already injured by ozone. Fujiwara (10, 11) found that the number of clusters was markedly reduced after grapevines were exposed to continuous fumigations of SO_2 at 0.065, 0.13, or 0.26 ppm. In some cases, the reduction in cluster number was not directly related to the severity of foliar symptoms. A large varietal difference in injury and damage was reported (10, 11).

Acid Deposition

That injury to fruit crops can be caused by SO_2 and its interaction with O_3 suggests potential implications of acid deposition on vegetation (20). The effect of acid deposition alone or in combination with ozone on fruit crops is not clear. One of the processes critical to the production of fruits, is the reproductive cycle which is initiated the year prior to harvest in the mature tree. Therefore, air pollutants may influence the fruiting process at the time of flower initiation in year one or following fertilization, fruit set, fruit development, and maturation stages in year two. The bloom period is of brief duration in the spring but could be very vulnerable to external influences such as acid rain.

Mature grapevines in the vineyard were sprayed with simulated acidic rain solutions ranging from pH 2.5 to pH 5.5 both as acute treatments at anthesis and chronically throughout the growing season (7). Few foliar lesions on leaves were visible at pH 2.75. In contrast, many leaf lesions as well as decreased fruit soluble solids were observed at pH 2.5. Solutions of pH 2.75 at anthesis reduced pollen germination in four of the nine varieties tested. However, fruit set was reduced in only one of these. Two varieties had reduced berry soluble solids with chronic weekly sprays at pH 2.75. There was no evidence that acid rain in combination with ambient ozone increased oxidant stipple as occurs when ozone and SO_2 are combined. Grape yields were not influenced by acid rain treatments. There was no evidence that acid rain at ambient pH levels (pH 3.3 to pH 5.8) had negative effects on grape production or fruit quality.

In a study of the effects of acid rain on apple tree productivity and fruit quality, Forsline et al. (8) sprayed mature trees with simulated acid rain solutions in the pH range of 2.5 and 5.5 at full bloom. Necrotic lesions developed on apple petals at pH 2.5 with slight injury appearing at pH 3.0 and pH 3.5. Apple foliage had no acid rain lesions at any of the pH levels tested. Pollen germination was reduced at pH 2.5 in the variety 'Empire'. Slight fruit set reduction at pH 2.5 was observed in 'McIntosh'. The incidence of russetting on 'Golden Delicious' fruits was ameliorated by the presence of rain-exclusion chambers but was not affected by acid rain. Even at the lowest pH levels no detrimental effects of simulated acid rain were found on apple tree productivity and fruit quality when measured as fruit set, seed number per fruit, and fruit size and appearance.

Marketability of fruits may also be impaired by acid rain, as evidenced by russetting of 'Golden Delicious' apple fruits (4). However, it is rainfall alone, whether acidic or not, that induces russetting (3). Blanpied (2) examined acid rain (pH 3.3 and pH 4.2) as a possible causal agent in the decline of calcium in apple leaf samples analyzed over several years in New York. Although simulated acid rain leached calcium from apple leaves, the leaves recovered the lost calcium within one week after treatment stopped.

In a greenhouse study (9), apomictic apple seedlings (*Malus hupenhensis*) were treated weekly with simulated acid rain solutions ranging from pH 2.25 to pH 7.0. Necrotic lesions developed on leaves at pH 2.25 and pH 2.50 immediately after the first application at the eight node stage. Following the ninth weekly application on seedlings with 23 to 26 nodes, lesions developed at pH levels up to 3.25. At final destructive harvest, 20% of the leaf area at pH 2.25 and 8% of the leaf area at pH 2.50 were injured. Significant growth reduction occurred at these two pH levels. Regression analysis indicated extensive growth inhibition at < pH 3.0, no growth inhibition at pH 3.5, some inhibition between pH 4.5 and pH 5.6, and normal growth at pH 7.0 in comparison to the unsprayed control. Growth was negatively correlated with lesion formation. Relative growth rates were reduced only at pH 2.25 and pH 2.50 as were unit leaf rates. Lesion development continued to increase on the basal leaves through the sixth weekly application but leveled off during the final applications. Negative correlations of photosynthesis rate to lesion percentage and dry weight to lesion percentage were observed.

During the period 1971 to 1981, average acidity of rain has increased from an average of pH 5 to pH 4 in many areas of northern Europe and eastern United States. This change has been attributed to increased combustion of fossil fuels, resulting in wide dispersal of sulfur and nitrogen oxides which are converted to corresponding strong acids in the atmosphere (27).

Conclusions from Research in the Northeast

1. Even with the possibility of ambient rains as low as pH 3.0, no conclusive evidence of a negative effect on the productivity of apple trees and grapevines was found when treated with these levels of acid rain at the flowering stage or later.

2. At pH 2.5 and pH 2.75 reduction in sugars in grapes was the only economically significant effect measured.

3. Threshold levels of acid rain appear to be variety specific.

4. In the vegetative stage, apple seedlings have a threshold of growth reduction between pH 2.5 and pH 2.75, which is closely related to leaf injury.

5. Correlations between acid rain effects on deciduous fruit crops in the Northeast and citrus in Florida are not clear but studies on this important Florida crop are warranted as a precautionary measure.

Air Pollution Research Needs on Agricultural Systems in Florida

Effects studies should be undertaken on agricultural systems and related to data collected from monitoring stations in all major agricultural areas of Florida.

Potential toxic effects and threshold levels should be determined for plant growth, appearance, and yield, as well as other components of productivity such as seed viability and quality, for all common crops in Florida. Visual symptoms of phytotoxicity should be characterized for these crops. Toxic doses should be quantified to determine chronic vs. acute exposures (low levels of prolonged duration vs. brief exposure to high levels). Research should also be conducted to determine stages of sensitivity in the life cycle of each plant as well as the season of highest sensitivity.

Evaluations of differences in crop and variety tolerance are necessary, with emphasis on research to develop improved varieties through plant breeding.

Research should also be conducted to determine possible antagonistic/synergistic interactions with individual farming practices and to devise remedial measures to reduce losses. Finally, the economic impacts of potential losses due to air pollution injury should be identified as well as to evaluate any cost-savings to farmers by atmospheric contributions of N and S.

Studies Underway and Those Needed for a Physiological Disorder Analysis

Limited research has been directed at determining to what extent atmospheric acid deposition will influence tree cropping systems, particularly citrus production. No definitive data are available on which to predict the degree of atmospheric deposition that would influence citrus tree productivity or fruit quality. However, physiological responses of citrus trees to ozone, SO_2 and SO_3, rain and soil pH, toxic minerals, etc. should be related to responses of other plants (5, 6, 12, 13, 14, 21, 22, 23, 24) either more sensitive or less sensitive. Knowing these responses will allow the construction of physiological responses in a hierarchical manner that may influence citrus productivity, fruit quality, and tree life. Secondly, and allied to this, can strategies be found to offset any deleterious effects should they occur?

Form and transport of gases, liquids and solids will determine the type of plant response elicited. Some of the main natural and pollutant gases, liquids, and solids that interface with plant chemical systems are listed in Table 1. Interactive components on both roots and shoot responses must be considered. For example, lowering the pH of the soil by acidic components may change Al^{+++} from an insoluble or bound form to a soluble species. Copper toxicity could occur in older citrus groves where it is applied in a disease control program (1).

Sensitivity of Tissue. From the standpoint of physiological responses, a chart of sensitive tissues versus resistant tissues would be similar to Table 2. As the cell starts to divide, it becomes more sensitive to changes. Concurrently, the basic structural units of the cell are also more sensitive. Very critical is the consideration of the interactive interface between plants and gases and liquids. This interaction indicates that the primary root and shoot tissues with their resultant ontogenic development should be influenced first, i.e., leaves, root, and shoot primary tissues.

Table 1. Natural and Expected Pollutants that Interface with Plants

ATMOSPHERE:*	Gases:	
	Primary:	N_2, CO_2, O_2, SO_x, NO_x, H_2O, CL_2, others
	Secondary:	H_2O_2, PAN, NH_x, S_xH, others
	Particulates and Aerosols:	
	H_2O, H_2SO_4, HNO_3, dusts, bacteria, pollen	
SOIL:	Gases:	
	CO_2, SO_2, O_2, H_2O, NH_3, RNO_x, RN_xH_x, C_xH_x	
	Minerals:	
	Soluble:	solvent (H_2O) essential beneficial toxic non-essential, beneficial or toxic
	Non-available:	insoluble non-toxic toxic
	Organisms:	
	Parasitic, Pathogenic, Pests, Symbiotics Non-interactive	

*The Equilibrium between gases, liquids, particulates and aerosols is very much dependent on radiant energy and the spectrum of that energy in the troposphere.

Table 2. Sensitivity Analysis of Interface between Plants and Environmental Factors that May Lead to a Response

SENSITIVITY ANALYSIS:		
Insensitive →→→→→→→→→	Moderate →→→→→→→→→→→→	Sensitive
Cellular		
Cells w/sec. walls	Resting cells	Dividing/expanding cells
Structures		
Secondary plant structures	------------	Primary plant structures (i.e. meristems, flowers)

Measurable resultants (Research needs). Based on the reaction of other tree crops to atmospheric deposition, a brief list of parameters of tree growth and development that may be influenced by atmospheric deposition is found in Table 3. Also listed are the estimated times it will take to parameterize the plant responses and determine if atmospheric deposition has an influence on citrus tree production and fruit quality.

These areas need to be investigated so that we may better understand the long-range effect(s) of acid rain deposition on one of our major tree crops of Florida.

Table 3. Physiological Responses that Could Indicate a Problem with Atmospheric Deposition

MEASURABLE RESULTANTS:	MEANINGFUL OBSERVATION PERIOD
Leaves:	
fewer, smaller, chlorotic, bronzed	(1 year)
Branches:	
fewer, dead terminals	(1-5 years)
Flowers:	
fewer flowers set fruit	(1 year)
Fruit:	
fewer & smaller fruit with blemishes	(1-2 years)
Roots:	
less absorptive primary roots	(1-3 years)
uptake of toxic minerals	(2-5 years)
Shorter-Tree Life	(1-30 years)
More Diseases Related to Stress	(1-30 years)

Literature Cited

1. Biggs, R. H. Unpublished.

2. Blanpied, G. D., 1979. Effect of Artificial Rain Water pH and Calcium Concentration on the Calcium and Potassium in Apple Leaves, HortScience 14 (6):706-708.

3. Creasy, L. L., 1980. The Correlation of Weather Parameters with Russet of Golden Delicious Apples under Orchard Conditions, Journal of the American Society for Horticultural Science 105:735-738.

4. Creasy, L. L., and Swartz, H. J., 1981. Agents Influencing Russet on Golden Delicious Apple Fruits, Journal of the American Society for Horticultural Science 106:203-206.

5. Facteau, T. J., and Rowe, K. E., 1981. Response of Sweet Cherry and Apricot Pollen Tube Growth to High Levels of Sulfur Dioxide, Journal of the American Society for Horticultural Science 106: 77-79.

6. Feder, W. A., 1981. Bioassaying for Ozone with Pollen Systems, Environmental Health Perspectives 37:117-123.

7. Forsline, P. L., Musselman, R. C., Dee, R., and Kender, W. J., 1983. Effects of Acid Rain on Grapevines, American Journal of Enology and Viticulture 34 (1):17-22.

8. Forsline, P. L., Musselman, R. C., Kender, W. J., and Dee, R. J., 1983. Effects of Acid Rain on Apple Tree Productivity and Fruit Quality, Journal of the American Society for Horticultural Science 108 (1):70-74.

9. Forsline, P. L., Dee, R. J., and Melious, R. E., 1983. Growth Changes of Apple Seedlings in Response to Simulated Acid Rain, Journal of the American Society for Horticultural Science (in press).

10. Fujiwara, T., 1970. Sensitivity of Grapevines to Injury by Atmospheric Sulfur Dioxide, Journal of the Japanese Society for Horticultural Science 39:219-223.

11. Fujiwara, T., 1975. Studies on the Development of Injury Symptoms Caused by Sulfur Dioxide at Low Level Concentrations in Plants and the Diagnosis of Injury, Central Research Institute of the Electric Power Industry, Japan. Bulletin of the Agricultural Laboratory No. 74001.

12. Haines, B., Stefani, M., and Hendrix F., 1980. Acid Rain: Threshold of Leaf Damage in Eight Plant Species from a Southern Appalachian Forest Succession, Water, Air, and Soil Pollution 14: 403-407.

13. Jacobson, J. S., and van Leuken, P., 1977. Effects of Acidic Precipitation on Vegetation, Proceedings of the Fourth International Clean Air Congress, Tokoyo, Japan. pp. 124-126.

14. Kawamata, S., 1975. Effects on Air Pollution to Fruit Tree Field Injury of Japanese Pear as Affected with the Industrial Smog. Bulletin of the Tokoyo Agricultural Experiment Station No. 8, Tokoyo Agricultural Experiment Station, Tachikawa, Tokoyo.

15. Kender, W. J., and Forsline, P. L., 1983. Remedial Measures to Reduce Air Pollution Losses in Horticulture, HortScience (in press).

16. Kender, W. J., and Carpenter, S. G., 1974. Susceptibility of Grape Cultivars and Selections to Oxidant Stipple, Fruit Varieties Journal 23 (3):59-61.

17. Kender, W. J., and Shaulis, N. J., 1973. Air Pollution Injury in New York Vineyards, New York State Horticultural Society Proceedings 118:70-73.

18. Kender, W. J., and Shaulis, N. J., 1976. Vineyard Management Practices Influencing Oxidant Injury on 'Concord' Grapevines, Journal of the American Society for Horticultural Science 101: 129-132.

19. Kender, W. J., Taschenberg, E. F., and Shaulis, N. J., 1973. Benomyl Protection of Grapevines from Air Pollution Injury, HortScience 8:396-398.

20. Kender, W. J., and Forsline, P. L., 1982. The Effects of Acid Rain on Fruit Crops, Proceedings of the New York State Symposium on Atmospheric Deposition, January 7-8, 1982, Ithaca, NY. pp. 47-49.

21. Lee, J. J., Neely, G. E., and Perrigan, S. C., 1980. Sulfuric Acid Rain Effects on Crop Yield and Foliar Injury, EPA-600/3-BO-016. U. S. EPA Environmental Research Laboratory, Corvallis, OR.

22. Lee, J. J., Neely, G. E., Perrigan, S. C., and Grothaus, L. C., 1981. Effects of Simulated Sulfuric Acid Rain on Yield, Growth and Foliar Injury of Several Crops, Environmental and Experimental Botany 21:171-185.

23. Lee, J. J., and Weber, D. E., 1979. The Effects of Simulated Acid Rain on Seedling Emergency and Growth of Eleven Woody Species, Forest Science 25:393-398.

24. Murdy, W. H., and Ragsdale, H. L., 1980. The Influence of Relative Humidity on Direct Sulfur Dioxide Damage to Plant Sexual Reproduction, Journal of Environmental Quality 9:493-496.

25. Musselman, R. C., Kender, W. J., and Crowe, D. E., 1978. Determining Air Pollutant Effects on the Growth and Productivity of Concord Grapevines Using Open-top Chambers, Journal of the American Society for Horticultural Science 103:645-648.

26. Musselman, R. C., Hertzberg, W. M., Kender, W. J., and Rosen, P. M., 1979. Relationship of Concord Grapevine Growth and Productivity to Sulfur Dioxide Exposure, Department of Pomology & Viticulture, New York State Agricultural Experiment Station, Geneva, NY. Final Report for Energy Impact Associates Contract. 38 pp.

27. Musselman, R. C., Shaulis, N. J., and Kender, W. J., 1980. Damage to Grapevines by Fossil Fuel Wastes and Pollutants, Search Agriculture (No. 3), New York State Agricultural Experiment Station, Geneva, NY. 19 pp.

28. Shaulis, N. J., Kender, W. J., Pratt, C., and Sinclair, W. A., 1972. Evidence for Injury by Ozone in New York Vineyards, HortScience 7:570-572.

29. Shertz, R. D., Kender, W. J., and Musselman, R. C., 1980. Effects of Ozone and Sulfur Dioxide on Grapevines, Scientia Horticulturae 13:37-45.

30. Shertz, R. D., Kender, W. J., and Musselman, R. C., 1980. Foliar Response and Growth of Apple Trees Following Exposure to Ozone and Sulfur Dioxide, Journal of the American Society for Horticultural Science 105:594-598.

EFFECTS OF ATMOSPHERIC DEPOSITION ON FORESTED ECOSYSTEMS IN FLORIDA -- SUGGESTED RESEARCH PRIORITIES

Henry L. Gholz
School of Forest Resources and Conservation
University of Florida, Gainesville, FL 32611

Abstract

The predicted increases in SO_x and NO_x emissions have generated research into the effects of altered atmospheric deposition on aquatic ecosystems and soil properties. However, direct effects on forest productivity and alterations of the role of vegetation in forest nutrient cycling (and, hence, soil properties and water quality) have not been addressed. This appears to be a result of: (1) logistical difficulties in assessing changes in forests, especially those that may become apparent only in the long-term, and (2) a plethora of potential research topics, but, to counter this, (3) a lack of cases where direct effects of increased acid deposition on forest productivity have been documented.

In this paper, I suggest a scheme for prioritizing research on the effects of altered atmospheric deposition on forests in Florida. In addition to the need for better estimates of atmospheric inputs into forests, there are 3 broad areas of research emphasized: effects on runoff quality, forest productivity and forest structure.

Introduction

The study of the effects of pollutants on a terrestrial ecosystem, by logical necessity, should build upon the wealth of information already available on productivity, nutrient cycling, soil chemistry, hydrology, pest management, etc. However, there are major gaps in our knowledge and understanding which currently prevent us from being able to predict many of the potential effects of altered atmospheric chemistry on the structure and function of forested ecosystems. Forests, by nature of their massiveness and longevity, impose major constraints on our efforts to learn. Also, although we have an abundance of knowledge about the functioning of certain forest types, because their dominant species are of major economic importance, for other forest types we have little information. Examples of the latter include watersheds of many aquatic systems important for sport fishing, recreation, and water consumption.

Natural forests covered 85% of Florida, according to Davis (1). Although most of these forests have been cut, and perhaps replaced by other forest types (including citrus groves and orchards), the aerial extent of forests in Florida is not much reduced. The dominant forest type is the pine flatwoods, a level, wet area, with soils dominated by poorly drained Spodosols and Ultisols (2). This type is now heavily dominated by commercial slash pine (*Pinus elliottii*) plantations. Dryer sites, dominated by *Pinus clausa*, *Pinus palustris* or mixed hardwoods (*Quercus* spp.), are the other major "upland" forest type. A diversity of forested wetlands occupies the "lowlands" either as ponds or as transitional forests between upland types and aquatic or non-forested swamp systems.

The literature on effects of acid deposition on forested ecosystems throughout the world has been reviewed recently by Johnson et al. (3), Pinkerton (4), and Johnson (this volume). Although public speculation abounds, there is currently very little documentation of a direct connection between increased acid deposition and specific forest responses.

In this paper I will identify potentially important research areas, and will attempt to prioritize them for the purpose of establishing an integrated view of acid deposition effects on Florida forests.

Overview

Three broad research areas can be identified pertaining to acid deposition and forests in Florida:

Runoff Quality -- effects of acid deposition in forests on the quality of water leaving these systems and entering "downstream" ecosystems.

Forest Productivity -- effects of acid deposition on the primary production, including economic yields, of forests.

Forest Structure -- effects of acid deposition on plant and animal species densities (or carrying capacities) and diversities.

In all three areas, all forest types in Florida are of concern, although in the case of forest productivity the commercial pine forests are of obvious specific interest. Most research should focus on forests in the northern 1/2 to 2/3 of the state, because, except for local acute cases, marine influences should dominate inputs to forests further south.

A fourth area, not addressed in detail here, must be the improved monitoring of atmospheric chemistry, description of processes involved, and estimation of magnitudes of inputs of water and elements into forests. Recent evidence indicates that forest canopies may be efficient "scavengers" of water, nutrients and pollutants in the form of aerosols, fog droplets, gasses, and dry particulates. For example, fog droplets may have very high H^+ and other ion concentrations compared to those of rainfall (5, 6, 7, 8). Lovett et al. (8) illustrate that without taking into account fog droplet impaction, inputs into a subalpine forest in New Hampshire of N, P and major cations are greatly underestimated (Table 1), undermining current views of nutrient cycling and retention in those forests. The high humidity surrounding canopies of forests in Florida may result in similar underestimates of inputs here. These problems must be dealt with in order to realistically interpret potential effects of altered atmospheric chemistry.

Runoff Quality

Potential alterations of surface and groundwater runoff from forested watersheds should be of primary concern. Research must be coordinated with work in upland, flatwood, wetland, and aquatic ecosystems, as these are all joined in a drainage catena, with the outputs from one

Table 1. Annual ion deposition at Mount Moosilauke, N.H. (from Lovett et al. 1983)

Ion	Cloud Deposition ($kg\ ha^{-1}\ yr^{-1}$)	Bulk Precipitation ($kg\ ha^{-1}\ yr^{-1}$)
H^+	2.4	1.5
NH_4^+	16.3	4.2
Na^+	5.8	1.7
K^+	3.3	2.1
SO_4^{-2}	275.8	64.8
NO_3^-	101.5	23.4

compromising the inputs to the next. One approach is to analyze soil buffering capacity and anion/cation mobility in soil cores for the various areas in response to added acid. This would help identify potentially "leaky" forest/soil types. However, this approach can only provide an index, as, for example, nutrients may also be leached in greater amounts from the canopy and forest floor, changing inputs to the soil and perhaps further altering soil leaching processes.

In addition, certain elements are of particular interest, as they have demonstrated toxic effects on fish (Al^{+++}), or other particularly negative characteristics. As almost all the Spodosols have high Al concentrations (Ref. 2, Table 2), this indicates a need for more research in Al mobility in soils. In order to develop predictive capabilities, controls on the reactions that release nutrients (and toxins) from "unavailable" or immobile forms in the forest floor and soil, into "available" or mobile forms must be understood. Such processes include mineralization and/or immobilization of nutrients from litter, soil organic matter, and mineral cation exchange sites, and nitrification.

Table 2. pH and Al concentrations (weak double acid extractable) in the soil under an unfertilized 14-year-old slash pine plantation

Horizon	pH	Al (ppm)
A1	3.9	39
A2	4.5	17
Bh	4.4	1530 (range 500-2380)
Bt	4.8	254

Forest Productivity

A connection between increased acid deposition and forest growth has been sought by many investigators because of the potentially great economic and ecological impacts such a change could have. All the areas of interest in the previous section are central to understanding effects on forest productivity as well. However, there are many potential alterations that would not directly affect water quality, and which also may need to be considered over a long time period -- at least one rotation in pine plantations.

The focus of research in this area must be the important connections between soil fertility, environmental factors and tree growth in our highly important commercial forest areas. Many of these sites already are very deficient in nutrients and suffer either extreme excesses or deficiencies in available water. There are four primary research needs: (1) How does acid deposition affect reactions that make nutrients available for plant uptake in the short-term (decomposition, mineralization, nitrification)? Increases in these reaction rates could lead to more rapid nutrient cycling and increase in forest productivity, or to more rapid leaching losses of nutrients from the ecosystems. (2) Does acid deposition affect the storage of organic matter and nutrients in the long-term? (3) Does acid deposition affect the physiological functions of the trees directly (i.e., is stomatal control altered)? (4) Does acid deposition alter the balance of beneficial and harmful microorganisms associated with trees? This last area includes such direct effects as cuticular wax erosion and potential increases in foliar pathogen infections (would such erosion lead to a greater fusiform rust infection of oak or pine leaves?), and indirect effects, such as destruction of mycorrhizae through increased Al concentrations in the rhizosphere.

In all cases, a keyword is "stress". Currently, our best indication of increased stress on a forest is either reduced tree growth or increased mortality. In both cases, damage is assessed retroactively. Therefore, research objectives must include both an identification of immediate alterations in forest functioning, and an effort to understand in the long-term how these alterations may interact to change forest productivity and consumer dynamics.

Certain stresses can be overcome through fertilization, drainage, tree improvement and site preparation. Current forest management practices may, therefore, eliminate or reduce negative aspects of increased acid deposition. There is the possibility, especially when NO^{-}_{3} is the major anion, for increased acid deposition to act as a fertilizer. However, little is known about the S cycle in forests, and it may be equally plausible for S additions to exaggerate widespread N and/or P deficiencies.

The mycorrhizal/fine root connection is an extremely vulnerable target for alterations in the rhizosphere environment, especially in our often P limited systems. For example, Ulrich (9) proposed that the dieback of *Abies alba* in northern Europe is the result of drought stress induced by Al^{+++} damage to fine roots in Spodosol soils chronically exposed to acid deposition.

Forest Structure

Forests are not monocultures, except perhaps in very dense old field plantings, or where herbicides are heavily used. Instead, many species interact, and perhaps compete, in a milieu of limiting resources. Forests in Florida, of all types, contain a variety of plant species and support an abundance and diversity of animal life, although only a few species may be of major economic importance. It is our legislative mandate, as a result of the National Forest Management Act of 1976 and the Endangered Species Act of 1975, if not an intrinsic and central goal of all land management, to maintain this diversity of life.

Pollutants can act to alter natural competitive balances and selectively reduce or eliminate certain plant or animal species (10). Therefore, it is important to identify not only susceptible species of commercial importance, but to anticipate changes in overall community structure. Furthermore, changes in plant communities could alter populations of game animals, cattle grazing in forests, and insects.

Research Priorities for Florida

Obviously, detailed analyses of all the areas mentioned above is not possible, so where might we start? Work in certain areas in certain forest types is already underway (e.g., Volk, Mansell papers in this volume and Ref. 11, 12, 13). My suggestions, based on my discussions with many of you participating in this workshop and others, and on my own assessment of the state-of-the-art, follow.

1. Enlarge monitoring efforts. Develop better techniques for quantifying *in situ* inputs of substances from the atmosphere into forest systems.

2. Establish means for quantifying the movement of ions in groundwater-dominated forest soils, as lysimetric methods are of limited application in much of the flatwoods. The application of simplified soil water flow models combined with multiple-ion exchange theory should be explored.

3. Determine effects of altered solution chemistry on soil reactions (decomposition, mineralization/immobilization, nitrification).

4. Establish S cycling patterns for selected Florida forests. How do N, P and S interact, and how mobile is SO_4^{--} in the soil solution?

5. Determine the potential for leaching of Al^{+++} (and/or other potentially harmful substances) from forest soils exposed to more acid inputs.

6. Determine the sensitivity of the fine root/mycorrhizal complex to acidification and potentially elevated concentrations of Al^{+++}.

7. Make judicious use of simulation models to help predict potential long-term effects of increased acid deposition on forest structure and productivity and on the balance of specific plant, insect, microbial and animal populations in forests.

The most progress would be in the context of an integrated program contrasting the structure and function of typical Florida forest types and their relative sensitivity to increased acid deposition. In all phases, the connectedness among upland, flatwood, wetland, and aquatic ecosystems should be emphasized.

Literature Cited

1. Davis, J.H. 1967. General map of natural vegetation of Florida. Agric. Expt. Stn. Institute of Food and Agricultural Sciences, Univ. Fla., Gainesville.

2. Smith, F.B., R.G. Leighty, R.E. Caldwell, V.W. Carlisle, L.G. Thompson, Jr., and T. C. Mathews. 1967. Principal soil areas of Florida, a supplement to the general soil map. Institute of Food and Agricultural Sciences Bulletin 717. Agricultural Experiment Station, Univ. Fla., Gainesville.

3. Johnson, D.W., J. Turner, and J.M. Kelly. 1982. The effects of acid rain on forest nutrient status. Water Resourc. Res. 18: 449-461.

4. Pinkerton, J.E. 1981. Acidic deposition and its effects on forest productivity--a review of the present state of knowledge, research activities, and information needs. Tech. Bull. 110, National Council of the Paper Industry for Air and Stream Improvements, Inc. N.Y., N.Y.

5. Falconer, P.D., and J.A. Kadelecek. 1980. Cloud chemistry and meteorological research at Whiteface Mountain: Summer 1979. Atmospheric Science Research Center, Publ. 748, Albany, N.Y.

6. Lindberg, S.E., R.C. Harriss, and R. R. Turner. 1982. Atmospheric deposition of metals to forest vegetaion. Science 215:1609-1610.

7. Waldman, J.M., J.W. Munger, D.J. Jacob, R.C. Flagan, J.J. Morgan, and M. R. Hoffmann. 1982. Chemical composition of acid fog. Science 218:677-680.

8. Lovett, G.M., W.A. Reiners, and R.K. Olson. 1983. Cloud droplet deposition in subalpine balsam fir stands: hydrological and chemical inputs. Science 218:1303-1304.

9. Ulrich, V.B. 1981. Eine okosystemare Hypothese uber die Ursachen des Tannensterbesn (*Abies alba* Mill.). Forstwissenshaftliches Centralblatt 100:228-236

10. Shugart, H.H., S.B. McLaughlin, and D.C. West. 1980. Forest models: their development and potential applications for air pollution effects research. pp. 203-214 *in* P.R. Miller (ed.) Proceedings of Symposium on Effects of Air Pollutants on Mediterranean and Temperate Forest Ecosystems. Proc. S.W. For. and Range. Expt. Stn., Berkeley, CA.

11. Gholz, H.L., and R.F. Fisher. 1982. Organic matter production and distribution in slash pine (*Pinus elliottii*) plantations. Ecology 63: 1827-1839.

12. Gholz, H.L., R.F. Fisher, and W.L. Pritchett. Nutrient dynamics of managed pine ecosystems in north Florida. (ms. in review.)

13. Riekerk, H. 1983. Impacts of silviculture on flatwoods runoff, water quality, and nutrient budgets. Water Res. Bull. 19:73-79.

ION EXCHANGE COUPLED WITH CONVECTIVE-DISPERSIVE TRANSPORT OF CATIONS DURING "ACID RAIN" INFILTRATION IN SOIL: A REVIEW

R. S. Mansell
Department of Soil Science
University of Florida, Gainesville, FL 32611

Abstract

During periods of rainfall, infiltrating rainwater displaces the electrolyte solution initially present in pores within the soil matrix. As the rainwater penetrates the soil profile, convection or mass flow results in the transport of multiple species of ions in the mobile liquid phase of the soil. The convective transport of these ions is coupled with hydrodynamic dispersion which mixes the displaced and displacing solutions and with ion exchange which selectively retards the velocities for different species. If the rainwater is acidic, basic cations such as Ca^{2+} and K^+ can be desorbed from electrostatically-charged soil colloids and possibly transported to underlying groundwater. Thus exchange sites in the soil provide a chemical buffer against excessive leaching of important basic cations. The effectiveness of this buffer depends upon the cation exchange capacity of the soil, the ratio of the number of "constant charge" exchange sites to the number of "variable-charge" sites, ionic composition of the acid rain relative to that of the initial soil solution, relative ionic strengths for both solutions, the degree of base saturation in the soil initially, selectivity exchange coefficients for the ions involved, the liquid flow velocity, and the dispersion coefficient. Recent transport models utilize multispecies ion exchange to describe cation movement through aquifers when the cation exchange capacity is constant. Experimentally-observed features such as multiple ion fronts separated by plateaus have been successfully described by those models. Convective-dispersive models which use multispecies ion exchange are needed to investigate the influence of acid rain upon leaching losses of cations in soils when the cation exchange capacity decreases with increasing acidity.

Introduction

In humid climates, basic cations are continually leached from soil (Black 1960) as rainwater infiltrates the soil surface and penetrates downward through the soil profile. This leaching process occurs naturally as hydrogen ions from carbonic and soluble organic acids replace a portion of the basic cations during water flow through the soil. Pure rainwater in equilibrium with atmospheric CO_2 (Reuss 1975) has an acidic pH of 5.65. Consequently soils of humid regions are generally acid with substantial quantities of exchangeable hydrogen and aluminum. The degree of base saturation of these soils is typically less than 100%, and Ca^{2+}, Mg^{2+}, and K^+ are the principal exchangeable bases.

During the past decade, public and scientific sectors have become concerned over acidification of rainwater due to anthropogenic (emissions from automobiles and coal-burning power generators) inputs of HNO_3 and

H_2SO_4 acids. One of the most likely effects of acid precipitation upon unlimed and unfertilized soils (Reuss 1980) is acceleration of the leaching of basic cations. However, cation leaching by either acid rain or naturally produced acids will not necessarily result in (Johnson et al. 1982) an equivalent decrease in base saturation or an equivalent increase in soil acidification. Only moderately acid soils with low cation exchange capacity CEC are generally considered to be susceptible (Johnson et al. 1982) to acidification by acid rain. Soils with high CEC have a high storage capacity for basic cations and are buffered against large changes in pH. In soils with high CEC the composition of the soil solution will be controlled by the degree of base saturation on exchange sites.

The extent to which cations are leached from a given soil during infiltration with acid rain is highly dependent (Johnson et al. 1982) upon the mobility of anions associated with the acid. Mobility of individual anions is inversely related to the selectivity (affinity for adsorption) sequence

$$PO_4^{3-} > SO_4^{2-} >> NO_3^- > Cl^- > OH^- .$$

Thus the exchange of anions between the soil solution and the exchange complex has a controlling influence upon the leaching of basic cations. Of the three most prevalent anions -- Cl^-, NO_3^-, and SO_4^{2-} -- in acid rain, only SO_4^{2-} has limited mobility due to specific adsorption and chemisorption by sesquioxides in soil. Both Cl^- and NO_3^-, however, can undergo nonspecific or physical adsorption by anion exchange sites in the soil. Most soils in humid regions are amphoteric (Arnold 1978, Uehara and Gillman 1981) since they contain an intimate admixture of constant electrostatic charge (crystalline clay minerals) and variable-charge (Fe and Al oxides and organic matter) colloids. Lowering the pH of the soil solution by acid rain infiltration should tend to increase the anion exchange capacity AEC and thereby decrease the mobility of anions such as SO_4^{2-}. On the other hand, acid rain should decrease the CEC and thus increase cation mobility. Thus soils containing a high content of sesquioxides should be less susceptible to cation leaching during acid rain infiltration than coarse textured quartz sands.

The process of cation leaching during infiltration of acid rain can be thought of as miscible displacement (Diggar and Nielsen 1976) where the infiltrating rainwater displaces the solution initially present in the porous soil matrix. As the rainwater penetrates into the soil profile, mass flow (convection) moves the mobile solution phase thus, transporting multiple species of cations and anions. As the mass flow occurs, hydrodynamic dispersion mixes the displacing and displaced solutions to produce a mobile transition zone that tends to widen with time. Ions moving with the solution phase undergo exchange with counter ions present on soil exchange sites distributed along the walls of pores.

If we assume that soil systems consist of 3 discrete phases--solution, exchanger, and nonexchangeable--the process of cation leaching can be portrayed by the accompanying schematic diagram (Figure 1). Although each of these phases contains multispecies of ions, the immobile exchanger and nonexchangeable phases generally (Black 1960) contain much larger quantities of basic cations than found at any given moment in the

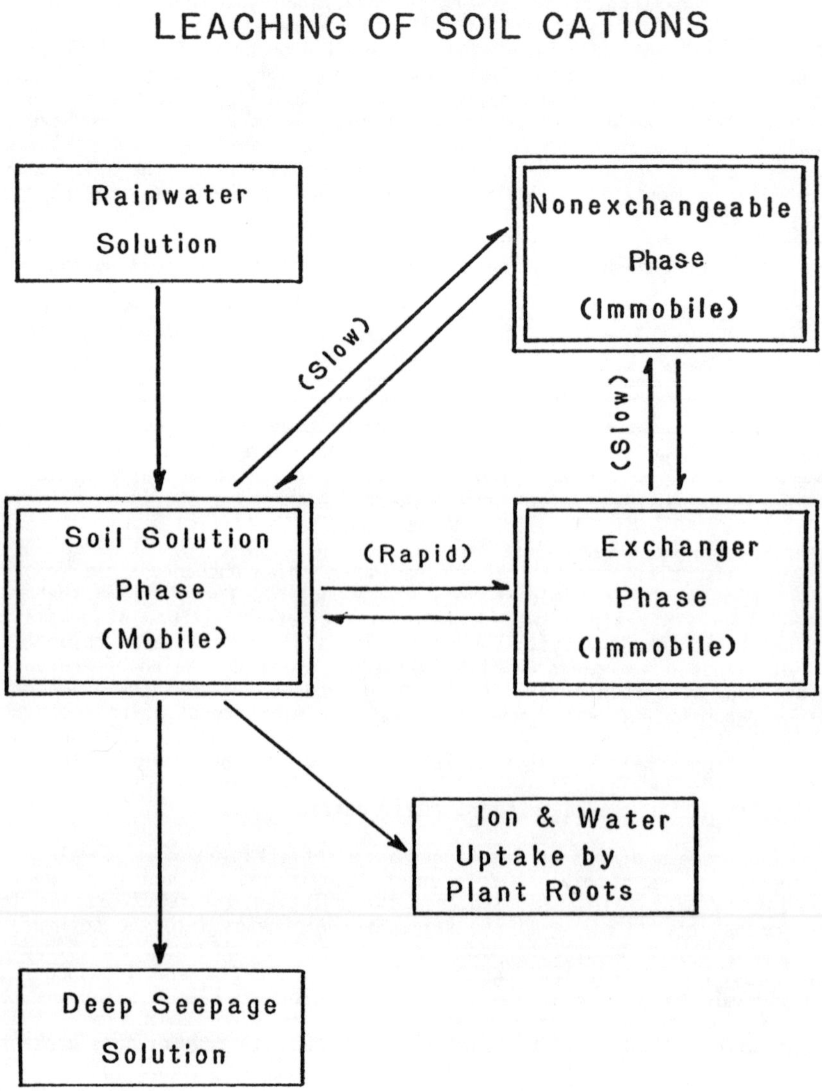

Fig. 1 Leaching of soil cations

mobile solution phase. Cations held in the exchanger phase are held by physical or nonspecific adsorption and undergo rapid exchange with counterions in the solution phase. The nonexchangeable phase contains chemisorbed or specifically adsorbed cations which undergo very slow transfer with counterions in the solution phase. During rainwater infiltration, the ionic composition and component concentrations in the soil solution will undergo change which will result in a rapid transfer of ions between the exchanger and solution phases. This transfer is similar in many respects to multicomponent chromatography. Removal of water and nutrient cations such as Ca^{2+} and K^+ by plant roots will also cause perturbations in the solution phase which will also induce ion exchange. Deep seepage loss of cations with water into underlying groundwater occurs as a result of the complex process referred to as cation leaching. One can readily observe that ion exchange, more precisely multispecies ion exchange, is an extremely important process that largely determines the extent of cation leaching in soils.

Helfferich (1962) has defined ion exchange as the mass transfer of electrostatically held counterions balancing the surface charge of the exchanger phase with other ions in the solution phase. Ion exchange is characteristically 1) reversible, 2) instantaneous, 3) stoichiometric, and 4) selective among different competing ions. Since ion exchange reactions occur rapidly, cation concentrations in the solution phase (Freeze and Cherry 1979) can be expected to be in local exchange equilibrium even when the solution is moving. Following any changes in composition of the solution phase due to rainfall, the equilibrium cation concentrations in solution will depend upon 1) concentrations of all major cations in the displacing rainwater, 2) concentrations of all major cations in the solution initially present in the soil pores, and 3) mole fractions of all adsorbed cations initially present in the exchanger phase. Even if local exchange equilibrium is maintained at all times, continual movement of rainwater through soil can be accompanied (Freeze and Cherry 1979) by a gradually changing chemistry of the solution phase.

Cation selectivity for exchange sites is largely determined by the lyotropic series where the affinity for adsorption occurs as the sequence

$$Al^{3+} \gg Ca^{2+} > Mg^{2+} \gg NH_4^+ > K^+ > Na^+ > H^+.$$

Cations with the smaller hydrated radii are preferred (Freeze and Cherry 1979) over other ions having the same valence. Divalent ions are generally preferred over monovalent ions, and trivalent ions are preferred over divalent ions. Cations such as Ca^{2+} with high affinity for exchange sites will be expected to have smaller mobility than ions such as Na^+ which have low affinity.

For simple binary ion exchange where the exchanger phase is initially saturated with cation species #j and a solution comprised of cation species #i, Vanselow selectivity coefficients can be determined (Sposito 1981)

$$K_{ij} = \left(\frac{\overline{C}_i^*}{C_i} \right)^{\nu_j} \Big/ \left(\frac{\overline{C}_j^*}{C_j} \right)^{\nu_i} \qquad [1]$$

where C_i, C_j are the activities of ions #i and #j respectively in solution, ν_i, ν_j are valences for ions #i and #j respectively, and $\bar{C}^*_i$, $\bar{C}^*_j$ are the mole fractions of ions #i and #j respectively in the exchanger phase. For the case of homovalent binary exchange and constant ionic strength I of the solution phase, the magnitude of the selectivity coefficient indicates the relative preference of exchange sites for the two cations:

K_{ij}	Ion Selectivity
> 1.0	preference for ion #i
< 1.0	preference for ion #j
= 1.0	no preference

The selectivity coefficient K_{ij} when Mg^{2+} represents ion #i and Ca^{2+} represents ion #j typically (Freeze and Cherry 1979) occurs in the range from 0.6 to 0.9 indicating that Ca^{2+} is adsorbed preferentially to Mg^{2+}. Quite often, the parameter K_{ij} is referred to as Vanselow's selectivity function (Freeze and Cherry 1979) since it is not a true constant but is a function of the composition of the exchanger phase. The dependence of K_{ij} is the same (Sposito 1981) as the ratio of the activity coefficients for the exchanger phase since

$$K_{ij} = K \frac{\left(f_j\right)^{\nu_i}}{\left(f_i\right)^{\nu_j}} = K \frac{\left(\bar{C}_j / \bar{C}^*_j\right)^{\nu_i}}{\left(\bar{C}_i / \bar{C}^*_i\right)^{\nu_j}} \qquad [2]$$

where $\bar{C}_i$, $\bar{C}_j$ are the activities of species #i and #j in the exchanger phase and K is the exchange equilibrium constant.

Principles of ion exchange chromatography can be utilized to provide useful insight into convective-dispersive transport of multiple species of ions through soil. In order to utilize multicomponent chromatography we will assume that steady liquid flow with an average pore water velocity υ is maintained through a water-saturated soil column of length L and that ion exchange is the only active chemical reaction in the soil. The effluent is analyzed with time for concentrations of each ion i for M species so that breakthrough curves (Plots of equivalent fractions C^*_i of each ion versus the number of pore volumes τ of effluent) can be determined for each species.

In order to mathematically describe multispecies ion transport for such a system (Rubin and James 1973) a set of M transport equations is required where one equation is needed for each ion i

$$\frac{\partial C_i}{\partial t} = D\frac{\partial^2 C_i}{\partial x^2} - \upsilon \frac{\partial C_i}{\partial x} - \frac{\rho}{\theta} \frac{\partial \bar{C}_i}{\partial t} \qquad [3]$$

where D is the hydrodynamic dispersion coefficient, ρ is soil bulk density, and θ is the volumetric water of the soil. The first term on the right hand side of [3] represents the dispersion or mixing process as a displacing solution moves into the soil, the second term represents the convection process which moves the ions with the mobile solution phase, and the third term determines the rate of ion exchange of ion i between the exchanger and solution phases. For simple homovalent binary exchange the exchange term can be written (Valocchi et al. 1981) as

$$\frac{\rho}{\theta}\frac{\partial \bar{C}_i}{\partial t} = \frac{\rho}{\theta}\frac{\partial \bar{C}_i}{\partial C_i}\frac{\partial C_i}{\partial t} \qquad [4]$$

where

$$\bar{C}_i = 1\Big/\left(\frac{1}{K_{ij}C_i} + \frac{1}{(1-1/K_{ij})}\right) \qquad [5]$$

and

$$\frac{\partial \bar{C}_i}{\partial t} = 1\Big/K_{ij}C_i^2\left(1 - \frac{1}{K_{ij}} + \frac{1}{K_{ij}C_i}\right)^2 \qquad [6]$$

If the cation exchange capacity $\bar{C}_T$ is kept constant and the total solution normality C_T is kept constant, then a single equation of the type given by equation [3] can be written for the binary exchange case

$$\frac{\partial C_1^*}{\partial t} = \frac{D}{R}\frac{\partial^2 C_1^*}{\partial x^2} - \frac{\upsilon}{R}\frac{\partial C^*_1}{\partial x} \qquad [7]$$

where $R = R(C_1^*)$ is a retardation function given by

$$R = 1 + \frac{\rho}{\theta}\frac{\bar{C}_T}{C_T}\left\{\frac{1}{K_{12}C^{*2}_1\left(1-\frac{1}{K_{12}} + \frac{1}{K_{12}C_1^*}\right)^2}\right\} \qquad [8]$$

Thus ion #1 will move through the soil with a variable velocity $\upsilon_1 = \upsilon/R$ which depends upon the selectivity coefficient K_{12} and the solution phase concentration C_1^*. If $K_{12} > 1$ the velocity υ_1 will increase as C_1^* increases (Valocchi et al. 1981) and if $K_{12} < 1$ the velocity will decrease as C_1^* increases. For the simplist case where $K_{12} = 1$, the retardation

$$R = 1 + \rho\ \bar{C}_T/\theta C_T$$

is constant and the velocity of the ion will not change with changes in C_1^*. From equations [8] and [9] one can readily see that the velocity of a given ion will be most highly retarded for soils with large values of $\bar{C}_T$ or CEC. If the value of the selectivity coefficient K_{12} is large then the velocity of the ion will be retarded even more. Thus binary exchange retards the movement of an ion through the soil and gives a delayed breakthrough of the ion in the effluent from a soil column. Biggar and Nielsen (1963), Lai and Jurinak (1971), and Persaud and Wierenga (1982) have presented cation breakthrough curves for binary soil exchange systems which demonstrate the retardation effect of the exchange process.

When multispecies ion exchange occurs during convective-dispersive transport, adsorption of each cation species i from the solution phase is highly dependent (Helfferich 1962, Helfferich and Klein 1970, and Rachinski 1965) upon the mole fractions of all cations in both the solution and exchanger phases. This mutually dependent adsorption results in a very complex form of the ion exchange term given in equations [3] and [4] since for each species i

$$\bar{C}_i^* = \bar{C}_i^* \left(C_1^*, C_2^*, \cdots C_m^*, \bar{C}_1^*, \bar{C}_2^*, \cdots \bar{C}_m^*\right). \qquad [9]$$

Rubin and James (1973) however have published complex but useable forms of equation [4] that can be used for multispecies exchange.

When a multicomponent solution is displaced through a soil column, cations with higher affinities for exchange sites will move more slowly through (Rachinski 1965) the soil than species with lower affinities. Helfferich and Klein (1970) conclude that for any given composition of the solution phase, ion species with higher affinities will have lower average velocities. The average velocity for each individual ion will vary with time depending upon how the concentrations of all other accompanying species in the solution phase vary with time. Helfferich (1962) has shown that when an ion exchanger is being equilibrated with a multicomponent solution in a batch study, the concentration of some ion species in solution may go through a maximum or a minimum before reaching the equilibrium concentration. This occurs because rates of adsorption may be quite different for various counter ions.

For multicomponent chromatography Helfferich and Klein (1970) have shown that a system with M cation species will produce M-1 coherent ion concentration fronts moving through the exchange column and these fronts are separated by plateau zones. If for example M=3, a single intermediate plateau zone will separate a slower, upstream front from a faster, downstream front that is associated with changes in the solution normality C_T will travel at the pore water velocity.

Valocchi et al. (1981) used a model based upon multicomponent chromatography to simulate the simultaneous transport of 3 cations - Ca^{2+}, Mg^{2+}, and Na^+ - and a common anion Cl^- during groundwater flow in an aquifer. Treated wastewater with much lower solution normality than the original groundwater was injected into the aquifer. The order of selectivity for the different cations was $Ca^{2+} > Mg^{2+} > Na^+$. For that tertiary exchange system, the model successfully predicted breakthrough curves for all three cations in well water downstream from the injection site. Breakthrough curves for both of the divalent cations reflected the influence of two fronts separated by a plateau. In the plateau region, Mg^{2+} was concentrated in excess of the injection solution but the Ca^{2+} concentration was much less than the injection solution. The difference in the Ca^{2+} and Mg^{2+} breakthrough curves reflect the mutually dependent adsorption and desorption processes that occurred simultaneously for three cations with different affinities for the exchange sites. Valocchi et al. (1981) postulated that many multispecies exchange and transport problems may be approximated by only using two or three major components.

Conclusion

Convective-dispersive transport models which include multispecies ion exchange offer a potential means for describing sequential leaching of cations from soils during infiltration of acid rain. Such models must however be modified to include the influence of acid rain upon the soil CEC due to variable charge colloids, specific adsorption of anions such as SO_4^{2-}, and chemical transformations such as the form of aluminum. Such models will require evaluation by carefully comparing computed cation leaching with actual data from soil columns during miscible displacement with "acid rain" solutions.

Literature Cited

1. Arnold, P.W. 1978. Surface-electrolyte interactions. In the Chemistry of Soil Constituents edited by D. J. Greenland and M. H.B. Hayes, John Wiley, New York.

2. Biggar, J. W. and D. R. Nielsen. 1963. Miscible displacement:V. Exchange processes. Soil Sci. Soc. Proc. 27:623-627.

3. Biggar, J. W. and D. R. Nielsen. 1976. Mechanisms of chemical movement in soils. In Agrochemicals In Soils, edited by A. Banin and U. Kafkafi, Pergamon Press, New York. pages 213-227.

4. Black, C. A. 1960. Soil-Plant Relationships. John Wiley and Sons, Inc. See pages 102-158.

5. Freeze, R. A. and J. A. Cherry. 1979. Groundwater. Prentice-Hall, Inc., Englewood Cliffs, New Jersey. Seepages 127-134.

6. Helfferich, F. 1962. Ion Exchange McGraw-Hill Book Company, Inc. New York.

7. Helfferich, F and G. Klein. 1970. Multicomponent Chromatography. Marcel Dekker, New York.

8. Johnson, Dale W., J. Turner, and J. M. Kelly. 1982. The effects of acid rain on forest nutrient status. Water Resour. Res. 18:449-461.

9. Lai, Sung-Ho and J. J. Jurinak. 1971. Numerical approximation of cation exchange in miscible displacement through soil columns. Soil Sci. Soc. Amer. Proc. 35:894-899.

10. Persaud, N. and P. J. Wierenga. 1982. A differential model for one-dimensional cation transport in discrete homoionic ion exchange media. Soil Sci. Soc. Amer. J 46:482-490.

11. Rachinski, V. U. 1965. The General Theory Of Sorption Dynamics and Chromatography. Translated from Russian, Consultants Bureau, New York.

12. Reuss, John O. 1975. Chemical/Biological Relationships Relevant To Ecological Effects of Acid Rainfall. EPA-660/3-75-032, U.S. Environmental Protection Agency. 47 pages.

13. Reuss, J. O. 1980. Simulation of soil nutrient losses resulting from rainfall acidity. Ecological Modelling 11:15-38.

14. Rubin, Jacob and R. V. James. 1973. Dispersion-affected transport of reacting solutes in saturated porous media:Galerkin Method applied to equilibrium-controlled exchange in undirectional steady water flow. Water Resour. Res. 9:1332-1356.

15. Sposito, Garrison. 1981. The Thermodynamics of Soil Solutions. Oxford Press, New York.

16. Uehara, Goro and G. Gillman. 1981. The Mineralogy, Chemistry, and Physics of Tropical Soils With Variable Charge Clays. Westview Press, Boulder, Colorado.

17. Valocchi, A. J., R. L. Street and Paul V. Roberts. 1981. Transport of ion-exchanging solutes in groundwater: Chromatographid theory and field simulation. Water Resour. Res. 17:1517-1527.

EFFECTS OF ACID PRECIPITATION ON A TYPIC QUARTZIPSAMMENT SOIL IN FLORIDA [1/]

Bob G. Volk and Gerald E. Byers
EPA/NCSU Acid Precipitation Program
University of Florida, Gainesville, FL 32611

Abstract

The magnitude of leaching of nutrients from the soil surface to the water table and to a nearby drainage system depends on many factors unique to the site. The sensitive deep sands of Florida present an extreme case of low nutrient status and high water flux to great depth. In addition, water table characteristics, both physical and chemical, influence the horizontal flow vis-a-vis the vertical flows. Investigations using artificial acid treated irrigation water (70:30, H_2SO_4:HNO_3) at pH 3.0, 3.7, and a 4.7 control over a two year period at a lake site in north central Florida suggest a relatively high and rapid loss of nutrients from the soil surface to the water table 4 m below under acid irrigation conditions. Considering the low CEC (3.5 me/100g), the low base saturation (4%) and the high hydraulic conductivity (74 cm/hr), the implication of effects in the rooting zone of natural vegetation is clear. The magnitude of leaching also depends on the nature of the ion, the depth being considered and the acid treatment. Readsorption of ions in the soil profile was shown to occur which further complicates the overall assessment of the real effect of acid rain. Most of the applied water, regardless of the amount, is lost from the soil site to the water table after 48 hours. Thus plant nutrients could potentially be rapidly and completely lost from the soil. Lateral movement in the groundwater to the lake likewise is very fast due to the high hydraulic conductivity.

Mounding of water was observed to occur at the water table surface. There was essentially little non-vertical flow to the water table, and horizontal wave movement towards the lake is rapid, despite the essentially flat water table extending more than 50 meters from the lake. The effect of the acid treatments from a very large rain event could be observed in the groundwater near the lake within 24 hours.

Laboratory studies on sealed soil columns under saturated water flow conditions at pH 3.0, 3.7, and 4.7 (70:30, H_2SO_4:HNO_3) show ion movement characteristics very similar to those observed in the field. The effect of acid rain is not long lasting and the soils recover the normal soil leachate chemistry within 48 hours. Calcium was the dominant nutrient in

1/ "Although the research described in this article has been funded in part by the U.S. Environmental Protection Agency, through a cooperative agreement with North Carolina State University, it has not been subjected to EPA's policy review and therefore does not necessarily reflect the views of the Agency."

the system and showed the greatest magnitude of loss. Magnesium, however, although present at 25% of the Ca, exhibited the greatest percentage loss from the rooting zone. Potassium showed lower but continual loss. Sulfate adsorption was clearly observed. Aluminum showed increased concentration in soil leachates with the more acid treatments.

Introduction

The impact of acid precipitation on terrestrial environments may result in chemical and biological alterations. Chemical alterations may consist of leaching of essential alkaline cations with subsequent eventual decrease in soil fertility and productivity as well as release of heavy metals and aluminum impacting groundwater and surface water environments (Babich et al., 1980; Cronan et al., 1978; Wainwright, 1980b).

The magnitude of leaching of nutrients from the soil surface to the water table and to a nearby drainage system depends on many factors unique to the site. The sensitive deep sands of Florida present an extreme case of low nutrient status and high water flux to great depth. In Florida approximately 35% of the northern half of the state is covered by quartzipsamments. In fact the largest acreage of citrus production in Florida is on this type of soil. Acid precipitation could remove significant quantities of bases and micronutrients from these soils. Investigators have noted the leaching of nutrients from soils treated with acid precipitation (Strayer and Alexander, 1981; Bache, 1980; Wilson, 1980).

Losses of bases from soils used for forest production would be particularly damaging to forest productivity in that basic cations are not applied to forest plantations. Johnson et al. (1980), in an examination of the growth rates on a New Jersey Hapludult and a Quartzipsamment, have suggested that acid precipitation may be having a deleterious effect on tree growth.

The purpose of this investigation was to examine the effects of acid precipitation on the movement of selected elements and on changes in chemical properties of a Typic Quartzipsamment soil in a field study.

Acidic Deposition Effects on Agricultural Crops

Recent research has shown that acid deposition has the potential to affect agricultural crops. Presently, the knowledge of effects is primarily based on experimental studies with plants grown under controlled conditions and on evidence from natural systems (Cowling, 1980b). Direct effects of acid deposition on agricultural crops under natural conditions have not yet been clearly documented; however, experimentation now underway may provide such evidence.

Agricultural crop or vegetation effects may be divided into direct or indirect categories (Cowling, 1980a, 1980b; Tamm and Cowling, 1976).

Direct Effects on Vegetation

As outlined by Newman et al. (1982), the following effects have been hypothesized or documented under experimental conditions:

1. Development of foliar lesions, galls, and necrotic zones (Shriner and Johnston, 1981; Irving, 1978; Evans and Curry, 1979; Evans et al., 1977a, 1977b; Ferenbaugh, 1976; Wood and Bormann, 1974);

2. Accelerated erosion of leaf cuticle and protective coatings (Tamm and Cowling, 1976; Shriner, 1976);

3. Interference with normal functioning of guard cells (Tamm and Cowling, 1976);

4. Poisoning of plant cells after diffusion of acidic substances through stomata or cuticle (Cowling, 1980a);

5. Disturbance of normal metabolism and growth processes such as photosynthesis and respiration (Ferenbaugh, 1976);

6. Alteration of leaf and root exudation processes (Tamm and Cowling, 1976);

7. Interference with reproductive processes such as pollen viability and seed germination (Lee and Weber, 1976);

8. Synergistic interactions with other stress factors such as gaseous pollutants (Irving and Miller, 1978); and

9. Inhibition of bud and seedling growth (Matziris and Nakos, 1977; Lee and Weber, 1976).

Jacobson (1980) has stated that herbaceous plants are more sensitive than deciduous trees, which are more sensitive than conifers. The most sensitive foliar stage is immediately preceding maximum leaf development.

Indirect Effects on Vegetation

According to Newman et al. (1982), indirect effects on vegetation and soil are thought to be capable of inducing significant effects on ecosystem function.

Reported or hypothesized indirect effects include (Newman et al., 1982):

1. Accelerated leaching of substances from foliar organs;

2. Increased susceptibility to environmental stresses such as water;

3. Changes in symbiotic associations including lichens, mycorrhizae, and nitrogen-fixing organisms (Shriner and Johnston, 1980); and

4. Reduced microbial activity and litter decomposition.

Agricultural Crop Effects

Experimental studies with plants grown under controlled (or semicontrolled) conditions have demonstrated that visible foliar symptoms and yield responses can be produced on certain crops (Table 1). The effects

Table 1. Simulated Acid Precipitation Effects on Vegetation or Yields of Selected Agricultural Crops

Crop	pH at Which Effect Occurs: Yield decrease (fresh weight)	pH at Which Effect Occurs: Foliar damage[a]	Source
Bean (red kidney)	none	3.2[b]	Shriner & Johnson 1981
Beet	3.0	3.5	Lee et al. 1980
Broccoli	3.0	3.5	
Cabbage	none	3.0	
Carrot	4.0	3.0	
Cauliflower	none	3.5	
Lettuce	none	3.5	
Lettuce	4.0(2.7)[c]	---	Evans et al. 1982
Mustard greens	4.0	3.5	Lee et al. 1980
Onion	none	none	
Pea	none	3.5	
Pepper	3.5[d]	4.0	
Potato	4.0	3.5	
Radish	3.5	3.5	
Soybean (Amsoy)	4.1	---	Evans et al. 1983
Soybean (Lee)	none	3.2[b]	Shriner & Johnson 1981
Soybean (Williams)	none	---	Evans et al. 1983
Spinach	none	3.5	Lee et al. 1980
Strawberry	4.0[e]	3.0	
Swiss Chard	none	4.0	
Tobacco	---	3.5	
Tomato	3.5[e]	3.5	

a 10% of plants showed leaf injury
b 75% reduction in number of nodules
c pH in parentheses notes when dry weight was affected
d Significant increase in yield at pH 3.5, no significant difference at pH 3.0
e Significant increase in yield at this and lower pH's

of acid rain are variable and at any one pH depend upon plant species, variety, soil and fertility, length of acid rain irrigation, the number of irrigations and whether control rain is applied between the acid rain irrigation events. Shiver and Johnson (1981) used pH 6.0 as the control rain whereas all others used a control rain at pH 5.6. Lee et al. (1980) applied acid rain three times a week and watered with control rain as needed. This control rain could ameliorate the effects of the acid irrigation. In contrast, Evans et al. (1983) allowed no additional rain on the plots. The differences which cultivars may exhibit is shown by Evans et al. (1983) for soybeans. Shriner and Johnston (1981) showed no difference in yield with the "Lee" cultivar of soybeans or with the red kidney beans but did show a reduced number of nodules in both cases.

Materials and Methods

Location and Soil Type

The study site was located on the south side of Lake McCloud, an ultra-oligotrophic lake in the Ordway Preserve, three miles east of Melrose, Florida. The preserve is property of the University of Florida. The primary soil of the lake watershed and the experimental plots was Candler, a Typic Quartzipsamment. Chemical analysis of this soil is presented in Table 2.

Table 2. Selected Chemical and Physical Characteristics of the Candler Soil (0-15 cm)

Sand	Silt	Clay	Organic Carbon	pH	Exchangeable Cations Ca	Mg	K	Na	H	CEC	Base Saturation
-----------	%	-----------			-----	cmol (p^+)/kg	---	---	---	---	%
97.0	1.9	1.1	0.73	4.0	0.06	0.027	0.01	0.023	3.32	3.44	3.73

Plot Description

The 1981 field plot design consisted of two transects, each including three 75 m² plots. Transects were placed parallel to each other and approximately 30 meters apart projecting perpendicular to the lake. Plots within each transect were 30 meters apart, with the plot nearest the lake approximately 30 meters from the lake edge and above the grass line. Distances between transects and plots within transects were chosen to minimize aerial drift between treatments. In 1982 an additional transect was added. Figure 1 illustrates plots, transects, and lake locations.

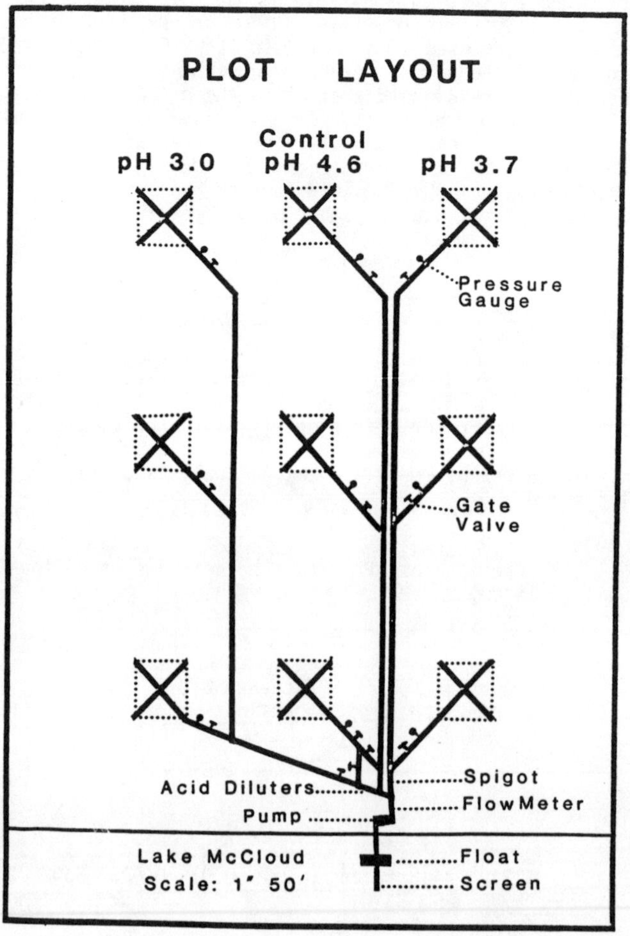

Fig. 1 Field experimental design showing relationship of individual plots and pH treatment transects

Plot Irrigation and Treatments

The first series of irrigations was conducted from March 1981 to July 1981. Fifty cm total of irrigation water was applied in 2.54 cm increments during that period. The irrigation events were evenly spaced. Simulated acid rain treatments consisted of acidified lake water at pH 3.7 and untreated lake water at pH 4.6 (control). Using a liquid chemical injector (Chemilizer Products, Inc., Clearwater, FL) on the transect main irrigation lines, a mixture of 70:30 H_2SO_4:HNO_3 was added to lake water to bring the pH to 3.7. One treatment was applied per transect. Irrigations were made using a 10 h.p. centrifugal pump at 65 psi head pressure and at 40 psi plot pressure. Four impact sprinklers were mounted per plot (one at each corner) with each sprinkler swinging ¼ arc and delivering 7.95 Lpm. Flowmeters, pressure gauges, a conductivity bridge and pH meter were used to monitor water flow and chemistry during irrigation events.

In 1982, the field plot acid rain treatment was modified and a pH 3.0 transect was added consisting of three plots. The second year irrigation events consisted of 18 irrigation applications (approximately 6 cm/irrigation) for a total of 100 cm irrigation from March 1 to August 31, 1982.

Soil Chemistry

Suction tubes with a porous fritted glass disc were placed at 15, 45, and 75 cm depths for the 1981 irrigation. Five tubes were placed at each depth in each plot. The 45 cm depth tubes were eliminated for the 1982 irrigation. Suction was drawn on each tube after each irrigation event and soil solutions were collected and analyzed for conductivity, pH, K, Ca, Mg, NH_4, NO_3, SO_4, P, and Al by appropriate atomic absorption or colorimetric techniques. Soil samples were collected at various depths from the plots and analyzed for pH and extractable cations by neutral normal ammonium acetate.

Results and Discussion

Soil pH's were analyzed at three depths before and after acid rain irrigations for the 1981 and 82 irrigation periods (Table 3). Data are averages of all plots receiving similar treatments. The pH 4.7 (or control) lake water irrigation had no effect on soil pH at any depth, which is not surprising since the untreated soil and lake pH are almost the same. For the pH 3.7 irrigations no soil pH change could be detected below the 7 cm depth. Shallow soil depth samples (0-7 cm) were collected for the 1982 irrigation since it was determined that any change in soil pH would only occur at the very surface of the soil. A decrease of approximately 0.3 pH units occurred for the 0-7 cm depth with the pH 3.7 treatment.

Table 3. Effects of Acid Rain on Soil pH over Two Years

Soil depth (cm)	pH 4.7				pH 3.7				pH 3.0	
	Control				Irrigation					
	B*		A**		B		A		B	A
	81	82	81	82	81	82	81	82	82	82
0- 7		4.73		4.77		4.87		4.60	4.83	4.43
0-15	4.73	4.73	5.20	5.00	4.88	4.87	5.13	4.83	4.83	4.63
15-75	4.80	4.53	5.33	4.73	5.30	4.57	5.20	4.63	4.70	4.53

* Before irrigation
** After irrigation

Data is only available for 1982 with the pH 3.0 treatment. Decreases in soil pH after acid rain irrigations were noted at all soil depths with the largest (0.4 units) occurring at the 0-7 cm depth.

Figures 2 and 3 show comparisons of data collected from 1981 and 1982 as well as the effects of acid rain pH on the soil solution Ca and Mg levels. Soil solution samples were collected from the suction tubes after each acid rain irrigation and analyzed. The horizontal axis on the figures of acid rain represents the cumulative amount of irrigation water applied. Each line represents 16 and 17 irrigation events for 1981 and 1982, respectively.

Levels of Ca and Mg found in the soil solution were generally slightly higher in 1981 for the control and pH 3.7 treatments when compared to 1982. This could be due to the fact that higher levels of Ca and Mg were present in the soil initially and the acid leaching in 1981 slightly lowered extractable levels for 1982. Another cause could be that higher natural rainfall occurred in 1982 thus slightly diluting soil solution Ca and Mg levels.

In all cases, the acid irrigation treatments of pH 3.7 and 3.0 increased the concentrations of Ca and Mg in the soil solution sampling tubes over the control irrigations. At the 15 cm sampling depth, the highest levels of Ca and Mg in the solution occurred early in the acid rain irrigations sequence (15-25 cm). The pH 3.0 treatment in 1982 clearly illustrates the rapid increase in Ca or Mg levels in soil solution followed by a gradual decrease in levels with additional irrigations. The rather sudden decrease in Ca or Mg concentrations in the pH 3.0 treatment at the 15 cm depth and at the 25 cm acid rain irrigation level may have been due to a rather heavy natural rain (4 cm) that occurred between irrigations and sampling thus diluting the sample collected.

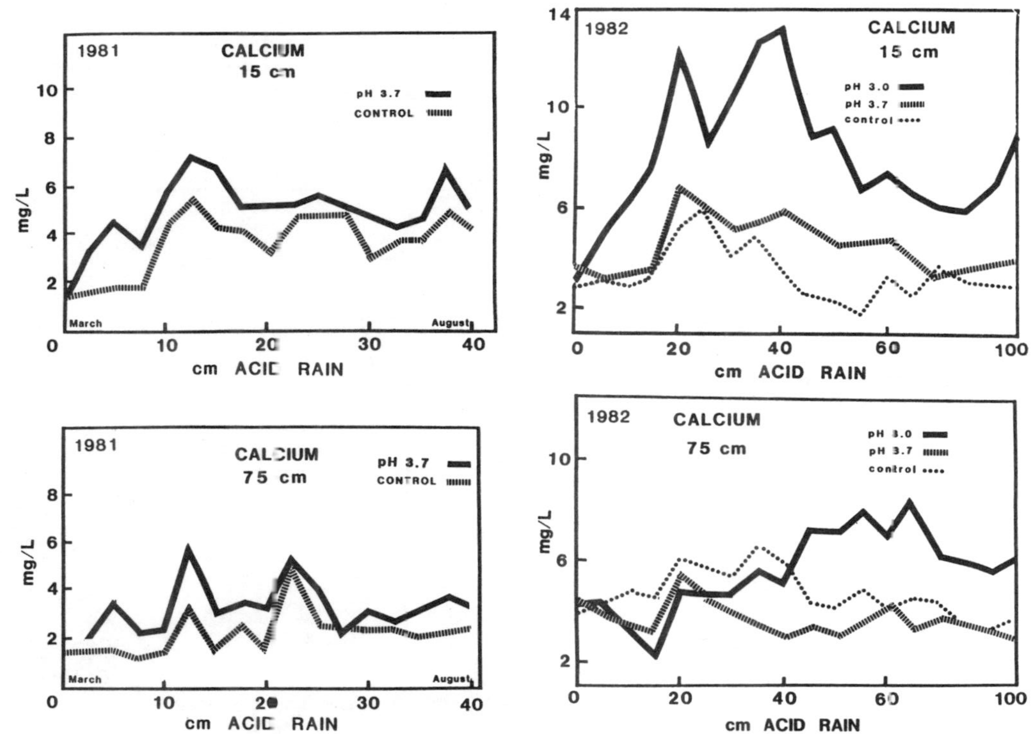

Fig. 2 Effects of acid rain irrigations at pH 3.0, 3.7, or 4.7 (control) on the level of Ca in soil solution samplers at 15 or 75 cm soil depths under field conditions for 1981 and 1982

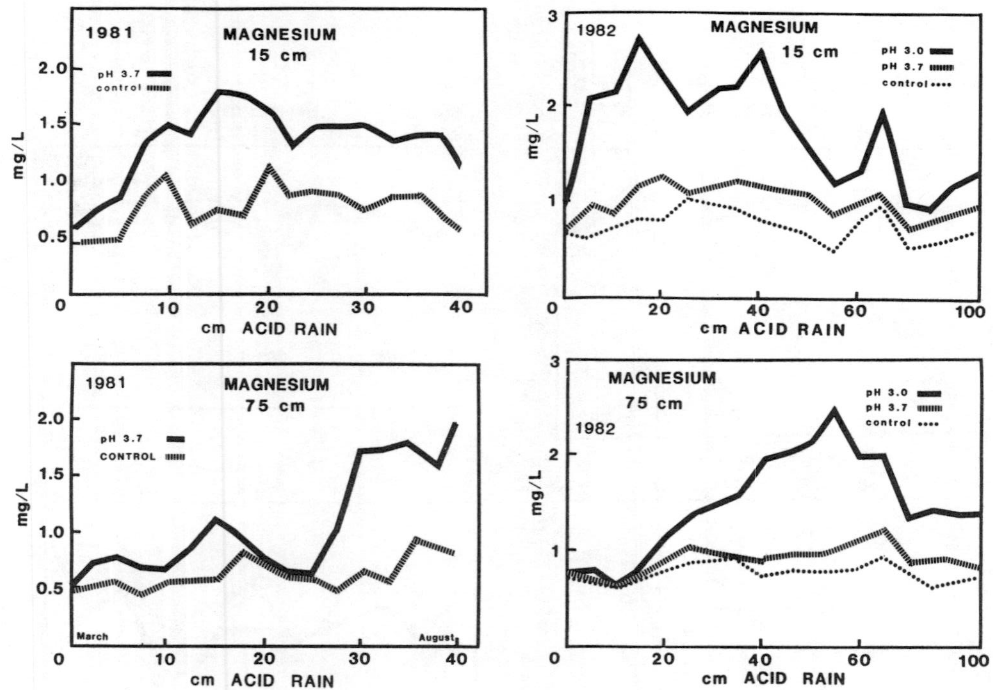

Fig. 3 Effects of acid rain irrigations at pH 3.0, 3.7, or 4.7 (control) on the level of Mg in soil solution samplers at 15 or 75 cm soil depths under field conditions for 1981 and 1982

Soil solution samples collected at 75 cm show a different elution pattern for Ca and Mg compared to the 15 cm depth. The graphs indicate a slow rise in Ca and Mg levels to a maximum level for the pH 3.0 treatment of 8 and 2.5 mg/L, respectively. The 75 cm 1981 graph for Ca does not show the gradual rise, however only 40 cm of acid irrigation were applied to the plots whereas 100 cm of acid irrigation were applied in 1982. At the 40 cm level of acid rain in 1982, the pH 3.0 treatment begins to show a steady increase in Ca concentration in the soil solution followed by a slow decrease.

In contrast to Ca, Mg levels in the pH 3.7 treatment for 1981 began to rise after 25 cm of precipitation and continued to increase. The 1982 data for Mg at 75 cm provides a clearer view of changes in Mg concentration with acid irrigations. For the pH 3.7 treatment a consistently higher but small difference (0.5 mg/L) is noted when compared to control Mg concentrations. The pH 3.0 treatment shows a gradual rise in Mg levels peaking at 2.5 mg/L then gradually decreasing.

It must be noted that 1982 was the first year for acid treatments at the pH 3.0 level whereas for the control and pH 3.7 treatments, it was the second year of irrigation. Since significant quantities of Ca or Mg had already been leached from the pH 3.7 treatments the first year, second year graphs may show substantially smaller differences between the control and pH 3.7 treatments.

In contrast to Ca and Mg, the graphs for K (Figure 4) show that the acid irrigation treatments in 1982 of pH 3.0 and 3.7 do not increase levels of K in the soil solution samplers at 15 or 75 cm compared to the control irrigation. Apparently the control irrigation at pH 4.7 is sufficiently acid to remove K from the exchange complex of the soil. This fact, plus the nature of monovalent K ions, causes lower pH irrigations to have little or no additional effect on K concentration in the soil solution.

Aluminum is an important element when considering toxic or detrimental effects of an element to plant roots and aquatic organisms. More specifically, concentrations of 0.2 to 0.4 mg/L Al^{+3} soil solution could be toxic to plant roots. Figure 5 illustrates Al^{+3} levels in the soil solution samplers at the various acid irrigation treatments. A maximum of 0.2 mg/L Al were found in the pH 3.7 treatment and greater than 5 mg/L were determined in the pH 3.0 irrigation. The Al level in the soil solution was found to be increasing in the pH 3.0 treatment after 100 cm of acid irrigation was applied. The movement pattern for Al at the 15 cm depth was considerably different than for other elements at this depth. Very little difference in Al levels for the different acid irrigations was noted until approximately 40 cm of acid rain had been applied. At this point the pH 3.0 treatment showed a rapid rise in Al levels. The rapid rise in Al levels in the soil solution with the pH 3.0 treatment can probably be explained by examining the change in pH of the soil solutions collected (Figure 6). At the 40 cm acid rain level the pH of the soil solution for the pH 3.0 treatment rapidly falls to below 5.0 and remains approximately level. At pH's below 5, Al becomes much more mobile

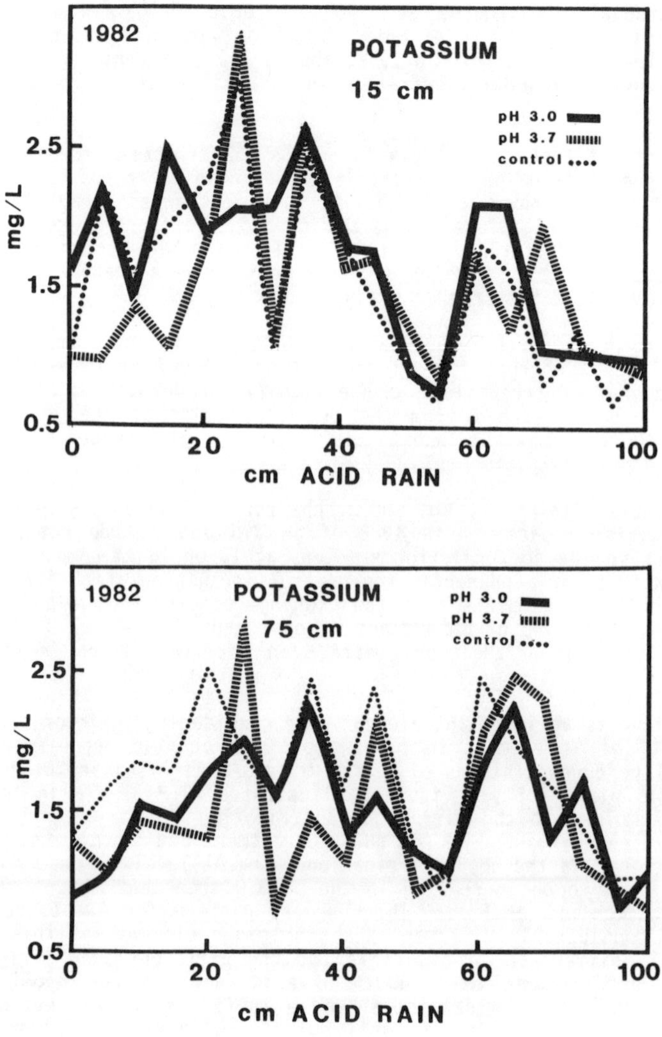

Fig. 4 Effects of acid rain irrigations at pH 3.0, 3.7, or 4.7 (control) on the level of K in soil solution samplers at 15 or 75 cm soil depths under field conditions

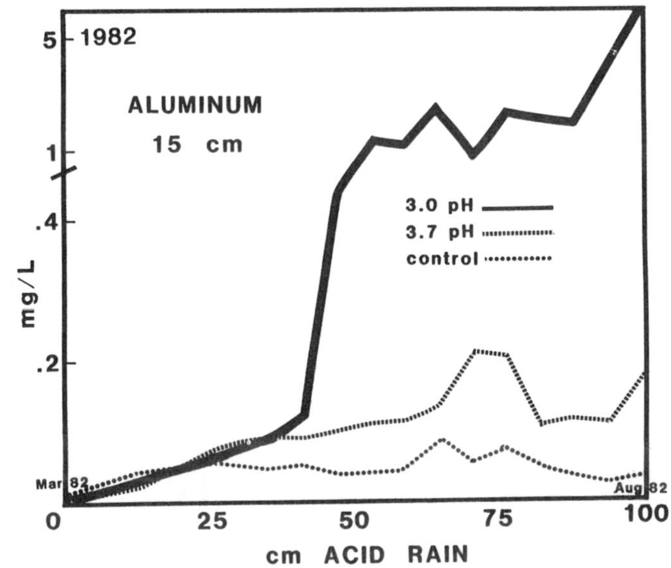

Fig. 5 Effects of acid rain irrigations at pH 3.0, 3.7, or 4.7 (control) on the level of Al in soil solution samplers at 15 or 75 cm soil depths under field conditions

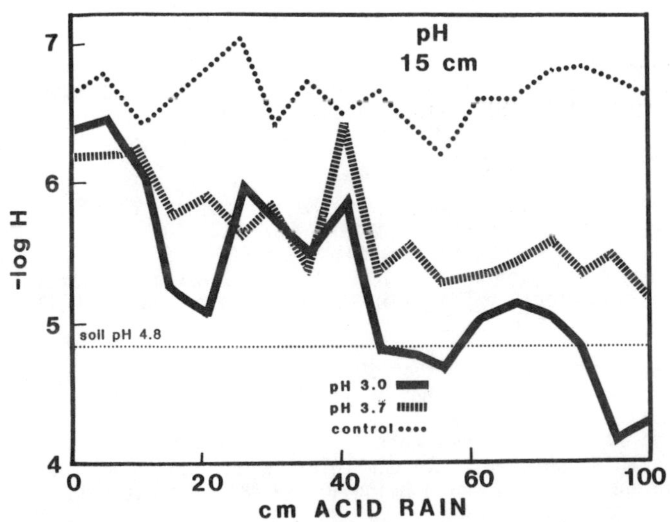

Fig. 6 Effects of acid rain irrigations at pH 3.0, 3.7, or 4.7 (control) on pH levels in soil solution samplers at 15 or 75 cm soil depths under field conditions

in the soil and can be removed from the exchange complex. This exchangeable Al is apparently causing the much higher levels of Al shown in Figure 5 with the pH 3.0 treatment. The further decline in solution pH (close to 4.0) for the 3.0 treatment at the 90-100 cm in acid rain level (Figure 6) corresponds with another increase in Al^{+3} in the soil solution at the same irrigation pH and cm of acid rain.

Effects of acid rain irrigations on changes in soil solution pH (Figure 6) show the control remains approximately constant at pH 6.7 over the entire 100 cm irrigation. The pH 3.7 irrigation treatment shows a slow decline from pH 6.2 to 5.2 of the soil solution over the entire period of application of acid rain. For the pH 3.0 irrigation treatment soil solution pH declines more steeply and falls from 6.4 to 4.3. If additional acid rain at pH 3.7 were to be added to these plots, it is anticipated that a sharp rise in Al levels in the soil solution would occur when the soil solution pH declined to values less than 5.

The major anion measured in the soil solution tubes for the field irrigation treatments was SO_4 (Figure 7). Sulfate was the major anion in the natural rain and was added in the sulfuric:nitric acid simulated acid rain treatments. Thus the magnitude of SO_4 concentration in the soil and its movement or possible adsorption could have an important effect on the movement of the cations.

Figure 7 shows the movement of SO_4 during the irrigation period at two depths. Sulfate at the 15 cm depth increased rapidly from the level added in the lake water and by the 40 cm application was slightly above the input concentration of 41 mg/L for the pH 3.0 treatment. There was then a sharp decline and the SO_4 stabilized in soil solution at approximately 25 mg/L. This decrease in part could be explained by three heavy rain events during this time in the irrigation sequence.

Sulfate in the soil solution for the pH 3.7 and control treatments approached the input concentrations of 14 and 7 mg/L, respectively. Note that the dotted horizontal lines (Figure 7) are the input irrigation SO_4 concentrations for pH 3.0, pH 3.7 and control (lake pH 4.7), and are 41, 14 and 7 mg/L, respectively.

Sulfate at the 75 cm depth for the pH 3.0 treatment initially remained lower until 40 cm of irrigation where there occurred a sharp rise to 25 mg/L. Although the SO_4 concentration at 75 cm depth never reached the input concentration of 41 mg/L, it did follow a similar pattern as SO_4 at the 15 cm depth. The pH 3.7 treatment also never reached the input concentration of 14 mg/L.

These results indicate that SO_4 at low levels is relatively mobile in the upper 15 cm of this sandy soil. There is obviously a capacity for some initial adsorption. At the 75 cm depth, approximately 65% of the total added SO_4 at the higher concentration (41 mg/L) was adsorbed by the soil above this depth, with lesser relative adsorption at the lower SO_4 concentrations (14 and 7 mg/L) being 40 and 38%, respectively.

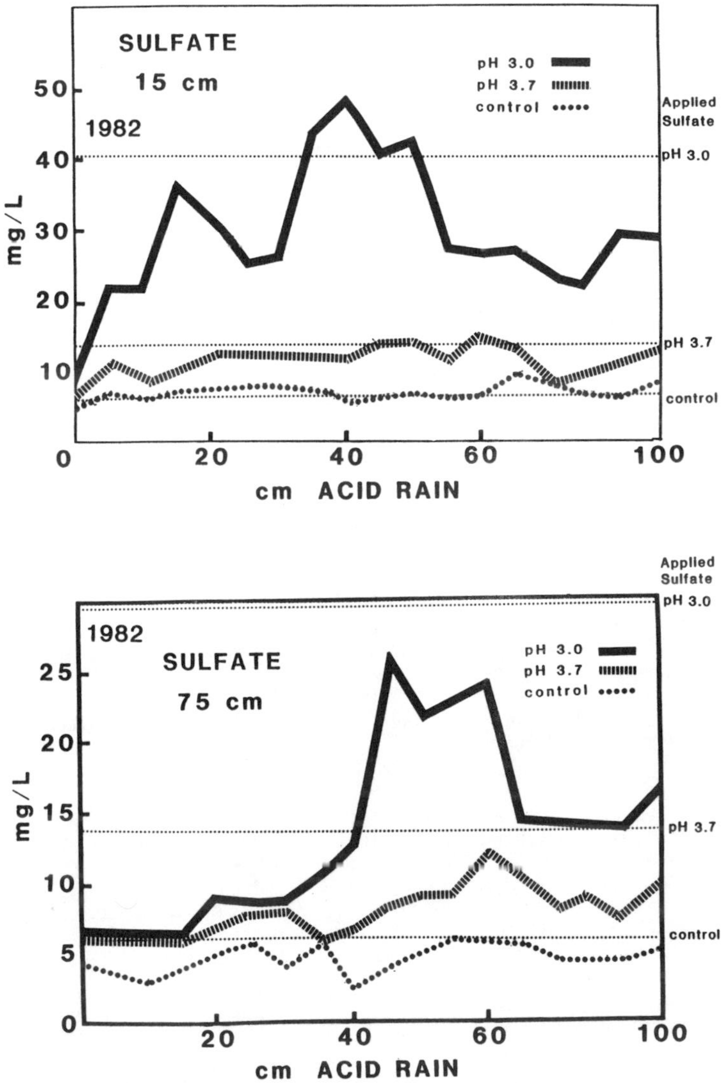

Fig. 7 Effects of acid rain irrigations at pH 3.0, 3.7, or 4.7 (control) on the level of SO_4 in soil solution samplers at 15 or 75 cm soil depths under field conditions

It is not unreasonable to project that a higher percentage of the SO_4 would be adsorbed at even greater depths. Thus movement of cations could be impeded by the SO_4 adsorption mechanism, since SO_4 could act as a bridging mechanism in cation retention, even in soils of low clay and organic matter content.

Research Needs in Florida

It is anticipated that acid deposition or acid rain problems will increase in Florida over the next decade. Direct effects on soil chemical changes including increased leaching of plant nutrients and movement of Al in the soil solution need considerably more research in field and laboratory studies. Effects of acid deposition on different types of soils under different vegetation needs more research. Initial studies have indicated that potential problems exist with increases in acid deposition but the magnitude of the problem can not yet be quantified.

Literature Cited

Babich, H., D. L. Davis, and G. Stotzky. 1980. Acid precipitation: Causes and consequences. Environment 22:6-41.

Bache, B. W. 1980. The acidification of soils. In: T. C. Hutchinson and M. Havas (Eds.), Effects of Acid Precipitation on Terrestrial Ecosystems. New York: Plenum Press. 183-202.

Cowling, E. B. 1980a. Acid precipitation and its effects on terrestrial and aquatic ecosystems. In: J. Kneis and P.J. Licy (Eds.), Aerosols: Anthropogenic and Natural Sources and Transport. Annals, New York Academy of Science 338:540-555.

Cowling, E. B. 1980b. A Status Report on Acid Precipitation and its Ecological Consequences as of December 1980. North Carolina State University, Raleigh, North Carolina.

Cronan, C. S., W. A. Reiners, R. C. Reynolds, Jr., and G. E. Lang. 1978. Forest floor leaching: Contributions from mineral, organic, and carbonic acids in New Hampshire subalpine forest. Science 200:309-311.

Evans, L. S., and T. M. Curry. 1979. Differential responses of plant foliage to simulated acid rain. Amer. J. Bot. 66:953-962.

Evans, L. S., A. J. Francis, and G. S. Raynor. 1977a. Acid Rain Research Program Annual Progress Report. Brookhaven National Laboratories, Upton, New York.

Evans, L. S., N. F. Gmur, and F. DaCosta. 1977b. Leaf surface and histological perturbations of leaves of *Phaseolus vulgaris* and *Helianthus* after exposure to simulated acid rain. Amer. J. Bot. 64:905.

Evans, L., N. Gmur, and D. Mancini. 1982. Effects of simulated acid rain on yields of *Raphanus sativa*, *Lactuce sativa*, *Triticum aestivum*, and *Medicago sativa*. *Env. Exp. Bot.* 22:445-453.

Evans, L., G. Kendrey, and K. Leivin. 1983. Productivity of field-grown soybeans exposed to simulated acid rain. National Acid Precipitation Assessment Program Effects Research Review, N.C. State University, February 21-25.

Ferenbaugh, R. W. 1976. Effects of simulated acid rain on *Phaseolus vulgaris* L. (Fabaceae). Amer. J. Bot. 63:283-288.

Irving, P. M. 1978. Induction of Visible Injury in Chamber-Grown Soybeans Exposed to Acid Precipitation. RER Division Annual Report, ANL-78-65-III, Argonne, Illinois.

Irving, P., and Miller, J. E. 1978. The Effects of Acid Precipitation Alone and in Combination with Sulfur Dioxide on Field-Grown Soybeans. RER Division Annual Report, ANL-78-65-III, Argonne, Illinois.

Jacobson, J. S. 1980. Experimental studies on the phytotoxicity of acidic precipitation: The United States experience. *In*: T. C. Hutchinson and M. Havas (Eds.), Effects of Acid Precipitation on Terrestrial Ecosystems. Plenum Press, New York.

Johnson, A. H., T. G. Siccama, D. Wang, R. S. Turner, and T. H. Baringer. 1981. Recent changes of the growth rate in the New Jersey Pinelands: A possible effect of acid rain. J. Environ. Qual. 10:427-430.

Lee, J. J., G. E. Nealy, and S. C. Perregan. 1980. Sulfuric Acid Rain Effects on Crop Yield and Foliar Injury. United States Environmental Protection Agency. Report No. EPS-600/3-80-016.

Lee, J. J., and D. E. Weber. 1976. A Study of the Effects of Acid Rain on Model Forest Ecosystems. Paper 76-25.5, 69th Annual Meeting, Air Pollution Control Association, Portland, Oregon.

Matziris, D. I., and G. Nakos. 1977. Effect of simulated "acid rain" on juvenile characteristics of Aleppo pine *(Pinus halepensis* Mill.). Forest Ecol. Management 1:267-272.

Newman, J. R., S. W. Fletcher, C. D. Hendry, C. D. Pollman, L. C. Carter, R. M. Matter, and J. R. Brice. 1982. A Literature Review of the Ecological Material Effects of Acid Deposition. Florida Acid Deposition Study, Environmental Science and Engineering, Inc., Gainesville, FL ESE No. 81-631-600.

Shriner, D.S. 1976. Effects of simulated rain acidified with sulfuric acid on host-parasite interactions. *In*: L. S. Dochinger and T. A. Selinga (Eds.), Proceedings of the First International Symposium on Acid Precipitation and the Forest Ecosystem, pp. 919-925, USDA Forest Service General Technical Report NE-23.

Shriner, D., and J. Johnston. 1981. Effects of simulated acid rain on nodulation of leguminous plants by Rhizobium spp. Env. Exp. Bot. 21:199-209.

Strayer, F. R., and M. Alexander. 1981. Effect of simulated acid rain on glucose mineralization and some physicochemical properties of forest soil. J. Environ. Qual. 10:460-465.

Tamm, C. O., and E. B. Cowling. 1976. Acidic precipitation and forest vegetation. In: Proc. First International Symposium on Acid Precipitation and the Forest Ecosystem, p. 854, USDA Forest Service, Gen. Tech. Report NE-23, Upper Darby, Pennsylvania.

Wainwright, M. 1980. Man-made emissions of sulphur and the soil. Intern. J. Environmental Studies 14:279-288.

Wilson, S. O. 1980. Actual and Potential Effects of Acid Precipitation on a Forest Ecosystem in the Adirondack Mountains. New York State Energy, Research, and Development Authority. 242 pp.

Wood, T., and F. H. Bormann. 1974. The effects of an artificial acid mist upon the growth of Betula alleghaniensis, Britt. Envir. Poll. 7:259.

SENSITIVITY OF FLORIDA LAKES TO ACIDIC PRECIPITATION

Daniel E. Canfield, Jr., H. L. Schramm,
J. V. Shireman and W. T. Haller
Center for Aquatic Weeds
University of Florida, Gainesville, FL 32611

Abstract

To assess the vulnerability of Florida lakes to damage by acidic precipitation, data from a survey of 165 lakes was used to determine pH, total alkalinity, calcium hardness and calcite saturation index values. Mean lake pH ranged from 4.1 to 8.9. Mean total alkalinity concentrations ranged from 0 to 204 mg/l as $CaCO_3$ and mean calcium hardness concentrations ranged from 1 to 215 mg/l as $CaCO_3$. Total alkalinity averaged below 10 mg/l as $CaCO_3$ in 49% of the sampled lakes. Calcite saturation index values were greater than 4 in 36% of the sampled lakes. Based on these data, Florida has a large number of lakes that are potentially vulnerable to damage by acid precipitation. However, data on chlorophyll a concentrations, zooplankton abundance and recent fisheries data suggest Florida lakes may not be as sensitive as the alkalinity and calcite saturation indices suggest. Low phosphorus and nitrogen concentrations are primarily responsible for reduced animal and plant populations in acidic Florida lakes. Consequently, predictions of future impacts or trends based on current sensitivity indices should be regarded with caution.

Introduction

The State of Florida has over 7,700 natural and artificial lakes (Florida Board of Conservation 1969). These lakes, which range in size from small 0.4 ha ponds to the 181,000 ha Lake Okeechobee, are an extremely valuable natural resource because they provide habitat for a diverse flora and fauna as well as water for many of man's domestic, industrial, agricultural and recreational activities. The acidity of rainfall in Florida, however, has increased over the past 25 years and annual average pH values below 4.7 now occur over the northern three quarters of the state (Brezonik et al. 1980). Although pH values in Florida's rainfall are not as low as those reported in other regions, Brezonik et al. (1980) found that the annual deposition of hydrogen ions in northern Florida is 33 to 50% of the deposition in the northeastern United States and sulfate deposition rates are 50 to 90% of the deposition rates at Hubbard Brook, New Hampshire. Recently, Crisman and Brezonik (1980), Crisman et al. (1980) and Beaver and Crisman (1981) have used data from a synoptic survey of 10 to 20 lakes to demonstrate that phytoplankton, zooplankton and benthic invertebrate communities are reduced in acidic Florida lakes. Because Florida has a number of poorly buffered lakes and sandy soils of low cation exchange capacity (Shannon and Brezonik 1972; Carlisle et al. 1978), the sensitivity of Florida lakes to acidic precipitation could be extremely high. The number of potentially sensitive lakes in Florida, their geographical location and their potential biological response to reductions in pH and alkalinity, however, are unknown.

During 1979 and 1980, we conducted a survey of 165 Florida lakes to determine their chemical and trophic state characteristics in relation to regional geology. Although the study was not specifically designed to determine the sensitivity of Florida lakes to damage by acidic precipitation, data collected during the survey can be used to provide a limited assessment of lake sensitivity. In this paper, our objectives are to document which geographical areas of Florida may contain lakes vulnerable to reductions in pH and total alkalinity by acidic precipitation and what the biological response may be if reductions in pH and total alkalinity occur.

Methods

Between August 1979 and September 1980, 165 Florida lakes were sampled 3 times to determine their pH, total alkalinity, calcium hardness, total phosphorus, total nitrogen, chlorophyll *a* and color levels. Zooplankton were sampled twice between February and September. All study lakes were chosen to represent lakes located on the major physiographic and geologic regions. Lake location, physiographic region and geologic formation are given in Canfield (1981). At each lake, surface water (0.5 m) and zooplankton samples were collected from 3 midlake locations. On larger lakes such as Lake Okeechobee, samples were collected from near-shore, open-water areas. Zooplankton were collected by making vertical tows with a 80 μm Wisconsin net. Water samples were placed on ice and zooplankton were preserved with 10% rose bengal stained formaldehyde prior to being transported to the laboratory for analysis.

At the laboratory, pH was measured by use of an Orion Model 601A pH meter calibrated against buffers at 4.0, 7.0, and 10.0. Total alkalinity was determined by titration with 0.02 N sulfuric acid (A.P.H.A. 1976). All samples were titrated to a pH of 4.5. Because the equivalence point occurs at pH > 4.5 in low alkalinity samples and many Florida lakes have low total alkalinities, reported alkalinities are greater than true alkalinities in many lakes. Calcium hardness was determined by titration with HexaVer chelating reagent (Hach Chemical Company 1975) using microburets. CalVer calcium indicator (Hach Chemical Company 1975) was used for endpoint determination. Calcium concentrations were calculated from calcium hardness data. Total phosphorus concentrations were determined by using the procedures of Murphy and Riley (1962) with a persulfate digestion. Total nitrogen concentrations were determined by using a modified Kjeldahl technique (Nelson and Sommers 1975). Color concentrations were determined by use of the platinum-cobalt method and Nessler tubes (A.P.H.A. 1976). The concentration of plankton algae in each lake was estimated by measuring chlorophyll *a* concentrations. For chlorophyll *a* analysis, a measured volume of lake water was filtered through a Gelman type A-E glass fiber filter. Filters were stored over dessicant and frozen until analyses could be completed. Chlorophyll *a* concentrations were determined by using the methods of Richards with Thompson (1952) and Yentsch and Menzel (1963). Chlorophyll *a* values were calculated by using the equations of Parsons and Strickland (1963). Corrections for phaeophytin were not made. Zooplankton abundance was determined by counting all copepods, cladocerans, and rotifers in a 5 ml aliquot from each sample.

A dissecting microscope and zooplankton counting wheel were used. Net efficiency was assumed to be 100% thus zooplankton counts represent relative abundances. Calcite saturation indices (CSI) were calculated for each lake by:

$$CSI = p(Ca^{2+}) + p(ALK) - p(H^{+}) + pk \quad (1)$$

where $p(x) = -\log(x)$, $pk = +2$, (Ca^{2+}) = mol/l, ALK = eq/l and H^{+} = eq/l (Conroy et al. 1974; Kramer 1976; Galloway et al. 1978). All chemical values used in calculating the CSI values were lake averages.

Results and Discussion

During our lake survey (Table 1), average pH values in the lakes ranged from 4.1 to 8.9. Mean total alkalinity concentrations ranged from 0 to 204 mg/l as $CaCO_3$ and mean calcium hardness concentrations ranged from 1 to 215 mg/l as $CaCO_3$. Average pH values were below 5.0 in 10% of the lakes while 20% had values between 5.0 and 6.0 and 30% had average pH values between 6.0 and 7.0 (Table 2). Total alkalinity concentrations averaged below 10 mg/l as $CaCO_3$ in 49% of the study lakes. An additional 29% of the lakes had an average total alkalinity concentration between 10 and 40 mg/l as $CaCO_3$ (Table 3). Average calcium hardness concentrations were below 10 mg/l as $CaCO_3$ in 41% of the lakes and 33% had values between 11 and 40 mg/l as $CaCO_3$ (Table 4). Based on these data, the majority of Florida lakes can be classified as acidic, low alkalinity, softwater lakes despite the presence of underlying limestone formations (Puri and Vernon 1964).

Although there is considerable variability in measured lake pH, total alkalinity, and calcium hardness values, the distribution of acidic and alkaline lakes in Florida is not random. There is a general increase in pH, total alkalinity and calcium hardness levels as one moves from the northwest to southeast and from inland highlands to lowlands (Figures 1, 2, 3). Although the lowest values occur in northern Florida where rainfall is acidic (Brezonik et al. 1980), the northwest to southeast gradient increases abruptly along the boundary between panhandle Florida and peninsular Florida. This corresponds to a change from Florida's northern physiographic zone to the central physiographic zone (White 1970) and to a change from the North Gulf Coast sedimentary province, which consists mainly of clastic sediments, to the Florida Peninsula sedimentary province, which consists primarily of carbonate, phosphate, and anhydrite sediments (Pressler 1947; Puri and Vernon 1964). Central Florida has the greatest variability in lake pH, total alkalinity, and calcium hardness levels because the region has a diverse geology and physiography (Puri and Vernon 1964; White 1970; Canfield 1981). In this region, discontinuous highlands form subparallel ridges which are generally separated by broad valleys (White 1970). Sinkhole lakes, which are generally above the piezometric surface and have small watershed to lake volume ratios are common on the ridges while broad shallow lakes, which are generally below the piezometric surface, are common on the valley floors (White 1970).

Table 1. Average pH, total alkalinity, calcium hardness, calcite saturation index (CSI), total phosphorus, total nitrogen, chlorophyll *a*, color, total zooplankton, and Secchi disc values for the Florida lakes sampled between August 1979 and September 1980. B for Secchi disc is a bottom reading

LAKE	pH	TOTAL ALKALINITY (mg/l as $CaCO_3$)	CALCIUM HARDNESS (mg/l as $CaCO_3$)	CSI INDEX	TOTAL PHOSPHORUS (mg/m^3)	TOTAL NITROGEN (mg/m^3)	COLOR (Pt-Co units)	CHLOROPHYLL *a* (mg/m^3)	TOTAL ZOOPLANKTON (organisms/l)	SECCHI DISC (m)
Lake Agnes	6.9	6	13	3	55	860	77	19	153	0.9
Alligator Lake	7.4	38	37	1.2	384	1700	35	50	1140	0.5
Alligator Lake	5.7	2	9	5	15	570	47	4	68	1.6
Lake Apopka	8.9	131	96	-1.3	166	4600	37	156	1480	0.2
Lake Arbuckle	7.0	10	24	2.3	49	850	112	19	332	0.8
Ariana Lake	8.0	35	36	0.7	21	700	7	20	165	1.0
Lake Arietta	7.0	6	14	2.8	16	360	8	8	80	2.9
Lake Ashby	6.9	13	17	2.5	18	580	146	12	167	0.8
Lake Baldwin	8.4	99	86	-0.6	39	630	11	22	233	1.5
Bear Lake	4.5	2	1	7.1	16	340	30	11	85	1.7
Blue Cypress Lake	7.7	31	44	0.9	67	750	110	5	81	1.1
Lake Broward	5.5	1	7	5.2	4	170	4	2	67	5.7
Lake Buffam	5.4	1	9	5.2	56	380	37	3	183	1.1
Lake Butler	5.5	1	20	4.8	11	240	3	1	123	6.0
Lake Butler	6.0	3	7	4.4	24	440	63	11	112	1.3
Carr Lake	6.1	4	3	4.7	15	440	15	7	157	2.0
Charles' Bay	4.9	2	1	6.3	14	470	45	3	284	1.5
Cherry Lake	5.9	2	5	4.8	27	290	4	5	244	2.3
Lake Clinch	6.7	4	10	3.4	12	310	8	7	116	2.2
Compass Lake	5.2	2	3	5.8	6	160	1	1	36	4.9
Lake Conway	7.5	29	32	1.2	13	400	3	4	95	3.7
Corn Landing Lake	7.0	27	30	1.8	13	570	82	3	271	1.3

Table 1. Continued

LAKE	pH	TOTAL ALKALINITY (mg/l as $CaCO_3$)	CALCIUM HARDNESS (mg/l as $CaCO_3$)	CSI INDEX	TOTAL PHOSPHORUS (mg/m³)	TOTAL NITROGEN (mg/m³)	COLOR (Pt-Co units)	CHLOROPHYLL *a* (mg/m³)	TOTAL ZOOPLANKTON (organisms/l)	SECCHI DISC (m)
Crescent Lake	6.3	4	11	3.7	14	410	15	3	125	3.0
Crescent Lake	6.8	20	39	2.1	30	1100	250	28	149	0.6
Crews Lake	7.1	21	25	1.8	13	710	45	4	285	1.5
Lake Crosby	5.7	2	5	4.9	17	420	30	5	107	1.9
Crystal Lake	5.7	2	2	5.5	6	120	0	1	24	8.1
Cypress Lake	7.8	22	25	1.2	131	1800	57	78	381	0.4
Dead Lake	6.1	12	12	3.5	15	310	68	4	32	2.0
Lake Deaton	7.1	25	14	2.0	29	2100	23	37	224	0.5
Deer Point Lake	6.8	23	19	2.2	9	180	62	2	13	2.1
Lake Dexter	7.2	51	100	0.8	115	990	136	18	268	0.7
Lake Dias	6.7	9	20	2.7	53	640	133	6	70	1.0
Dinner Lake	8.0	37	36	0.6	9	460	3	4	99	2.9
Lake Disston	5.2	3	7	5.2	10	830	383	2	65	0.4
Lake Dora	8.9	120	93	-1.3	90	3100	431	24	822	0.4
Lake Dorr	5.2	2	5	5.5	20	370	92	4	80	1.2
Lake Down	6.5	3	23	3.4	8	310	3	1	55	6.2
Dunford Pond	4.9	1	2	6.5	7	220	6	1	52	5.0
Eagle Lake	7.6	29	31	1.1	29	930	13	25	71	0.8
Lake Eaton	7.0	27	38	1.7	55	1100	416	3	212	0.4
Lake Ellen	4.9	1	3	6.2	12	320	97	3	57	1.5
Lake Eustis	8.7	104	82	-0.9	46	1800	19	65	595	0.4
Lake Fairview	8.0	52	55	0.3	15	450	5	3	82	4.8
Gap Pond	5.1	1	2	4.0	12	310	4	2	96	4.5
Lake Geneva	6.4	3	9	4.0	13	200	3	2	68	3.3

Table 1. Continued

LAKE	pH	TOTAL ALKALINITY (mg/l as $CaCO_3$)	CALCIUM HARDNESS (mg/l as $CaCO_3$)	CSI INDEX	TOTAL PHOSPHORUS (mg/m^3)	TOTAL NITROGEN (mg/m^3)	COLOR (Pt-Co units)	CHLOROPHYLL a (mg/m^3)	TOTAL ZOOPLANKTON (organisms/l)	SECCHI DISC (m)
Lake Gentry	6.1	3	9	4.2	19	610	80	5	82	0.8
Lake George	8.4	59	111	-0.5	56	1400	88	48	309	0.7
Lake Gibson	6.4	10	13	3.1	269	690	69	6	93	0.8
Lake Griffin	8.6	96	74	-0.8	83	2200	23	73	502	0.4
Hampton Lake	5.5	2	6	5.2	13	410	30	6	145	2.2
Lake Harney	7.5	47	130	0.5	99	1100	103	12	169	0.8
Lake Harris	8.5	95	78	-0.6	24	1100	15	21	459	0.8
Lake Hart	5.9	4	10	4.2	19	1100	183	4	77	0.6
Lake Hatchineha	7.3	24	28	1.5	46	1200	129	17	530	0.6
Lake Howard	8.9	41	42	-0.5	52	2000	27	105	208	0.3
Lake Huntley	6.0	3	10	4.3	22	410	13	6	252	1.6
Hurricane Lake	5.7	3	1	5.6	11	320	16	3	91	3.5
Lake Iamonia	5.6	3	2	5.2	7	430	29	8	130	2.2
Lake Iola	7.6	27	27	1.3	15	560	10	8	57	2.9
Lake Istokpoga	6.9	11	22	2.4	45	990	92	31	341	0.5
Lake Jackson	5.3	3	9	4.9	14	270	6	3	222	2.8
Lake Jackson	6.5	5	5	3.8	16	440	9	4	173	4.0
Lake Jackson	6.4	4	3	4.3	14	360	10	3	24	2.7
Lake Jessamine	7.8	44	43	0.7	16	610	5	5	183	3.4
Lake Jessup	8.4	72	101	-0.5	347	2800	46	157	1200	0.3
Johns Lake	6.5	5	16	3.3	27	580	18	7	489	1.0
Lake Josephine	5.6	3	13	4.6	24	520	52	11	219	0.8
Juniper Bay	5.5	2	1	5.9	20	390	17	8	144	2.4
Karick Lake	6.0	9	5	4.1	19	240	20	4	91	2.9

Table 1. Continued

LAKE	pH	TOTAL ALKALINITY (mg/l as $CaCO_3$)	CALCIUM HARDNESS (mg/l as $CaCO_3$)	CSI INDEX	TOTAL PHOSPHORUS (mg/m³)	TOTAL NITROGEN (mg/m³)	COLOR (Pt-Co units)	CHLOROPHYLL a (mg/m³)	TOTAL ZOOPLANKTON (organisms/l)	SECCHI DISC (m)
Lake Kerr	6.1	3	17	4.0	13	220	15	2	93	3.3
Kingsley Lake	7.0	10	13	2.6	11	280	6	2	43	4.2
Lake Kissimmee	7.9	22	27	1.0	42	1300	53	29	327	0.7
Lawne Lake	7.6	59	69	0.5	172	1100	62	18	245	0.9
Lake Lindsey	6.0	4	10	4.0	41	2200	44	40	996	0.8
Little Crooked Lake	5.4	2	9	5.1	54	1000	138	12	284	0.3
Little Red Water Lake	5.8	2	7	4.8	34	474	28	16	44	0.7
Lochloosa Lake	7.4	23	23	1.6	36	1200	61	32	187	0.7
Lake Lotela	7.3	14	5	2.6	17	330	6	3	54	2.6
Lake Louisa	4.9	2	8	5.6	9	870	230	3	71	0.7
Lake Louise	6.4	5	9	3.6	23	630	42	7	214	1.8
Lake Lowery	6.2	5	13	3.7	22	1100	79	10	185	1.4
Lowry Lake	5.1	1	1	6.5	13	90	1	2	57	3.8
Lake Maggiore	8.6	110	215	-1.2	77	2300	32	67	1500	0.3
Magnolia Lake	5.1	1	2	6.4	13	100	2	1	61	4.8
Lake Maitland	7.7	69	71	0.3	28	600	4	10	294	1.9
Lake Manatee	6.7	30	31	2.0	163	620	101	7	49	1.2
Lake Margaret	4.5	0	3	7.3	20	710	65	7	142	0.7
Lake Marian	7.8	22	26	1.1	75	1700	51	62	139	0.6
Lake Marion	7.3	29	34	1.4	22	810	37	11	430	1.0
Lake Mary Jane	5.1	3	9	5.1	18	1300	225	9	105	0.5
Lake McKenzie	4.7	1	1	6.9	4	180	9	1	46	5.3
Merial Lake	4.7	0	1	7.2	6	64	0	1	41	3.2
Merritts Mill Pond	8.1	96	97	-0.4	19	1500	2	1	7	B

Table 1. Continued

LAKE	pH	TOTAL ALKALINITY (mg/l as $CaCO_3$)	CALCIUM HARDNESS (mg/l as $CaCO_3$)	CSI INDEX	TOTAL PHOSPHORUS (mg/m^3)	TOTAL NITROGEN (mg/m^3)	COLOR (Pt-Co units)	CHLOROPHYLL a (mg/m^3)	TOTAL ZOOPLANKTON (organisms/l)	SECCHI DISC (m)
Lake Minnehaha	5.8	3	10	4.4	18	500	68	6	170	1.3
Lake Minneola	6.4	5	10	3.6	14	430	21	2	95	4.0
Lake Miona	7.2	25	31	1.6	15	1300	26	16	286	1.1
Mirrow Lake	4.9	1	2	6.6	3	110	3	1	93	7.1
Lake Monroe	7.6	45	108	0.4	91	1300	89	37	301	0.5
Moon Lake	6.7	8	14	2.9	14	660	28	3	127	2.7
Mountain Lake	6.3	12	12	3.2	16	820	123	7	349	1.3
Lake Munson	7.7	58	44	0.6	620	2900	51	145	768	0.4
Mystic Lake	6.7	10	8	3.1	12	550	13	4	401	B
Newnans Lake	6.8	14	14	2.6	52	1300	93	38	466	0.6
Ocean Pond	4.1	0	3	7.6	23	220	50	3	175	1.1
Ocheesee Pond	5.6	3	2	5.3	9	330	9	3	68	3.3
Lake Okahumpka	8.3	50	44	0.1	14	880	20	5	456	B
Lake Okeechobee	8.4	100	98	-0.7	105	1100	47	15	217	0.5
Orange Lake	7.2	18	18	2.0	31	1100	54	35	219	0.8
Osborne Lake	8.2	204	175	-1.1	140	1200	60	40	154	1.0
Otter Lake	4.5	2	11	5.8	29	490	220	3	152	0.9
Oyster Lake	6.5	19	97	2.0	34	550	210	4	142	0.9
Lake Padgett	7.3	23	31	1.6	14	530	15	2	92	3.6
Palestine Lake	5.0	1	3	6.4	20	290	37	4	332	1.1
Lake Panasoffkee	8.3	101	101	-0.6	17	580	38	4	277	B
Lake Parker	8.9	68	75	-0.9	79	1600	22	34	236	0.5
Lake Pasadena	6.4	6	11	3.5	27	800	55	9	317	1.2
Pate Pond	4.6	0	1	7.4	14	300	36	6	80	1.7

Table 1. Continued

LAKE	pH	TOTAL ALKALINITY (mg/l as $CaCO_3$)	CALCIUM HARDNESS (mg/l as $CaCO_3$)	CSI INDEX	TOTAL PHOSPHORUS (mg/m³)	TOTAL NITROGEN (mg/m³)	COLOR (Pt-Co units)	CHLOROPHYLL a (mg/m³)	TOTAL ZOOPLANKTON (organisms/l)	SECCHI DISC (m)
Lake Pierce	7.8	37	33	0.8	40	1200	28	23	466	0.8
Lake Placid	6.6	4	6	3.7	13	540	7	6	79	2.2
Lake Poinsett	7.7	58	146	0.1	49	1100	94	9	1030	0.6
Red Beach Lake	6.7	5	8	3.4	16	450	31	10	143	1.1
Reedy Lake	7.9	26	37	0.8	28	1100	8	27	227	0.7
Lake Rosalie	7.0	9	15	2.6	17	660	68	4	54	1.1
Round Lake	6.3	4	5	4.1	5	170	0	1	29	4.7
Lake Rousseau	7.3	92	95	0.5	48	460	70	2	10	2.1
Lake Rowell	7.4	24	56	1.2	66	720	47	12	166	0.8
Lake Sampson	6.9	10	30	2.3	34	520	67	3	83	2.0
Santa Fe Lake	6.0	3	5	4.5	16	360	33	5	90	2.2
Lake Sebring	5.7	2	8	4.7	112	690	133	6	67	0.4
Sellers Lake	4.9	1	2	6.6	4	260	3	2	109	5.8
Lake Seminole	6.8	20	16	2.4	44	510	20	11	35	0.5
Lake Seminole	0.8	90	93	-1.0	122	1900	27	65	723	0.3
Smith Lake	5.4	2	6	5.4	11	390	10	2	244	2.8
Lake Stanley	6.1	3	2	4.8	11	350	8	3	33	2.5
Lake Stella	7.0	16	42	1.9	13	460	12	3	130	4.1
Sun Lake	5.3	2	1	6.0	14	420	10	2	180	B
Lake Susan	5.1	2	9	5.3	14	840	200	9	104	0.9
Suwannee Lake	6.6	8	9	3.2	59	980	20	21	273	1.3
Lake Talquin	6.2	12	10	3.4	115	750	73	10	127	0.5
Lake Tarpon	6.8	16	62	1.9	39	640	50	4	50	1.5
Lake Thonotosassa	8.1	48	52	0.2	834	1500	82	58	101	0.7

Table 1. Continued

LAKE	pH	TOTAL ALKALINITY (mg/l as $CaCO_3$)	CALCIUM HARDNESS (mg/l as $CaCO_3$)	CSI INDEX	TOTAL PHOSPHORUS (mg/m³)	TOTAL NITROGEN (mg/m³)	COLOR (Pt-Co units)	CHLOROPHYLL a (mg/m³)	TOTAL ZOOPLANKTON (organisms/l)	SECCHI DISC (m)
Tiger Lake	7.3	33	13	1.7	43	780	67	16	152	0.7
Tiger Tail	8.9	66	52	-0.7	14	610	4	3	598	B
Lake Tohopekaliga	8.1	39	37	0.5	370	1700	53	70	308	0.4
East Lake Tohopekaliga	6.1	3	7	4.3	24	640	32	9	270	1.5
Townsend Pond	3.1	3	4	5.6	9	630-	120	6	196	1.3
Lake Trafford	8.4	111	82	-0.7	65	1300	48	28	921	1.0
Lake Tsala Apopka (Hernando Pool)	7.3	42	41	1.1	11	490	24	3	98	2.8
Turkey Pen Pond	4.7	1	1	7.3	5	68	1	1	76	4.7
Lake Underhill	7.6	58	72	0.5	53	780	10	38	136	0.8
Upper Myakka Lake	7.6	41	53	0.8	210	860	99	7	910	1.3
Lake Victor	5.7	6	6	4.5	12	290	15	3	74	2.7
Lake Virginia	7.9	57	59	0.2	31	520	3	17	78	1.6
Lake Wales	7.7	40	38	0.8	21	460	8	5	181	2.5
Lake Washington	7.8	62	101	0.1	32	1100	92	3	56	0.8
Watertown Lake	7.9	78	66	0.1	30	620	23	19	309	1.9
Lake Wauberg	7.8	21	16	1.4	80	2100	17	110	285	0.4
Lake Weir	7.1	12	10	2.5	12	640	3	8	128	2.3
Lake Weohyakapka	7.0	8	10	2.7	28	460	42	5	72	1.0
Western Lake	6.6	21	130	1.6	6	290	141	1	39	1.7
Wildcat Lake	4.8	1	2	6.7	8	190	18	1	68	3.2
Lake Wimico	6.6	21	20	2.5	28	490	113	4	51	0.5
Lake Winnimissett	6.8	7	33	2.5	7	260	4	1	52	5.8
Lake Yale	8.3	116	68	-0.5	14	680	7	10	380	1.4

Table 2. Frequency distribution and cumulative percentage of pH values in 165 Florida lakes

pH	Frequency	Cumulative Percentage
4.0 - 4.9	16	10
5.0 - 5.9	33	30
6.0 - 6.9	50	60
7.0 - 7.9	44	87
8.0 - 8.9	22	100

Table 3. Frequency distribution and cumulative percentage of total alkalinity concentrations (mg/l as $CaCO_3$) in 165 Florida lakes

Total Alkalinity	Frequency	Cumulative Percentage
0 - 10	81	49
11 - 20	17	59
21 - 40	30	78
41 - 80	22	91
81 - 160	14	99
>160	1	100

Table 4. Frequency distribution and cumulative percentage of calcium hardness concentrations (mg/l as $CaCO_3$) in 165 Florida lakes

Calcium Hardness	Frequency	Cumulative Percentage
0 - 10	68	41
11 - 20	26	57
21 - 40	28	74
41 - 80	22	87
81 - 160	19	99
>160	2	100

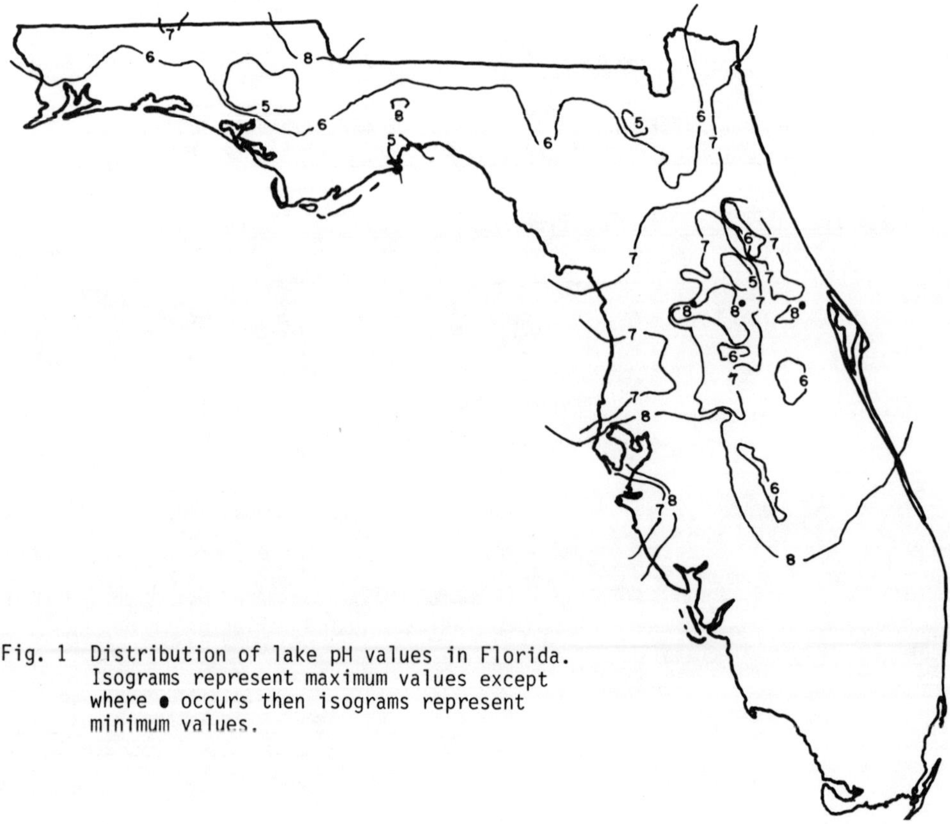

Fig. 1 Distribution of lake pH values in Florida. Isograms represent maximum values except where • occurs then isograms represent minimum values.

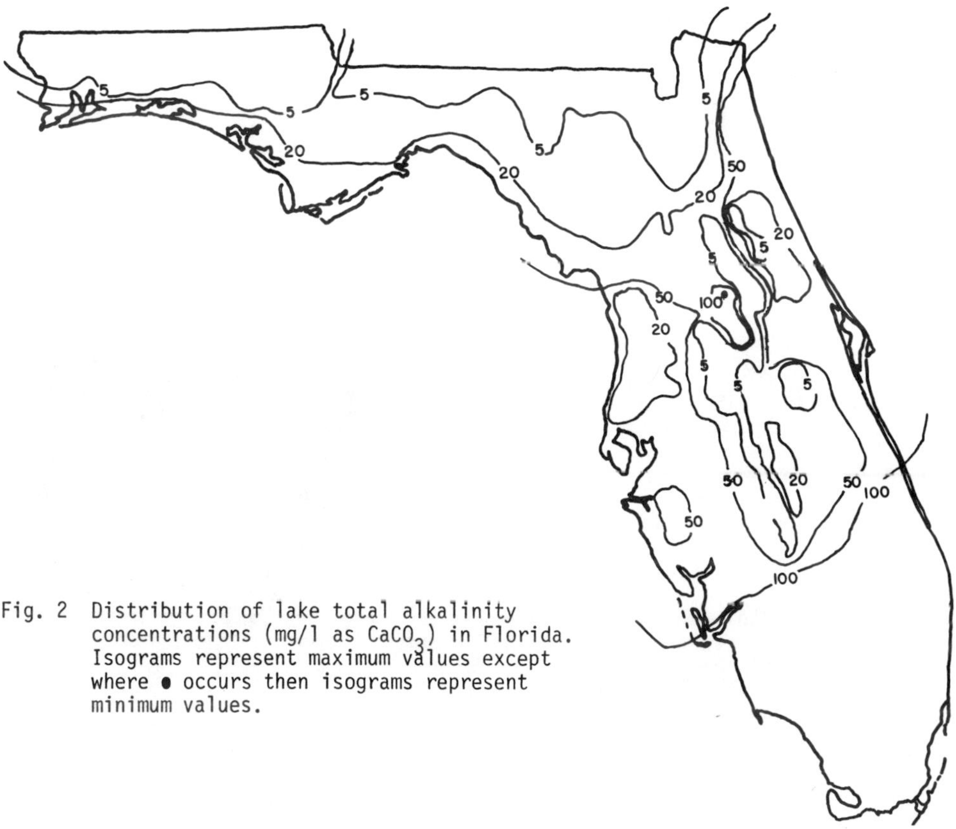

Fig. 2 Distribution of lake total alkalinity concentrations (mg/l as $CaCO_3$) in Florida. Isograms represent maximum values except where ● occurs then isograms represent minimum values.

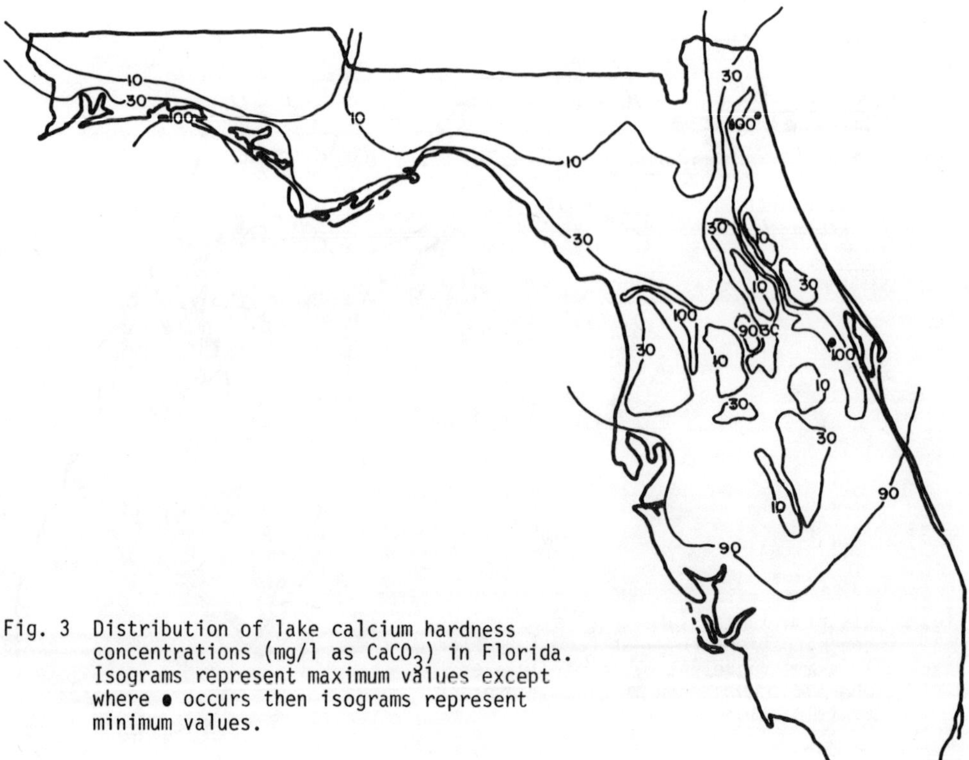

Fig. 3 Distribution of lake calcium hardness concentrations (mg/l as $CaCO_3$) in Florida. Isograms represent maximum values except where • occurs then isograms represent minimum values.

Prediction of geographical areas that may contain lakes sensitive to acidic precipitation are often based on bedrock geology, soil chemistry, or water chemistry (c.f. Haines 1981). A major criterion for assessing lake sensitivity fo acidic precipitation is a lake's total alkalinity as this reflects a lake's chemical capacity to neutralize acid inputs. Surface waters with a total alkalinity of 10 mg/l as $CaCO_3$ or less are often classified as highly sensitive to acidification and those with alkalinities between 10 and 20 mg/l as $CaCO_3$ are moderately sensitive (Altshuller and McBean 1979; Haines 1981). Using this criterion, 49% of the lakes we sampled would be classified as highly sensitive and 10% would be classified as moderately sensitive (Table 3). Conroy et al. (1974), Kramer (1976), and Galloway et al. (1978) suggested the calcite saturation index is a better index of neutralizing capacity. Glass and Loucks (1980) suggested lakes with CSI values less than 1 are located in terrain not susceptible to change or damage by acid loading. Lakes having CSI values between 1 and 2 may be susceptible to change and lakes with values between 2 and 3 are classified as probably susceptible to change. Lakes having CSI values between 3 and 4 are classified as susceptible to change and lakes having CSI values greater than 4 are classified as highly susceptible to change. For the lakes sampled in our survey, 36% had CSI values greater than 4, 11% had CSI values between 3 and 4, and 12% had CSI values between 2 and 3 (Table 5). By use of this criterion, only 35% of the lakes rather than 49% would be classified as unstable relative to acid loading and possibly vulnerable to damage by acidic precipitation.

Table 5. Frequency distribution and cumulative percentage of calcite saturation index (CSI) values in 165 Florida lakes

CSI Index	Frequency	Cumulative Percentage
<1	44	27
1 - 1.9	24	41
2 - 2.9	20	53
3 - 3.9	18	64
≥4	59	100

Although fewer lakes would be classified as highly sensitive by use of the CSI criterion, results obtained by use of either the CSI or alkalinity indices are in good agreement (Table 1). By use of either index, 59% of the sampled lakes would be classified as susceptible to change if acid loading increased. Both indices indicate sensitive lakes (c.f. Figures 1 and 2) are located primarily in north Florida but that there are sensitive geologic regions in peninsular Florida. Haines (1981), however, observed the accuracy with which lake sensitivity can be predicted by the CSI or alkalinity indices is unknown as factors other than the chemical buffering capacity of a lake can affect the response of a given lake to acid loading. As an example, he cited Schindler et al. (1980), who found the pH of Lake 223 in the Experimental Lakes Area, Canada decreased less than calculated when acid was added. Schindler et al. (1980) found sulfate reduction by bacteria in anoxic regions caused the generation of dissolved inorganic carbon. Mobilization of iron from the sediments during anoxic periods seemed to cause the dissolved organic carbon respired by decomposers to be accumulated as bicarbonate thus buffering against acidity. Hendrey et al. (1976), and Yan (1979) found phytoplankton biomass and production in acid and non-acid lakes were similar when phosphorus levels were similar.

For the lakes sampled in our survey, chlorophyll *a* concentrations and zooplankton abundance are lower in acid lakes (Table 6). There is a positive correlation between pH and logarithmically transformed chlorophyll *a* concentrations ($r = 0.61$; $P < 0.01$) and zooplankton concentrations ($r = 0.41$; $P < 0.01$). From these data, it is tempting to conclude reductions in pH lead to reductions in chlorophyll *a* and zooplankton concentrations in Florida lakes. However, the acid lakes also have low phosphorus and nitrogen concentrations; thus the question arises whether the low chlorophyll *a* and zooplankton concentrations are the result of low pH or the lack of nutrients. It is difficult to separate the effects of pH and nutrients as the variables are intercorrelated (Table 7). However, partial correlation coefficients between pH and chlorophyll *a* concentrations and pH and zooplankton concentrations with the effects of the variation of phosphorus and nitrogen eliminated are 0.17 and -0.02 respectively. This suggests pH has little effect on chlorophyll *a* and zooplankton concentrations in acid and non-acid lakes. Based on these data, we conclude phosphorus and nitrogen, not pH, are the factors responsible for the low chlorophyll *a* and zooplankton concentrations measured in the low pH lakes (Table 6).

Many of Florida's freshwater ecosystems have traditionally been characterized by diverse assemblages of both submerged and emergent aquatic vascular plants. Changes in aquatic macrophyte populations may have direct effects on fish populations. However, little information exists on the effects of pH on the composition and functioning of aquatic plant communities in Florida lakes. Submerged aquatic plant communities which have traditionally characterized low fertility, low pH lakes in northern Florida include the genera *Eleocharis*, *Myriophyllum*, *Sagittaria*, *Mayaca* and *Utricularia* among others. More alkaline waters in central and south Florida are now characterized by exotic *Hydrilla*, but prior to the introduction of *Hydrilla* the genera *Najas*, *Ceratophyllum*, *Vallisneria* and *Potamogeton* dominated the submerged flora (Blackburn et al. 1968).

Waters which are dominated by submerged macrophytes are characterized by widely fluctuating diurnal pH changes. Studies by Van et al. (1976) have shown that pH of *Hydrilla* infested water in lakes ranged from a high of 10 at dusk to a low of 7.1 at daybreak. Although the plant grows most profusely in more alkaline waters, laboratory studies showed that *Hydrilla* prefers to utilize free CO_2 in photosynthesis, but is capable of utilizing HCO_3^-. Optimal or maximum photosynthesis of *Hydrilla* occurs at pH of 3 to 6 coinciding with highest free CO_2 levels. Other macrophytes also grow in waters having a wide range of pH. Waterhyacinth (*Eichhornia crassipes*) is one of the major aquatic weeds of the world and grows profusely in waters of high fertility regardless of pH values between 4.0 and and 8.0 (Haller and Sutton, 1973). If these examples of submerged (*Hydrilla*) and floating (*Eichhornia*) aquatic macrophytes are indicative of the response of other plants in these groups, it appears that low pH alone will not limit their growth. More likely, it will be chemical complexes or changes in fertility levels.

Table 6. Average chemical and biological characteristics in Florida lakes for different pH ranges. Numbers in parentheses are the minimum and maximum values.

PARAMETER	pH				
	4.0-4.9	5.0-5.9	6.0-6.9	7.0-7.9	8.0-9.0
TOTAL ALKALINITY (mg/l as $CaCO_3$)	1 (0-4)	2 (1-6)	10 (3-30)	37 (8-92)	92 (39-204)
TOTAL PHOSPHORUS (mg/m^3)	12 (3.4-29)	20 (3.6-112)	33 (5.4-270)	67 (8.6-622)	126 (14-834)
TOTAL NITROGEN (mg/m^3)	319 (64-872)	460 (91-1300)	600 (170-2200)	960 (280-2900)	1700 (580-4600)
COLOR (mg/l as Pt)	53 (0-230)	60 (0-383)	64 (0-416)	42 (3-136)	34 (2-88)
SECCHI DISC (m)	2.8 (0.7-7.1)	2.4 (0.3-8.1)	1.9 (0.4-6.2)	1.5 (0.4-4.8)	0.6 (0.1-1.5)
CHLOROPHYLL *a* (mg/m^3)	3 (0.5-11)	5 (0.5-16)	8 (0.8-40)	23 (2-145)	53 (1-156)
TOTAL ZOOPLANKTON (organisms/l)	117 (41-332)	125 (24-284)	165 (13-996)	262 (10-1140)	529 (7-1500)

Table 7. Correlations (r) between pH and the logarithms of total phosphorus (TP), total nitrogen (TN), chlorophyll *a* (Chl *a*) and zooplankton (Zooplk) concentrations for 165 Florida lakes

	pH	TP	TN	Chl *a*	Zooplk
pH	1.00	0.54	0.67	0.62	0.41
TP		1.00	0.73	0.77	0.49
TN			1.00	0.85	0.62
Chl *a*				1.00	0.65
Zooplk					1.00

There have been no studies to determine the response of Florida's fish populations to reductions in pH, but there is evidence the fish can tolerate low pH and total alkalinity levels. For example, Ocean Pond, an oligo-mesotrophic lake (Table 1; total phosphorus = 23 mg/m^3; total nitrogen = 222 mg/m^3; chlorophyll a = 3.3 mg/m^3) had an average pH of 4.1, the lowest of any lake in our survey. Ocean Pond also had an average total alkalinity of 0, a calcium hardness concentration of 2.7 mg/l as $CaCO_3$ and a CSI index of 7.6 (Table 1). By use of either the alkalinity or CSI index, Ocean Pond, which is located in north Florida where rainfall is acidic, would be classified as one of Florida's most sensitive lakes. Yet, Ocean Pond supports a sustained sport fishery for largemouth bass (*Micropterus salmoides*), black crappie (*Pomoxis nigromaculatus*), bluegill (*Lepomis macrochirus*) and other centrarchids (Vaughn et al. 1978; Krummrich et al. 1979). In addition, the Florida Game and Fresh Water Fish Commission has successfully introduced the sunshine bass (*Morone chrysops* x *Morone saxatilis*) which further improved sport fishing (Vaughn et al. 1976). Although successful fish production at pH 4.1 might be due to low toxic metal (e.g., aluminum) levels or high color levels (Ocean Pond has a color value of 50 mg/l as Pt), Kautz (1981) has shown nutrients are a prime determinant of fish biomass in clear and colored Florida lakes. This suggests, similar to chlorophyll a, zooplankton and aquatic macrophytes, phosphorus and nitrogen concentrations not pH are the primary limiting factors.

Aquatic macrophyte abundance may also influence sportfish populations. Lake management procedures have been conducted to encourage growth of aquatic plants which in turn seems to stimulate productivity of sportfish populations (Wegener and Williams 1974). Generally, lakes which are characterized by having vegetated littoral regions of up to 30 or 40 percent of the surface area have been excellent sport fishing lakes and are characterized by balanced fish populations (Colle 1982). However, the necessity of aquatic macrophytes in lakes or a direct cause-effect relationship between macrophytes and sportfish production is still a matter of debate (Maceina and Shireman 1983). Comparison of condition factor of 300-400 mm total length (TL) largemouth bass (*Micropterus salmoides*) collected by electrofishing during August-December indicates significantly lower condition factor in Orange Lake (Alachua County) (Table 8). The significantly lower condition factor of largemouth bass in this intermediate pH lake is largely due to the abundance of aquatic vegetation. Dense *Hydrilla* covered 80% of the open water area of this 5142 ha lake. Colle and Shireman (1980) reported decreased condition factor of harvestable bass where *Hydrilla* exceeded 30% areal coverage. pH also does not significantly affect the distribution of largemouth bass and chain pickerel (*Esox niger*) in a diversely vegetated lake. Based on concomitant day and night electroshocking samples in 14 different habitats in Orange Lake, there was no significant relationship between abundance (catch/hour) of largemouth bass >150 mm TL or chain pickerel >200 mm TL and pH (Tables 9 and 10). Mean pH in these various habitats ranged from 5.3 - 7.8 during the May samples and 5.9 - 9.4 during the August samples.

We believe there are other factors that might reduce the potential sensitivity of Florida lakes to acidic precipitation. First, many Florida soils are naturally acidic and pH values below 4 are not uncommon

Table 8. Comparison of condition factor of 300-400 mm total length largemouth bass collected from four Florida lakes with different mean open water pH

Lake	Mean pH	Mean Condition	Sample Size	
Santa Fe Lake (Alachua County)	6.0	1.42	40	a[1/]
Newnans Lake (Alachua County)	6.8	1.47	23	a
Orange Lake (Alachua County)	7.2	1.36	13	b
Lake Henderson (Citrus County)	7.4	1.44	33	a

[1/] Means with same letter are not significantly different $p<.05$.

Table 9. Average (day and night) pH and catch/hour of largemouth bass >150 mm total length and chain pickerel >200 mm total length in various habitats in Orange Lake, Florida in May 1982

Habitat	pH		Catch/hour			
			Largemouth Bass		Chain Pickerel	
(Dominant macrophytes)	Value	Rank	Number	Rank	Number	Rank
Paspalidium (*Panicum*), *Hydrilla*	7.8	1	33.7	6	29.1	1
Hydrilla	6.6	2	20.0	11	20.0	2
Hydrilla	6.4	3	106.2	1	0.0	13.5
Hydrilla	6.2	4	33.8	5	7.5	12
Nuphar, *Hydrilla*	6.1	5	24.0	10	16.0	3
Polygonum, *Utricularia*, *Cabomba*	6.0	6.5	52.2	2	14.3	4
Paspalidium (*Panicum*), *Hydrilla*	6.0	6.5	39.1	3	11.1	10
Hydrilla	5.9	9	28.0	9	12.6	7
Nuphar	5.9	9	4.6	14	13.2	5
Nuphar	5.9	9	34.1	4	12.0	9
Polygonum	5.8	11	19.0	12	12.8	6
Hydrilla	5.7	12	29.7	7	12.2	8
Polygonum, *Nuphar*, *Utricularia*, *Cabomba*	5.6	13	29.4	8	10.1	11
Nuphar, *Hydrilla*	5.3	14	13.3	13	0.0	13.5
Spearman's Rank Correlation Coefficient (r)			.405		.424	

Table 10. Average (day and night) pH and catch/hour of largemouth bass >150 mm total length and chain pickerel >200 mm total length in various habitats in Orange Lake, Florida in August 1982

Habitat (Dominant macrophytes)	pH Value	pH Rank	Largemouth Bass Number	Largemouth Bass Rank	Chain Pickerel Number	Chain Pickerel Rank
Paspalidium (Panicum), Hydrilla	9.3	1	6.6	8	7.3	8
Polygonum, Nuphar, Utricularia, Cabomba	8.3	2	24.0	5	0.0	12
Hydrilla	8.0	3	0.0	12.5	0.0	12
Polygonum Utricularia, Cabomba	7.8	4	8.6	6	0.0	12
Paspalidium (Panicum), Hydrilla	7.5	5	36.4	3	14.3	5
Nuphar, Hydrilla	6.9	6	53.8	1	24.5	1
Hydrilla	6.7	7	0.0	12.5	15.0	4
Nuphar	6.6	8	33.4	4	12.0	6
Hydrilla	6.5	9	0.0	12.5	0.0	12
Hydrilla	6.4	10	4.0	9	9.5	7
Polygonum	6.1	11	3.3	10	0.0	12
Nuphar, Hydrilla	6.0	12.5	38.4	2	3.5	9
Nuphar	6.0	12.5	0.0	12.5	20.0	3
Hydrilla	5.9	14	8.0	7	24.0	2
Spearman's Rank Correlation Coefficient			.119		- .313	

in lower soil horizons (Carlisle et al. 1978). Because many Florida lakes receive the majority of their annual water budget from surface and subsurface runoff and Florida soils represent a large reservoir of hydrogen ions, a large percentage of the annual hydrogen ion loading rate will be due to hydrogen ions released from surrounding soils rather than from precipitation. Hydrogen ions originating from precipitation should dominate hydrogen ion loading rates only in those lakes that have small watershed to lake volume ratios and receive the majority of their annual water loading from direct precipitation. Second, Florida's fish fauna especially the sport fish fauna is composed of centrarchids. Unlike salmonids, which seem to be extremely sensitive to low pH, centrarchids tend to tolerate lower pH values. In Florida, the fish fauna has evolved in naturally acidic conditions thus reproduction and growth can occur even at pH 4.1 as is the case for Ocean Pond. Finally, damage from acidic precipitation has been documented primarily in the oligotrophic lakes of Scandinavia, Canada, and the eastern United States. In Florida, over 75% of the lakes are either mesotrophic or eutrophic (Canfield 1981). Because Hendrey et al. (1976), Yan (1979) and our data have shown phytoplankton biomass and zooplankton abundance are similar in acid and nonacid lakes with similar nutrient levels, we believe acidification probably will not alter primary production or zooplankton abundance in the majority of Florida lakes and thus will not adversely impact fish biomass or production as nutrients seem to be the prime determinant of fish biomass in Florida lakes. However, while the majority of Florida lakes probably are not vulnerable to major damage by acidic precipitation or at least their pH and alkalinity values could be reduced to the levels found in Ocean Pond without eliminating fish. This should not be construed to mean that there are no sensitive lakes in Florida because sensitive lakes probably do occur. These lakes are most likely Florida's small, clear, soft-water oligotrophic sinkhole lakes. Because these lakes often have small watershed to lake volume ratios, rainfall on the lake surface can represent a significant source of water. Many of these lakes are located in northern Florida where the acidity of rainfall has been increasing for the past 25 years. In our survey, pH in these lakes was near or below 5. However, if our survey is representative of other Florida lakes, these lakes probably represent 10 to 15% of the lakes in Florida.

Predictions concerning the future impact of acidic precipitation on Florida lakes, however, should be regarded with caution because there remains a great deal of uncertainty about the factors that control the sensitivity of lakes to acidification. Phytoplankton, zooplankton, and fish survive and reproduce in Florida lakes at pH values equal to or below those that have apparently caused damage to aquatic systems in northern regions. This apparent anomaly may result for several reasons including concentrations of toxic metals, complexation of toxic metals by organic ligands, different animal and plant species, long-term genetic adaptation to naturally acidic conditions, and the presence of sufficient nutrients. In Florida, as well as other regions, there is a paucity of good data on important variables such as fish populations, toxic metal concentrations, and sources of acid inputs. Models such as the Vollenweider nutrient loading model which have been used to evaluate the effect of changing nutrient loading rates on phosphorus (Canfield and Bachmann 1981) and nitrogen (Bachmann 1980) concentrations in lakes have not been developed to assess the effects of changing acid loading rates. Information on the possible interaction between acidification and eutrophication is also lacking, thus there is a need for the collection of good data which can be used to define general patterns of lake behavior. Until these problems are clarified, a great deal of uncertainty will be associated with any predictions concerning the impact of acidic precipitation on southern and northern aquatic ecosystems.

Literature Cited

Altshuller, A., and G. McBean. 1979. The LRTAP problem in North America: a preliminary overview. Report of the United States-Canada Research Consultation Group on the Long-Range Transport of Air Pollutants. United States Environmental Protection Agency, Research Triangle Park, North Carolina, USA.

American Public Health Association. 1976. Standard Methods for the Examination of Water and Wastewater. 14th ed., American Public Health Association, Washington, D.C.

Beaver, J. R., and T. L. Crisman. 1981. Acid precipitation and the response of ciliated protozoans in Florida lakes. Internationale vereinigung fur Theoretische und Angewandte Limnologie Verhandlungen 21:324-326.

Blackburn, R. O., P. F. White, and L. W. Weldon. 1968. Ecology of submerged aquatic weeds in South Florida canals. Weed Sci. 16:261-266.

Brezonik, P. L., E. S. Edgerton, and C. D. Hendrey. 1980. Acid precipitation and sulfate deposition in Florida. Science 208:1027-1029.

Canfield, D. E., Jr. 1981. Chemical and Trophic State Characteristics of Florida Lakes in Relation to Regional Geology. Final Report. Submitted to Florida Cooperative Fish and Wildlife Unit, University of Florida, Gainesville, Florida.

Carlisle, V. W., R. E. Caldwell, F. Sodek, III, L. C. Hammond, F. G. Calhoun, M. A. Granger, and H. L. Breland. 1978. Characterization Data for Selected Florida Soils. Soil Science Research Report Number 78-1, University of Florida, Gainesville, Florida.

Colle, D. E. 1982. Hydrilla - Miracle or migraine for Florida's sportfish. Aquatics 4(2):6.

Colle, D. E., and J. V. Shireman. 1980. Coefficients of condition for largemouth bass, bluegill, and redear sunfish in hydrilla-infested lakes. Transactions of the American Fisheries Society 109:521-531.

Conroy, N., D. S. Jeffries, and J. R. Kramer. 1974. Acid shield lakes in Sudbury, Ontario region. Proceedings 9th Canadian Symposium on Water Pollution Research. Canada 9:15-61.

Crisman, T. L., and P. L. Brezonik. 1980. Acid rain: threat to sensitive aquatic ecosystems. Proceedings 73rd Annual Meeting of the Air Pollution Control Association.

Crisman, T. L., R. L. Schulze, P. L. Brezonik, and S. A. Bloom. 1980. Acid precipitation: the biotic response in Florida lakes. Pages 296-297 _in_ Drabløs and Tollan (1980).

Florida Board of Conservation. 1969. Florida lakes. Part 3. Gazetteer. Div. Water Resour. Tallahassee, Florida.

Galloway, J., E. Cowling, E. Gorham, and W. McFee. 1978. A national program for assessing the problem of atmospheric deposition (acid rain). National Atmospheric Deposition Program, Report to the Council on Environmental Quality, Fort Collins, Colorado, USA.

Glass, G., and O. Loucks, editors. 1980. Impacts of Air Pollutants on Wilderness Areas of Northern Minnesota. United States Environmental Protection Agency. EPA-600/3-80-044, Duluth, Minnesota, USA.

Hach Chemical Company. 1975. Water and Wastewater Analysis Procedures. 3rd ed. Hach Chemical Company, Ames, Iowa.

Haines, T. A. 1981. Acidic precipitation and its consequences for aquatic ecosystems: A review. Transactions of the American Fisheries Society 110:669-707.

Haller, W. T., and D. L. Sutton. 1973. Effect of pH and high phosphorus concentration on growth of waterhyacinth. Hyacinth Contr. J. 11: 59-61.

Hendrey, G., K. Baalstrud, T. Traaen, M. Laake, and G. Raddum. 1976. Acid precipitation: some hydrobiological changes. Ambio 5:224-227.

Kautz, R. S. 1981. Effects of eutrophication on the fish communities of Florida lakes. Proc. Ann. Conf. S.E. Assoc. Fish & Wildl. Agencies 34:67-80.

Kramer, J. 1976. Geochemical and Lithological Factors in Acid Precipitation. United States Forest Service General Technical Report NE-23-611-618.

Krummrich, J., D. Jones, G. Byerley, and J. Chamberlain. 1979. 1978 Fish Management Report for the Northeast Florida Region. Florida Game and Fresh Water Fish Commission, Tallahassee, Florida.

Maceina, M. J., and J. V. Shireman. 1983. Influence of dense hydrilla infestation on Black Crappie growth. S.E. Association Game and Fish Comm. (In Press).

Murphy J., and J. P. Riley. 1962. A modified single solution method for the determination of phosphate in natural waters. Anal. Chim. Acta 27:31-36.

Nelson, D. W., and L. E. Sommers. 1975. Determination of total nitrogen in natural waters. J. Environ. Qual. 4:465-468.

Parson, T. B., and J. D. Strickland. 1963. Discussion of spectrophotometric determination of marine-plant pigments, with revised equations for ascertaining chlorophylls and carotenoids. J. Mar. Res. 21:155-163.

Puri, H. S., and R. O. Vernon. 1964. Summary of the geology of Florida and a guidebook to the classic exposures. Florida Geological Survey Special Publication No. 5.

Raddum, G. 1980. Comparison of benthic invertebrates in lakes with different acidity. Pages 330-331 in Drabløs and Tollan (1980).

Richards, F. A. with T. G. Thompson. 1952. The estimation and characterization of plankton populations by pigment analyses; II. A spectrophotometric method for the estimation of plankton pigments. J. Mar. Res. 11:156-171.

Schindler, D., R. Wagemann, R. Cook, T. Ruszczynski, and J. Prokopovich. 1980. Experimental acidification of Lake 223, Experimental Lakes area: background data and the first three years of acidification. Canadian Journal of Fisheries and Aquatic Sciences 37:342-354.

Shannon, E. E. , and P. L. Brezonik. 1972. Limnological characteristics of north and central Florida lakes. Limnology and Oceanography 17: 97-110.

Van, T. K., W. T. Haller, and G. Bowes. 1976. Comparison of the photosynthetic characteristics of three submersed aquatic plants. Plant Physiol. 58:761-768.

Vaughn, T. L., E. J. Moyer, D. Weaver, and K. Denson. 1976. 1975 Fish Management Report for the Northeast Florida Region. Florida Game and Fresh Water Fish Commission. Tallahassee, Florida.

Wegner, W., and V. Williams. 1974. Completion Report - Lake Tohopekaliga Drawdown Study. Florida Game and Fresh Water Fish Commission, O. J. Report F-29. Tallahassee, Florida.

White, W. H. 1970. The Geomorphology of the Florida Peninsula. Florida Department of Natural Resources Geological Bulletin No. 51.

Yan, N. 1979. Phytoplankton community of an acidified, heavy metal-contaminated lake near Sudbury, Ontario: 1973-1977. Water, Air and Soil Pollution 11:43-55.

Yentsch, C. S., and D. W. Menzel. 1963. A method for the determination of phytoplankton chlorophyll and phaeophytin by fluorescence. Deep Sea Res. 10:221-231.

PERSPECTIVES ON BIOTIC RESPONSES TO ACIDIFICATION IN FLORIDA LAKES

Thomas L. Crisman and Raymond W. Bienert, Jr.
Department of Environmental Engineering Sciences
University of Florida, Gainesville, FL 32611

Abstract

While generalizations regarding the response of biota to progressive acidification have been established for temperate lakes, comparable data for subtropical lakes are lacking. Based on: 1) a limnological survey of acidic Florida lakes and additional detailed two-year monitoring of the most acidic lake of this survey, and 2) *in situ* and laboratory experiments, an initial comparison of temperate and subtropical biotic responses to acidification is now possible. With the exception of reduced importance of diatoms and the absence of *Sphagnum*, autotrophic communities in acidic Florida lakes are similar to those of comparable temperate lakes. Major differences in the response of heterotrophic communities between the two climatic zones are apparent. Bacterial numbers are not reduced as noted for temperate lakes but approximate levels found in eutrophic lakes of the state. In addition, unlike temperate lakes, the species composition of zooplankton, benthic invertebrate, and fish communities does not appear markedly affected by acidification. Results from *in situ* enclosure and laboratory experiments suggest that major alterations in all autotrophic and heterotrophic communities will occur if the pH of Florida lakes drops below 3.5 - 3.8. Of most serious concern is the potential elimination of copepods and centrarchid fish. The implication of such biotic alterations on the functioning of lake ecosystems remains unknown.

Introduction

Acid rain was first described in 1852 by Robert Smith who attributed observed damage in Manchester, England to the burning of coal (5). It was not until the 1960's and 1970's that anthropogenic acid precipitation was recognized as a widespread phenomenon affecting Scandinavia, Canada and the eastern United States (20). Within the United States, acid rain has both expanded geographically and become more severe in the past three decades (8).

The susceptibility of a lake to acidification depends on its carbonate buffering capacity. The most sensitive lakes are those located in areas of igneous or metamorphic bedrock and having alkalinities less than 25 mg $CaCO_3$/L.

All levels of the trophic chain appear to be affected by the acidification of temperate lakes. Submergent macrophyte communities are often greatly simplified both with respect to species richness and abundance, and submergent species are replaced by *Sphagnum* moss (14). The importance of direct pH effects versus increased heavy metal availability and competitive exclusion by *Sphagnum* has not been adequately investigated.

Within the phytoplankton, species richness and diversity decline, and chrysophytes, pyrrhophytes, diatoms and green algae become the dominant components (35, 24, 34). While phytoplankton biomass may be reduced in oligotrophic acidic lakes, this may be more a reflection of phosphorus availability than direct pH effects (2, 22).

Lake acidification also has a pronounced impact on heterotrophic communities in temperate lakes. Heterotrophic bacterial decomposition rates appear to be reduced in acidic lakes (18) resulting in increased accumulation of coarse organic sediments (15). The species richness, diversity and total abundance of zooplankton are reduced with acidification (3), and rotifers often assume numerical dominance over crustaceans (25). It has been suggested that the importance of herbivorous zooplankton as a controlling factor for phytoplankton community structure is reduced in acidic lakes (22). As with zooplankton, the benthic invertebrate community is simplified, and major species and group replacements, including the loss of amphipods and gastropods, have been documented (23, 16).

The fish response to lake acidification is better documented than for any other biotic component. Loss of fish populations due to acidification has been reported from Norway (33), Sweden (1), Canada (17), and the United States (27). While direct pH effects may be expected, indirect effects of lowered pH on the availability of sodium, calcium (19), aluminum (11), and heavy metals (4) have been implicated as the primary reason for the loss of fish populations in acidic waters.

Acid Deposition in Florida and the Sensitivity of Lakes

Annual average rainfall pH (1978-1979) for the panhandle and the northern two-thirds of peninsular Florida is 4.7, while that of southern Florida is 5.0 - 5.4. Summer values are consistently lower than winter values and may be a reflection of the importance of local emissions (7).

The susceptibility of Florida lakes, as for temperate lakes, to acidification is a reflection of their carbonate buffering capacity and initial pH. The lake regions in the state most sensitive to acidification are those of Washington, southern Leon, Clay, Putnam, eastern Marion, and Highlands Counties (Figure 1). As an example of the potential magnitude of the problem, of 57 lakes surveyed in the Ocala National Forest during the winter of 1982-1983, 48 had pH values of 4.2 - 6.0 and alkalinities <5 mg $CaCO_3$/L (Crisman, Garren, Clarkson, unpublished data).

Limnological data for Florida lakes are sparse prior to 1970. Brezonik et al. (7) were able to compare current pH values for 11 acidic lakes in Clay and Putnam counties with published historical values. Of the 11, pH had declined progressively in 9 lakes since 1957 ranging from 0.1 to 0.6 pH units. The two lakes exhibiting a pH increase for the period, Geneva and Kingsley, have undergone pronounced residential development in their watersheds.

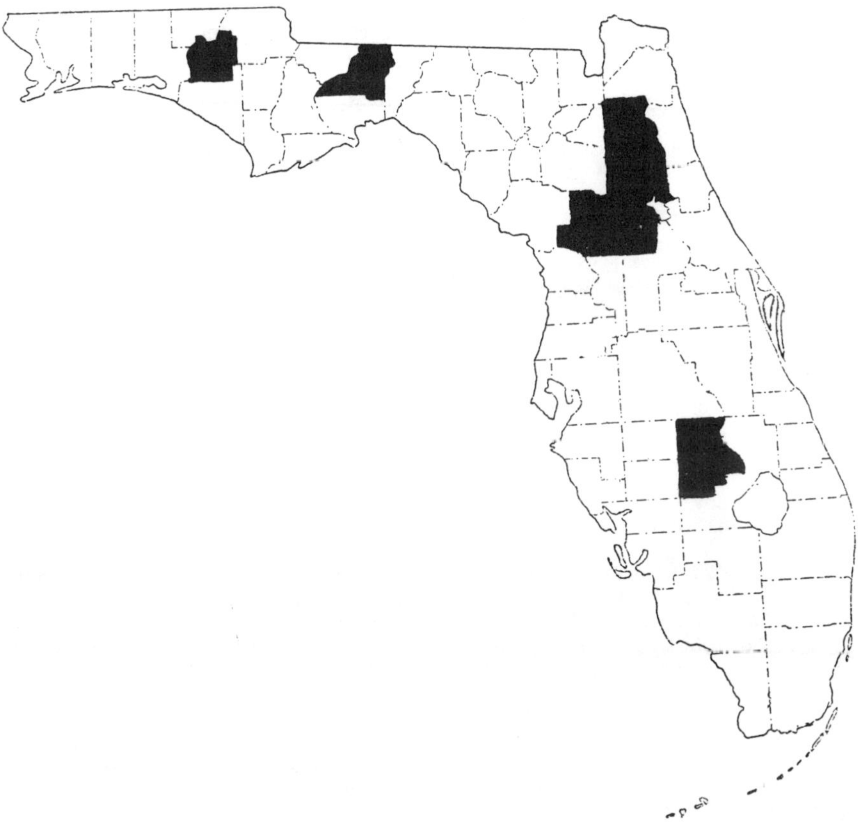

Fig. 1 Preliminary map of major lake districts in Florida sensitive to man-induced acidification

Finally, initial paleolimnological investigations have provided some insight into the duration of acidic conditions in Florida lakes and anthropogenic loading of heavy metals. Evidence from subfossil diatoms and cladocerans suggest that Lake Annie, a 40,000 year old oligotrophic lake (current pH 5.7) in Highlands County, has been acidic throughout most of its history (Crisman, unpublished data). In addition, analysis of heavy metal deposition rates in surficial sediment cores from seven Florida lakes (30) shows enrichment of cadmium, lead and nickel in the most recent sediments, presumably a result of enhanced atmospheric deposition.

Biotic Responses to the Acidification of Florida Lakes

Most of the past research effort in delineating the aquatic biotic response to the acidification of Florida lakes has concentrated on surveys of various biological and chemical parameters in a series of lakes representing a pH gradient. EPA initially funded P. L. Brezonik and T. L. Crisman (1977-1979) to survey 22 lakes, and work has continued with NADP funding (1980-1983) to conduct detailed biological and chemical monitoring and field experiments at Lake McCloud, the most acidic lake examined in the EPA survey (pH 4.6).

Bacteria. Water column bacteria concentrations in Lake McCloud (pH 4.6) during 1981 ranged from $0.3 - 2.7 \times 10^6$ cells/ml. Concentrations were always greater than summer values reported for temperate lakes of comparable trophic state (31). While bacteria data for subtropical lakes are generally lacking, Lake McCloud values are similar to those reported by Crisman and Scheuerman (9) for hypereutrophic Lake Brooker, Florida. Similarly, Boylen et al. (6) suggested that the acidification of Adirondack lakes has not appreciably reduced bacterial numbers.

Results from experimental enclosures in Lake McCloud suggest that acidification of Florida lakes below pH 3.6 may reduce bacterial numbers, but whether bacterial acitivity will also be lowered is unknown (10). Some authors (18) suggested that bacterial decomposition is reduced with the acidification of temperate lakes, while others (26) found no difference in decomposition rates.

Macrophytes. In a floristic survey of north Florida lakes, Garren (12) noted a general loss of submergent vegetation with increasing acidification. The vegetated littoral zone in the most acidic lakes is greatly restricted and consists of only a narrow shoreline fringe of emergent grasses and sedges. While the same trend has accompanied the acidification of temperate lakes (14), submergent vascular macrophytes are often replaced by a dense growth of *Sphagnum* moss. We suggest that the general loss of submergent vegetation with the acidification of Florida lakes may alter fish community structure via a loss of breeding habitat and increased susceptibility of juvenile fish to predation.

Phytoplankton. As in comparable temperate lakes, Schulze (28) found that phytoplankton abundance and chlorophyll levels generally declined in Florida lakes along a gradient of increasing acidity. While the importance of blue-green algae declines with increasing acidification in both temperate and subtropical lakes, acidic Florida lakes differ from comparable temperate lakes principally due to the near absence of diatoms in the former.

Schulze (28) suggested that the reduced abundance of phytoplankton in acidic Florida lakes may be a reflection of the general reduced availability of phosphorus in most acidic lakes. Subsequent *in situ* experiments at Lake McCloud, in which phosphorus and nitrogen were added to acidified (pH 3.7 and 4.0) and ambient (pH 4.6) experimental enclosures (Bienert, Ogburn, Crisman and Brezonik, unpublished data), demonstrated that in even the most acidic Florida lakes, phytoplankton abundance can be greatly enhanced through nutrient addition.

Zooplankton. While zooplankton abundance is greatly reduced with the progressive acidification of temperate lakes (3, 29), Schulze (28) failed to demonstrate a significant reduction in Florida lakes representing a comparable gradient of pH conditions (pH 4.6 - 6.6). In addition, the cladoceran *Daphnia* was not eliminated in the most acidic Florida lakes.

Further acidification of Florida lakes will likely cause a major alteration in zooplankton community structure. Results from a 96-hour bioassay experiment in which calanoid copepods (*Diaptomus*) collected from Lakes McCloud (pH 4.6) and Geneva (pH 6.2) were exposed to a gradient of pH conditions (3.7 - 6.2) suggested that acidification of Florida lakes to pH 3.8 will result in the elimination of this important zooplankton component (13). In addition, total copepod abundance was reduced and calanoid copepods eliminated in concurrent (1981) *in situ* experiments at Lake McCloud (pH 4.6) in which a water column enclosure was acidified to pH 3.6 (10). Similar multi-enclosure experiments in Lake McCloud during 1982 suggested that the entire copepod assemblage will be significantly reduced, if not eliminated, if Florida lakes are acidified to pH 3.6 (Bienert, Ogburn, Crisman and Brezonik, unpublished data). The impact on total ecosystem structure and productivity brought about by the potential elimination of these important algal grazers and fish food component is unknown.

Benthic Invertebrates. Neither the total biomass nor the diversity of benthic invertebrate assemblages in Florida lakes is significantly reduced with progressive lake acidification (7). In addition, with the exception of the elimination of mollusks in the most acidic lakes presumably reflecting an inability to form calcium carbonate shells, no major species replacements were noted. This is in marked contrast to observations in comparable temperate lakes where benthic community structure is greatly simplified, and species composition is altered dramatically (32, 21).

Fish. The most recent (1975-1980) fish data of the Florida Game and Fresh Water Fish Commission were compiled for 11 acidic lakes (pH 4.5 - 6.0) in the panhandle and north-central Florida (10). With the exception of brown bullhead, which was found only in Round Lake (pH 5.8), 11 fish species were common to all lakes. Thus, unlike comparable temperate lakes, no major species have been eliminated in acidic Florida lakes. In spite of a lack of species losses, the condition of most specimens collected from highly acidic lakes is considered poor.

Research Needs in Florida

With the exception of phytoplankton and macrophytes, the response of aquatic biota to the man-induced acidification of Florida lakes differs from that reported for comparable temperate systems. In part, this may be attributed to differences in the seasonal input of highly acidic water to lake ecosystems. For example, the greatest acidification of temperate lakes is associated with spring snow melt. Since most fish breeding occurs during this period, fish populations may be subject to severe pH stress during the most critical part of their life cycle. Florida fish populations, on the other hand, begin breeding during late winter and early spring, a period of low rainfall and thus low pH stress.

It is often difficult to separate direct and indirect pH effects as causal factors for observed biotic changes occurring during progressive lake acidification. Heavy metals and aluminum are mobilized with a reduction in pH, often reaching levels toxic to biota. In addition, alterations in one trophic level of an ecosystem can influence the structure and function of other trophic levels. Observed reductions in both phytoplankton and macrophytes with acidification may be strongly influenced by nutrient availability. The biomass of plant-dwelling benthic invertebrates may be reduced when submergent macrophytes are eliminated from acidic lakes thus directly affecting fish populations depending on benthos as food.

While we have not observed changes in Florida lakes as drastic as have been reported for comparable temperate systems, initial *in situ* acidification experiments suggest that further acidification will have a serious impact on the structure and function of aquatic ecosystems in the state. Future research in Florida should concentrate on the following questions:

1) At what pH will serious biotic effects be noted in Florida lakes? We must determine if clear water and colored lakes will be affected similarly.

2) Given projected estimates of human population growth in the state and associated needs for electric power, when will atmospheric loadings reach the level that will acidify lakes to the point that serious biotic changes will be expected?

3) How can we prevent the futher acidification of Florida lakes and in the future restore those lakes that have undergone serious biotic alterations?

Literature Cited

1. Almer, B., 1972. Forsurningens Inverkan Pa Fiskbestand I Vastkutsjoar. Information fran Sotuattens-Laboraotiret, Drottningholm, No. 12., 47 p.

2. Almer, B., Dickson, W., Ekstrom, C., and Hornstrom, E., 1978. Sulfur Pollution and the Aquatic Ecosystem, IN: Sulfur in the Environment. Part II. Ecological Impacts. Edited by J. O Nriagu. J. Wiley and Sons, Inc., New York, N.Y., pp. 271-311.

3. Almer, B., Dickson, W., Ekstram, C., Hornstrom, E., and Miller, U., 1974. Effects of Acidification of Swedish Lakes, Ambio., 3:30-36.

4. Beamish, R. J., 1976. Acidification of Lakes in Canada by Acid Precipitation and the Resulting Effects on Fishes, pp. 479-498. In: Proceedings of the First International Symposium on Acid Precipitation and the Forest Ecosystems. USDA Forest Service, Columbus. 1074 pp.

5. Boyle, R. H., and Boyle, R. A., 1983. Acid Rain, Nick Lyons/Schocken Books, New York. 146 pp.

6. Boylen, C. W., Schick, M. O., Roberts, D. A., and Singer, R., 1983. Microbiological Survey of Adirondack Lakes of Varying pH, Appl. Env. Microbiol. (In Press).

7. Brezonik, P. L., Hendrey, C. D., Jr., Edgerton, E. S., Schulze, R.L., and T. L. Crisman, 1980. Acidity, Nutrients, and Minerals in Atmospheric Precipitation over Florida: Deposition Patterns, Mechanisms, and Ecological Effects, Completion Rept. Grant no. 805560, U.S. EPA, Corvallis, Oregon.

8. Cogbill, C. V., 1976. The History and Character of Acid Precipitation in Eastern North America, pp. 363-370. IN: Proceedings of the First International Symposium on Acid Precipitation and the Forest Ecosystem. USDA Forest Service, Columbus. 1074 pp.

9. Crisman, T. L., and Scheuerman, P. R., 1982. Algal Management Through Lake Aeration: Annual Report for 1981, submitted to Florida Dept. of Natural Resources. 16 pp.

10. Crisman, T. L., Bienert, R. W., Foran, J. A., Gunn, M. A., Scheuerman, P. R., Gourlie, N., Garren, R. A., Binford, M. W., Ogburn, R. W., and Brezonik, P. L., 1982. Current and Projected Effects of Acid Precipitation on the Biota of a Softwater Florida Lake, annual report for the project "Ecological Effects of Acid Precipitation on a Sensitive Lake in Florida" funded by U. S. EPA-NCSU Grant APP-007-02-1980. 55 pp.

11. Dickson, W., 1978. Some Effects of the Acidification of Swedish Lakes, Verh. Internat. Verein. Limnol., 20:851-856.

12. Garren, R. A., 1982. Macrophyte Species Composition - Trophic State Relationships in Fourteen North Central Florida Lakes, M. S. Thesis, University of Florida, Gainesville. 89 pp.

13. Gourlie, N. M., 1981. The Effect of Low pH on Survival and Grazing Rates of *Diaptomus floridanus*, M. S. project, University of Florida, Gainesville. 30 pp.

14. Grahn, O., 1976. Macrophyte Succession in Swedish Lakes Caused by Deposition of Airborne Acid Substances, pp. 519-530. IN: Proceedings of the First International Symposium on Acid Precipitation and the Forest Ecosystem. USDA Forest Service, Columbus. 1074 pp.

15. Grahn, O., Hultberg, H., and Landner, L., 1974. Oligotrophication - A Self-Accelerating Process in Lakes Subject to Excessive Supply of Acid Substances, Ambio., 3:93-94.

16. Hagen, A., and Langeland, A., 1973. Polluted Snow in Southern Norway and the Effect of the Meltwater on Freshwater and Aquatic Organisms, Environ. Pollut., 5:45-57.

17. Harvey, H. H., 1975. Fish Populations in a Large Group of Acid Stressed Lakes, Verh. Internt. Verein. Limnol., 19:2406-2417.

18. Hendrey, G. R., Baalsrud, K., Traaen, T. S., Laake, M., and Raddum, G., 1976. Acid Precipitation: Some Hydrobiological Changes, Ambio., 5:224-227.

19. Leivestad, H., Hendrey, G., Muniz, I. P., and Snekvik, E., 1976. Effects of Acid Precipitation on Freshwater Organisms, pp. 87-111. IN: Impact of Acid Precipitation on Forest and Freshwater Ecosystems in Norway. FR 6/76. SNSF-project.

20. Likens, G. E., Wright, R. F., Galloway, J. N., and Butler, T. J., 1979. Acid Rain, Scientific American, 241:43-51.

21. Mossberg, P., 1979. Benthos of Oligotrophic and Acid Lakes. (In Swedish, summary in English), Inf. Inst. Freshw. Res., Drottningholm (9), 23 pp.

22. National Research Council Canada, 1981. Acidification in the Canadian Aquatic Environment: Scientific Criteria for Assessing the Effects of Acid Deposition on Aquatic Ecosystems, NRCC/CNRC Ottawa. 369 pp.

23. Okland, J., 1969. Distribution and Ecology of the Freshwater Snails (Gastropoda) of Norway, Malacologia, 9:143-151.

24. Schindler, D. W., and Holmgren, S. K., 1971. Primary Production and Phytoplankton in the Experimental Lakes Area, Northwestern Ontario and Other Low Carbonate Waters, and a Liquid Scintillation Method for Determining ^{14}C Activity in Photosynthesis, J. Fish. Res. Board Can., 28:189-201.

25. Schindler, D. W., and Noven, B., 1971. Vertical Distribution and Seasonal Abundance of Zooplankton in Two Shallow Lakes of the Experimental Lakes Area, Northwestern Ontario, J. Fish. Res. Board Can., 28:245-256.

26. Schindler, D., Wagemann, R., Cook, R., Ruszczynski, T., and Prokopowick, J., 1980. Experimental Acidification of Lake 223, Experimental Lakes Area: Background Data and the First Three Years of Acidification, Can. J. Fish. Ag. Sci., 37:342-354.

27. Schofield, C. L., 1976. Lake Acidification in the Adirondack Mountains of New York: Causes and Consequences, pp. 477-478. In: Proceedings of the First International Symposium on Acid Precipitation and the Forest Ecosystem. USDA Forest Service. 1074 pp.

28. Schulze, R. L., 1980. The Biotic Response of Acid Precipitation in Florida Lakes, M.S. Thesis, University of Florida, Gainesville. 146 pp.

29. Sprules, W. G., 1975. Midsummer Crustacean Zooplankton Communities in Acid Stressed Lakes, J. Fish. Res. Board Can., 32:389-395.

30. Thompson, D. M., 1981. Distribution of Heavy Metals in Selected Florida Lake Sediments, M.S. Thesis, University of Florida, Gainesville. 160 pp.

31. Wetzel, R. G., 1975. Limnology, W. B. Saunders Co., Philadelphia, PA. 743 pp.

32. Wiederholm, T., and Eriksson, L., 1977. Benthos of an Acid Lake, Oikos, 29:261-267.

33. Wright, R. F., Dale, T., Gjessing, E. T., Hendrey, G. R., Henriksen, A., Johannessen, M., Lysholm, C., and Muniz, I. P., 1976. Impact of Acid Precipitation on Freshwater Ecosystems in Norway, Water Air Soil Pollut., 6:483-499.

34. Yan, N. D., 1979. Phytoplankton Community of an Acidified Heavy Metal-Contaminated Lake near Sudbury, Ontario: 1978-1977, Water Air Soil Pollut., 11:43-55.

35. Yan, N. D., and Stokes, P. M., 1978. Phytoplankton of an Acidic Lake, and Its Responses to Experimental Alterations of pH, Environ. Conserv., 5:93-100.

AUTHOR INDEX